W9-BTG-973

HIPPOCRATES
VIII

LCL 482

HIPPOCRATES

VOLUME VIII

EDITED AND TRANSLATED BY
PAUL POTTER

HARVARD UNIVERSITY PRESS
CAMBRIDGE, MASSACHUSETTS
LONDON, ENGLAND
1995

Library of Congress Cataloging-in-Publication Data

Hippocrates.
(The Loeb classical library)
Vol. 8 with an English translation by Paul Potter.
ISBN 0–674–99531–7
1. Hippocrates — Translations into English.
2. Medicine, Greek and Roman.
III. Potter, Paul, 1944–   .   IV. Title.   V. Series.
PA3612.H65   1923   610   23–12030

Typeset by Chiron, Inc, Cambridge, Massachusetts.
Printed in Great Britain by St Edmundsbury Press Ltd,
Bury St Edmunds, Suffolk, on acid-free paper.
Bound by Hunter & Foulis Ltd, Edinburgh, Scotland.

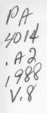

CONTENTS

INTRODUCTION

The ten treatises in this volume fall naturally into three groups.[1]

The works in the first group *Places in Man*, *Glands*, and *Fleshes* are monographs, each presenting a general account of the origin, structure, and function of particular parts of the human body, along with comments on disturbances of function and the origin of diseases. All three writers make effective use of observations drawn from common experience and from clinical practice, and of explanatory hypotheses deriving ultimately from contemporary philosophical thought, in constructing their systems.

The two books of *Prorrhetics* contribute to the art of prognosis, the prediction of disease outcome. *Prorrhetic I* is a collection of 170 short independent chapters which enumerate clinical signs and give their prognostic values in specific diseases. Individual cases are sometimes mentioned, and the writer occasionally poses questions. *Prorrhetic II* is a monographic treatment of the art of medical prediction, beginning with an account of its usefulness, its limits, and its frequent abuse, and then going on to a

---

[1] The individual works are analysed in more detail in their particular introductions.

detailed discussion of how the practitioner can best understand and profit from the signs that appear in a large selection of specific diseases.

The third group consists of five practical manuals on specific facets of Hippocratic surgical practice. *Physician* outlines professional behaviour, and gives instructions on setting up an office and on the performance of bandaging, incising, and cautery. *Use of Liquids* discusses the external application of fresh water, salt water, vinegar and wines. *Ulcers* lays down the principles and practice by which external lesions are to be evaluated and healed. *Haemorrhoids* and *Fistulas* summarize the pathology and treatment of haemorrhoids and condylomas, and of fistula in ano and its complications, respectively.

Manuscript Tradition

A = Parisinus Graecus 2253	XI c.
V = Vaticanus Graecus 276	XII c.
M = Marcianus Venetus Graecus 269	X/XI c.
I = Parisinus Graecus 2140	XIII c.
H = Parisinus Graecus 2142 Ha (older part)[2]	XII/XIII c.
Hb (newer part)	XIV c.
R = Vaticanus Graecus 277	XIV c.
Recentiores = approximately 20 manuscripts	XV/XVI c.

The *stemma codicum* appearing as Fig. 1 provides an overview of the interdependencies among the manuscripts containing the works in this volume. This is not, however,

[2] Folios 46, 49, 55–78 and 80–308; see A. Rivier, *Recherches sur la tradition manuscrite du traité hippocratique "De morbo sacro,"* Berne, 1962, pp. 97 ff.

2

Fig. 1: *Stemma Codicum*

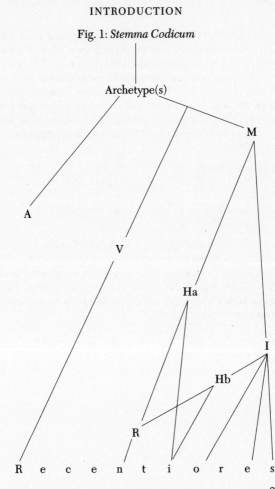

to suggest that the ten treatises share the same transmission: for in fact A, V and M, even in their original complete states, each contained only a selection of the Hippocratic Collection; furthermore M subsequently lost 72 leaves of text from between the folios now numbered 408 and 409, some time after I had been copied from it in the middle of the thirteenth century. For the treatises we are concerned with here, then, the independent witnesses are as follows:

Places in Man	AV	*Physician*	V
Glands	V	*Use of Liquids*	A
Fleshes	V	*Ulcers*	M
Prorrhetic I	I	*Haemorrhoids*	I
Prorrhetic II	I	*Fistulas*	I

Prorrhetic I and *II*, *Haemorrhoids* and *Fistulas* were all among the treatises contained in the section of M now lost.[3] They also happen to occupy the newer part of Parisinus Graecus 2142, Hb, which derives from I rather than from M directly (as Ha does), and thus possesses no independent textual authority.

[3] See H. Kuehlewein and J. Ilberg, *Hippocratis Opera Omnia*, Leipzig, 1894, vol. 1, xx.

BIBLIOGRAPHY

Editions, Translations and Commentaries

HIPPOCRATIC COLLECTION

Hippocratis Coi ... octoginta volumina ... per M. Fabium Calvum, Rhavennatem ... latinitate donata ..., Rome, 1525. (= Calvus)

Omnia opera Hippocratis ... in aedibus Aldi & Andreae Asulani soceri, Venice, 1526. (= Aldina)

Hippocratis Coi ... libri omnes ... [per Ianum Cornarium], Basel, 1538. (= Froben)

Vidus Vidius, *Chirurgia e Graeco in Latinum conversa ... nonnullis ... commentariis*, Paris, 1544. (= Vidius)

Hippocratis Coi ... opera ... per Ianum Cornarium ... Latina Lingua conscripta, Venice, 1546. (= Cornarius)

Hippocratis Coi ... viginti duo commentarii ... Theod. Zvingeri studio & conatu, Basel, 1579. (= Zwinger)

Stephan. Manialdus, *Hippocratis chirurgia ... graece ... latinitate ... commentar. illustr.*, Paris, 1619. (= Manialdus)

Magni Hippocratis ... opera omnia ... latina interpretatione & annotationibus illustrata Anutio Foesio..., Geneva, 1657–62. (= Foes)

Magni Hippocratis Coi opera omnia graece & latine edita ... industria & diligentia Joan. A. Vander Linden, Lei-

5

den, 1665. (= Linden)

Hippocratis Coi et Claudii Galeni Pergameni ... opera ... Renatus Charterius ... edidit, Paris, 1679. (= Chartier)

Hippocratis opera omnia ... studio et opera Stephani Mackii ..., Vienna, 1743–49. (= Mack)

J. F. K. Grimm, *Hippokrates Werke aus dem griechischen ...*, Altenburg, 1781–92. (= Grimm)

C. G. Kühn, *Magni Hippocratis Opera omnia ...*, Leipzig, 1825–27. (= Kühn)

Fr. Adams, *Genuine Works of Hippocrates translated from the Greek*, London, 1849. (= Adams)

E. Littré, *Oeuvres complètes d'Hippocrate*, Paris, 1839–61. (= Littré)

Ch. Daremberg, *Oeuvres choisies d'Hippocrate*, second edition, Paris, 1855. (= Daremberg)

F. Z. Ermerins, *Hippocratis ... reliquiae*, Utrecht, 1859–64. (= Ermerins)

C. H. Th. Reinhold, Ἱπποκράτης / Ψευδωνύμως Ἱπποκράτεια, Athens, 1865–67. (= Reinhold)

J. E. Petrequin, *Chirurgie d'Hippocrate*, Paris, 1877–78. (= Petrequin)

R. Fuchs, *Hippokrates, sämmtliche Werke. Ins Deutsche übersetzt ...*, Munich, 1895–1900. (= Fuchs)

J. L. Heiberg, *Hippocratis ... De medico ... De liquidorum usu ...*, Corpus Medicorum Graecorum I 1, Leipzig and Berlin, 1927. (= Heiberg)

R. Joly, *Hippocrate, ... De l'usage des liquides*, Budé VI(2), Paris, 1972. (= Joly)

R. Joly, *Hippocrate, Des lieux dans l'homme, Du système des glandes, Des fistules, Des hémorroïdes, ... Des chairs ...*, Budé XIII, Paris, 1978. (= Joly)

OTHER AUTHORS

I. E. Drabkin, *Caelius Aurelianus On Acute Diseases and On Chronic Diseases*, Chicago, 1950. (= Caelius)

Max Wellmann, *Pedanii Dioscuridis . . . De materia medica*, Berlin, 1906–1914. (= Dioscorides)

E. Nachmanson, *Erotiani Vocum hippocraticarum collectio*, Gothenburg, 1918. (= Erotian)

E. Nachmanson, *Erotianstudien*, Uppsala, 1917. (= Nachmanson)

C.G. Kühn, *Claudii Galeni Opera omnia . . .*, Leipzig, 1825–1833. (= Galen)

Ch. Daremberg and Ch. E. Ruelle, *Oeuvres de Rufus d'Ephèse*, Paris, 1879. (= Rufus)

General Works

Gerhard Baader and Rolf Winau (edd.), *Die hippokratischen Epidemien. Verhandlungen des Ve Colloque International Hippocratique*, Stuttgart, 1989. (= *Sudhoffs Archiv*, Beiheft 27)

Louis Bourgey, *Observation et expérience chez les médecins de la Collection hippocratique*, Paris, 1953.

L. Bourgey and J. Jouanna (edd.), *La Collection hippocratique et son rôle dans l'histoire de la médecine. Colloque de Strasbourg*, Leiden, 1975.

S. Byl, "Les dix dernières années (1983–1992) de la recherche hippocratique", *Centre Jean-Palerne: Lettre d'informations*, no. 22 (May 1993), pp. 1–39.

B. Celli, *Bibliografía Hipocrática*, Caracas, 1984.

Hans Diller, *Kleine Schriften zur antiken Medizin*, Berlin/New York, 1973.

Ludwig Edelstein, "Hippokrates", in *Paulys Real-Ency-*

clopädie der classischen Altertumswissenschaft, Supplement 6, Stuttgart, 1935. (cols. 1290–1345)

Gerhard Fichtner, *Corpus Hippocraticum. Verzeichnis der hippokratischen und pseudohippokratischen Schriften*, second edition, Tübingen, 1990.

Hellmut Flashar (ed.), *Antike Medizin*, Darmstadt, 1971. (Wege der Forschung, vol. 221)

M. D. Grmek (ed.), *Hippocratica. Actes du Colloque hippocratique de Paris*, Paris, 1980.

Mirko D. Grmek, *Les maladies à l'aube de la civilisation occidentale*, Paris, 1983.

Beate Gundert, "Parts and their Roles in Hippocratic Medicine", *ISIS*, vol. 83 (1992), pp. 453–465.

William Arthur Heidel, *Hippocratic Medicine. Its Spirit and Method*, New York, 1941.

Robert Joly, *Le niveau de la science hippocratique*, Paris, 1966.

Robert Joly (ed.), *Corpus Hippocraticum. Actes du colloque hippocratique de Mons*, Mons, 1977.

Jacques Jouanna, *Hippocrate*, Paris, 1992.

Pedro Laín Entralgo, *La medicina hipocrática*, Madrid, 1970.

Volker Langholf, *Medical Theories in Hippocrates. Early Texts and the 'Epidemics'*, Berlin/New York, 1990.

François Lasserre and Philippe Mudry (edd.), *Formes de pensée dans la Collection hippocratique. Actes du IVe Colloque International Hippocratique*, Geneva, 1983.

G. E. R. Lloyd (ed.), *Hippocratic Writings*, Harmondsworth, 1978.

Iain M. Lonie, *The Hippocratic Treatises "On Generation" "On the Nature of the Child" "Diseases IV". A Commentary*, Berlin/New York, 1981.

BIBLIOGRAPHY

J. A. López Férez (ed.), *Tratados hipocráticos. Actas del VIIe Colloque International Hippocratique*, Madrid, 1992.

G. Maloney and R. Savoie, *Cinq cents ans de bibliographie hippocratique. 1473–1982*, Quebec, 1982.

Gilles Maloney and Winnie Frohn, *Concordance des oeuvres hippocratiques*, 5 vols., Quebec, 1984. (Repr. Hildesheim, 1986)

Paul Potter, *Short Handbook of Hippocratic Medicine*, Quebec, 1988.

Paul Potter, Gilles Maloney and Jacques Desautels (edd.), *La maladie et les maladies. Actes du VIe Colloque International Hippocratique*, Quebec, 1990.

Wesley D. Smith, "Galen on Coans versus Cnidians", *Bulletin of the History of Medicine*, vol. 47 (1973), pp. 569–585.

Owsei Temkin, "Der Systematische Zusammenhang im Corpus Hippocraticum", *Kyklos*, vol. 1 (1928), pp. 8–43.

Antoine Thivel, *Cnide et Cos? Essai sur les doctrines médicales dans la Collection hippocratique*, Paris, 1981.

Supplementary bibliographical information is to be found in the introductions to the individual treatises.

NOTE ON TECHNICAL TERMS

The following terms require some explanation as they cannot be rendered simply and precisely into English.

ἀπόστασις/apostasis: the process of recovery from a disease is often associated with the collection and removal of morbid humours (peccant material) from the ailing part of the body, an apostasis. Cf. *Regimen in Acute Diseases* (*Appendix*) 39: "All diseases are resolved through either the mouth, the cavity, or the bladder; sweating is a form of resolution common to them all." The moment of this resolution by apostasis generally represents the disease's "crisis".

ἕλκος/helkos : the meaning of the term is wider than any single English term, and includes any discontinuity of tissue, whether internal or external, inflamed or livid, traumatic or spontaneous; possible translations include "sore", "ulcer", "wound" or simply "lesion". The Hippocratic treatise traditionally named *Ulcers* (περὶ ἑλκῶν) is in fact an account not of the pathological phenomenon "ulcer" in the strict dictionary sense, but rather of surface lesions of all types.

κοιλίη/cavity: generally the thorax and/or abdomen is meant, but more frequently the gastro-intestinal tract.

Anything a person feels to be "high up" or that involves nausea or vomiting is imagined to be in the "upper cavity", anything felt to be "low down" or that has a relationship to defecation is in the "lower cavity". The term "cavity" can also be applied to other hollows in the body, e.g. in bones.

φλεγμαίνειν/*phlegmainein*: either "to form phlegm", "to swell up", or "to become inflamed"; in many passages it is impossible to tell which of these three is meant.

φλέψ/vessel: generally a blood vessel, whether artery or vein, but occasionally some other tube such as the ureter.

PLACES IN MAN

INTRODUCTION

Erotian lists this work in his preface among the therapeutic works, and includes thirty-two words from it in his *Glossary*[1]; four of these glosses cite Bacchius of Tanagra, suggesting that the treatise was already considered Hippocratic by the third century B.C.[2]

Galen's *Glossary* contains several words referable to *Places in Man*, five with explicit references (ὡς) ἐν τῷ περὶ τόπων τῶν κατὰ ἄνθρωπον.[3] Galen also quotes many times from the treatise in his other writings, often however without naming his source.[4]

Caelius Aurelianus' (fifth century) Latin translation of Soranus' *Chronic Diseases* refers directly to ch. 22 of *Places in Man* in discussing the treatment of sciatica: "Hippocrates in his book *On Places* employs cupping without scarification. He also prescribes the drinking of hot drugs."[5] The continued visibility of *Places in Man* in the Middle Ages is attested by its citation by the Byzan-

---

[1] Erotian p. 9; see Nachmanson pp. 331–39.

[2] Erotian A58, Γ8, K18 and M9.

[3] Galen vol. 19, 74 ἀλαία φθίσις; 19, 103 θηρίον; 19, 107f. κάμμορον; 19, 114 κρημνοί; and 19, 118 λεπτά.

[4] For a detailed study of Galen's ambivalent attitude to *Places in Man* see Schubring pp. 61–70.

[5] Caelius pp. 920f.

tine excerptor of Rufus of Ephesus's *Names of the Parts of the Body*.[6]

Places in Man is organized as follows:

1: Introduction

2–8: Human anatomy and physiology, with particular emphasis on the brain and organs of sensation, the blood vessels and cords, and the bones and joints.

9–15: Fluxes and their causes: the seven cardinal fluxes from the head to the nostrils, ears, eyes, chest, spinal marrow, vertebrae, and joints.

16–40: The origin and treatment of specific diseases: e.g. pleurisy, internal suppuration, consumptions, sciatica, dropsies, fevers, jaundice, malignant ulcer, and angina. These chapters vary widely in form and content, lacking the more regular pattern (name; symptoms; course; prognosis; treatment) followed in *Affections*, *Diseases I–III* and *Internal Affections*.

41–46: General principles of medicine: e.g. the roles of "similars" and "opposites" in disease causation and treatment; the modes of action of medications; the meaning of "good luck" in medical practice.

47: Diseases of Women.

Throughout its somewhat irregular course, *Places in Man* exhibits a unity of viewpoint and purpose centred around a number of principles enunciated in its early chapters:

(i) The best medical treatment is that based upon an

[6] Rufus p. 235; see also Schubring pp. 70f.

understanding of how diseases arise in the body.

(ii) This understanding begins with a knowledge of normal structure and function, followed by an acquaintance with disease processes and mechanisms.

(iii) All parts of the body are interrelated and interdependent. Diseases often migrate from one part to another.

(iv) Fluxes of internal fluids play a central role in disease causation.

(v) Disease processes often commence from some excess or imbalance of temperature, moisture, or particular substance.

(vi) No simple rules exist by which a physician can know in each case how to act; variability of individual constitution, climate, and geography all affect health and disease.

Places in Man is included in all the standard collected editions and translations, including Zwinger, and has been the object of a number of special studies,[7] of which the two most recent are: K. Schubring, *Untersuchungen zur Überlieferungsgeschichte der hippokratischen Schrift "De locis in homine"*, Berlin, 1941[8] (= Schubring); and Robert Joly's Budé edition. Joly includes in his introduction a detailed study, with references, of the many questions that have occupied the attention of scholars regarding the treatise: unity of authorship; medical theories; philosophical parentage; relation to other Hippocratic works; school affiliation.[9]

It is on Joly's edition that mine for the most part depends.

[7] See Littré vol. 6, 275.   [8] This work is particularly useful for its discussion of the indirect transmission.   [9] Pp. 11–32.

ΠΕΡΙ ΤΟΠΩΝ
ΤΩΝ ΚΑΤΑ ΑΝΘΡΩΠΟΝ

VI 276
Littré

1. Ἐμοὶ δοκέει ἀρχὴ μὲν οὖν οὐδεμία εἶναι τοῦ σώματος, ἀλλὰ πάντα ὁμοίως ἀρχὴ καὶ πάντα τελευτή· κύκλου γὰρ γραφέντος ἀρχὴ οὐχ εὑρέθη. καὶ τῶν νοσημάτων ἀπὸ παντὸς ὁμοίως τοῦ σώματος, τὸ μὲν ξηρότερον, πεφυκὸς νόσους λάζεσθαι καὶ μᾶλλον πονέειν, τὸ δὲ ὑγρὸν ἧσσον· τὸ μὲν γὰρ ἐν τῷ ξηρῷ νόσημα πήγνυταί τε καὶ οὐ[1] διαπαύει, τὸ δ᾽ ἐν τῷ ὑγρῷ διαρρεῖ, καὶ τοῦ σώματος ἄλλοτε ἄλλο μάλιστα ἔχει, καὶ ἀεὶ μεταλλάσσον ἀνάπαυσιν ποιέει καὶ θᾶσσον παύεται, ὥστε οὐ πεπηγός.

Τοῦ δὲ σώματος τὰ μέλεα ἕκαστα τὸ ἕτερον τῷ ἑτέρῳ, ὁπόταν ἔνθα ἢ ἔνθα ὁρμήσῃ, νοῦσον παραυτίκα ποιέει, ἡ κοιλίη τῇ κεφαλῇ, καὶ ἡ κεφαλὴ τῇσι σαρξὶ καὶ τῇ κοιλίῃ, καὶ τἆλλα πάντα οὕτω κατὰ λόγον [ὥσπερ ἡ κοιλίη τῇ κεφαλῇ, καὶ ἡ κεφαλὴ τῇσί τε σαρξὶ καὶ τῇ κοιλίῃ].[2] ἡ γὰρ κοιλίη ὁκόταν ὑπεχώρησιν μὴ ποιέῃ τὴν μετρίην, καὶ ἐσίῃ ἐς αὐτήν, ἄρδει τῇ ὑγρότητι τὸ σῶμα τῇ ἀπὸ τῶν σιτίων τῶν προσφερομένων· αὕτη δὲ ἡ ὑγρότης ἀπὸ

PLACES IN MAN

1. Now in my opinion, there is no beginning point of the body, but rather every part is at the same time both beginning and end, in the same way that in the figure of a circle, no beginning point is to be found. Likewise, diseases arise from the whole body indifferently, although the drier component of the body is disposed to become ill and to suffer more, the moist component less. For whereas any disease that occupies a dry part is fixed and unremitting, one in a moist part flows somewhere else and generally occupies different parts of the body at different times; through constantly changing, it has interruptions and goes away sooner, and so it is not fixed.

Every part of the body, on becoming ill, immediately produces disease in some other part, the cavity[a] in the head, the head in the muscles and the cavity, and so on in the same way [as the cavity in the head and the head in the muscles and the cavity]. The cavity, for instance, when material enters it and it does not make a corresponding evacuation downwards, floods the body with moisture from the ingested food. This moisture, being

[a] See the note on technical terms above, following the general introduction.

---

[1] οὐ om. V.   [2] Del. Ermerins.

τῆς κοιλίης ἀποφρασσομένη ἐς τὴν κεφαλὴν ὡδοι-
πόρησεν ἀθρόη· καὶ ἐς τὴν κεφαλὴν ἐπὴν ἀφίκηται,
οὐ χωρευμένη ὑπὸ τῶν τευχέων τῶν ἐν τῇ κεφαλῇ,
ῥεῖ ᾗ ἂν τύχῃ, καὶ πέριξ τῆς κεφαλῆς, καὶ ἐς τὸν
ἐγκέφαλον [διὰ λεπτοῦ τοῦ ὀστέου]·[1] καὶ ἡ μὲν ἐν
τῷ ὀστέῳ ἐνδέδυκεν, ἡ δὲ περὶ τὸν ἐγκέφαλον διὰ
λεπτοῦ τοῦ ὀστέου· καὶ ἢν μὲν ἐς τὴν κοιλίην πάλιν
ἀφίκηται, τῇ κοιλίῃ νοῦσον ἐποίησεν· ἢν δ' ἄλλῃ πῃ
τύχῃ, ἄλλῃ νοῦσον ποιέει, καὶ τἆλλα οὕτως, ὥσπερ
278 τοῦτο, τὸ | ἕτερον τῷ ἑτέρῳ νοῦσον ποιέει· καὶ κάλ-
λιστον οὕτως εὐτρεπίζειν τὰ νοσεόμενα διὰ τῶν τὰς
νούσους ποιεύντων· οὕτω γὰρ ἂν κάλλιστα τὴν
ἀρχὴν τοῦ νοσεομένου τις ἰῷτο.

Τὸ δὲ σῶμα αὐτὸ ἑωυτῷ τωυτόν ἐστι καὶ ἐκ τῶν
αὐτῶν σύγκειται, ὁμοίως δὲ συνεχόντων,[2] καὶ τὰ
σμικρὰ αὐτοῦ καὶ τὰ μεγάλα καὶ τὰ κάτω καὶ τὰ
ἄνω· καὶ εἴ τις βούλεται τοῦ σώματος ἀπολαβὼν
μέρος κακῶς ποιέειν τὸ σμικρότατον, πᾶν τὸ σῶμα
αἰσθήσεται τὴν πεῖσιν, ὁποίη ἄν τις ᾖ, διὰ τόδε ὅτι
τοῦ σώματος τὸ σμικρότατον πάντα ἔχει, ὅσα περ
καὶ τὸ μέγιστον· τοῦτο δ' ὁποῖον ἄν τι πάθῃ τὸ σμικρότατον, ἐπαναφέρει πρὸς τὴν ὁμοεθνίην ἕκαστον
πρὸς τὴν ἑωυτοῦ, ἤν τε κακόν, ἤν τε ἀγαθὸν ᾖ· καὶ
διὰ ταῦτα καὶ ἀλγέει καὶ ἥδεται ὑπὸ ἔθνεος τοῦ σμι-
κροτάτου τὸ σῶμα, ὅτι ἐν τῷ σμικροτάτῳ πάντ' ἔνι
τὰ μέρεα, καὶ ταῦτα ἐπαναφέρουσιν ἐς τὰ σφῶν
αὐτῶν ἕκαστα, καὶ ἐξαγγέλλουσι πάντα.

2. Φύσις δὲ τοῦ σώματος, ἀρχὴ τοῦ ἐν ἰητρικῇ

turned away from the cavity, moves away all at once to the head, and when it arrives in the head, not finding room in the vessels there, flows anywhere it can, both all about the head and into the brain [through the fine bone]; some of it goes into the bone, some through the fine bone to around the brain. If it goes back to the cavity, it produces disease in the cavity; if it happens to go somewhere else, it produces disease there. Elsewhere it is just the same, one part produces disease in another. In view of this, it is best to set diseases right by means of the factors that produce them; for in this way you can best treat the source of diseases.[a]

The body is uniform throughout and is composed of the same things, connected in the same way, the small parts and the large, those lower down and those higher up. And if someone wishing to harm the body should take away some part, even the smallest, the whole body will experience the hurt, of whatever sort it be, because the smallest part of the body has everything that the greatest part has. Anything the smallest part experiences, either good or bad, it passes on to its related part, each to its own particular relative. For this reason, a person experiences both pain and pleasure from the smallest members of his body, since all parts are present in the smallest, and these smallest parts pass on their experience to each of their own, and inform them of everything.

2. The nature[b] of the body is the beginning point of

[a] See chapter 31 below.
[b] I.e. structure and function.

---

[1] Del. Ermerins.
[2] Potter: οὐκ ἐχόντων (-τος) AV.

21

λόγου· πρῶτον διατέτρηται ᾗ ἐσακούομεν· τὰ μὲν
γὰρ περὶ τὰ ὦτα πέριξ κενεά, οὐκ εἰσακούει ἄλλο ἢ
ψόφον καὶ ἰαχήν· ὅ τι δ᾽ ἂν διὰ τῆς μήνιγγος ἐς
τὸν ἐγκέφαλον ἐσέλθῃ, τοῦτο[1] διαφραδέως ἀκούεται
ταύτῃ· καὶ μόνη τρῆσις διὰ τῆς μήνιγγός ἐστι τῆς
περὶ τὸν ἐγκέφαλον περιτεταμένης. κατὰ δὲ τὰς
ῥῖνας τρῆμα μὲν οὐκ ἔνεστιν, σομφὸν δέ, οἷον
σπόγγια· καὶ διὰ τοῦτο διὰ πλείονος ἀκούει ἢ
ὀσφραίνεται· κατὰ πολὺ γὰρ σκίδναται ἡ ὀδμὴ τῆς
ὀσφρήσιος.

Καὶ ἐς τοὺς ὀφθαλμοὺς φλέβια λεπτὰ ἐς τὴν ὄψιν
ἐκ τοῦ ἐγκεφάλου διὰ τῆς μήνιγγος τῆς περιεχούσης
φέρονται· ταῦτα δὲ τὰ φλέβια τὴν ὄψιν τρέφουσι τῷ
ὑγρῷ τῷ καθαρωτάτῳ τῷ ἀπὸ τοῦ ἐγκεφάλου, ἐς ὃ
280 καὶ ἐμφαίνεται ἐν τοῖσιν | ὀφθαλμοῖσιν· ταῦτα δὲ τὰ
φλέβια καὶ ἀποσβεννύασι τὰς ὄψεις ὅταν ξηρανθῶ-
σιν. μήνιγγες δὲ τρεῖς εἰσιν αἱ τοὺς ὀφθαλμοὺς
φυλάσσουσαι, ἡ μὲν ἐπάνω παχυτέρη, ἡ δὲ διὰ
μέσου λεπτοτέρη, ἡ δὲ τρίτη λεπτὴ ἡ τὸ ὑγρὸν
φυλάσσουσα· τούτων ἡ μὲν ἐπάνω καὶ παχυτέρη
νοσέει[2] ἢν κωφωθῇ· ἡ δὲ διὰ μέσου ἐπικίνδυνος·
αὕτη καί, ὅταν ῥαγῇ, ἐξίσχει οἷον κύστις· ἡ δὲ τρίτη
ἡ λεπτοτάτη πάμπαν ἐπικίνδυνος, ἡ τὸ ὑγρὸν
φυλάσσουσα. μήνιγγες δὲ δύο εἰσὶ τοῦ ἐγκεφάλου·
ἡ λεπτὴ οὐκέτι ἰατὴ[3] ἐπὴν τρωθῇ.

3. Φλέβες δὲ περαίνουσι μὲν ἐς τὴν κορυφὴν διὰ
τῆς σαρκὸς ἔχουσαι πρὸς τὸ ὀστέον· φέρονται δὲ
διὰ τῆς σαρκός, δύο μὲν ἐκ τῆς κορυφῆς κατ᾽ ἰθὺ ᾗ

medical reasoning. Now first, the meningeal membrane is perforated at the point through which we hear, for the emptiness all around the ear only hears sounds and shouts, whereas what goes through the membrane into the brain is perceived distinctly. This is the only perforation through the membrane that covers the brain. In the region of the nostrils there is no perforation, but a kind of porousness, like a sponge; for this reason a person hears over a greater distance than he is able to smell, since odours are thoroughly broken up on their way to the organ of smell.

To the eyes: narrow vessels lead to the pupil from the brain through the membrane that encloses it. These vessels nourish the pupil with purest moisture from the brain, and it is in this moisture that the image appears in the eyes. The same vessels also extinguish the pupils if they dry out. There are three membranes protecting the eye: a thicker one on the surface, a thinner one at the intermediate position, a third fine one that protects the moisture. Of these, the thicker one on the surface becomes ill if it is dulled. The one in the intermediate position is vulnerable, and when it tears it protrudes like a bladder. The third and finest membrane is extremely vulnerable, i.e. the one that protects the moisture. Of the brain, there are two membranes; the fine one cannot be healed, when once injured.

3. There are vessels that have their terminus at the vertex of the head and that pass through the tissue close to the bone. Two run through the tissue in a straight line

---

[1] τοῦτο om. V. [2] Potter: νοῦσος AV.
[3] R: ἡ αὐτὴ AV.

αἱ ὀφρύες συγκλείονται καὶ τελευτῶσιν ἐς τοὺς καν-
θοὺς τῶν ὀφθαλμῶν,[1] μία δὲ ἀπὸ τῆς κορυφῆς ἐς τὴν
ῥῖνα φέρεται καὶ σχίζεται ἐς τὸν χόνδρον τῆς ῥινὸς
ἑκάτερον· ἄλλαι δύο φλέβες παρὰ τοὺς κροτάφους
φέρονται ἐν μέσῳ τῶν κροτάφων καὶ τῶν ὤτων, αἳ
πιέζουσι τὰς ὄψεις καὶ σφύζουσι αἰεί· μοῦναι γὰρ
αὗται οὐκ ἄρδουσι τῶν φλεβῶν, ἀλλ' ἀποτρέπεται
ἐξ αὐτῶν τὸ αἷμα·[2] τὸ δ' ἀποτρεπόμενον ἀποσυμ-
βουλεύει τῷ ἐπιρρέοντι· καὶ τὸ μὲν ἀποτρεπόμενον
βουλόμενον ἄνω χωρέειν,[3] τὸ δ' ἄνωθεν ἐπιρρέον
βουλόμενον κάτω χωρέειν, ἐνταῦθα ὠθεύμενά τε καὶ
ἀναχεόμενα πρὸς ἄλληλα καὶ κυκλούμενα, σφυγμὸν
παρέχουσι τοῖσι φλεβίοισιν.

Ἡ δὲ ὄψις τῷ ἀπὸ τοῦ ἐγκεφάλου ὑγρῷ τρέφεται·
ὅταν δέ τι τοῦ ἀπὸ τῶν φλεβῶν λάβῃ, τῇ ῥύσει
ταράσσεται, καὶ οὐκ ἐμφαίνεται ἐς αὐτὸ καὶ προκι-
νέεσθαι δοκέει ἐν αὐτῷ τοτὲ μὲν οἷον εἴδωλον ὀρνί-
θων, τοτὲ δὲ οἷον φακοὶ μέλανες, καὶ τἄλλα οὐδὲν
ἀτρεκέως κατ' ἀληθείην δύναται ὁρᾶν. ἄλλαι δύο
φλέβες ἐν μέσῳ τῶν τε ὤτων καὶ τῶν ἄλλων
282 φλε|βῶν, αἳ φέρονται ἐς τὰ ὦτα, καὶ πιέζουσι τὰ
ὦτα· ἄλλαι δύο φλέβες ἐκ τῆς συγκλείσεως τοῦ
ὀστέου ἐς τὰς ἀκοὰς φέρονται.

Αἱ δὲ κάτω τοῦ σώματος τετραμμέναι, δύο μὲν
φλέβες παρὰ τοὺς τένοντας τοῦ τραχήλου, φέρονται
δὲ καὶ παρὰ τοὺς σπονδύλους καὶ τελευτῶσιν ἐς τοὺς
νεφρούς· αὗται δὲ καὶ ἐς τοὺς ὄρχιας περαίνουσιν,
καὶ ὅταν αὗται πονήσωσιν, αἷμα οὐρέει ὤνθρωπος.

from the vertex to where the eyebrows meet, and end at
the corners of the eyes. One runs from the vertex to the
nose, and divides to each side of the nasal cartilage. Two
other vessels pass (from the vertex of the head) along the
temples, between the temples and the ears; these compress the eyes and are always throbbing, for they are the
only vessels that do not dispense their moisture, so that
the blood is turned back out of them; on turning back it
clashes with the new blood that is flowing to them; and
the blood that turns back tries to flow up the vessel, while
the new blood that is coming from above tries to flow
down the vessel, and both currents, thrust forward, pour-
ing out against one another, and moving in a circle, produce a pulsation in the vessels.

The pupil is nourished by moisture from the brain; if it
takes up anything from the vessels, it is disturbed by this
afflux, and the image does not appear normally in it, but
something seems to dance before it, sometimes like the
image of birds and sometimes like black specks, and otherwise it is not able to see anything clearly according to
reality. Two other vessels, located between the ears and
the pair of vessels mentioned above, pass to the ears and
press on them. Two other vessels run from the (sagittal)
suture to the organ of hearing.

Other vessels are directed down through the body.
Two vessels run beside the tendons of the neck and along
the vertebrae, to end at the kidneys; these also terminate
at the testes; and when they are afflicted, the person

---

[1] V: ἐς τοὺς ὀφθαλμοὺς τῶν κανθῶν A.

[2] V: φλέγμα A.

[3] ἄνω χ. Zwinger in margin: ἀποχωρέειν AV.

ἄλλαι δύο φλέβες ἀπὸ τῆς κορυφῆς φέρονται ἐς
τοὺς ὤμους, καὶ δὴ καὶ ὠμιαῖαι καλέονται. ἄλλαι
δύο φλέβες ἀπὸ τῆς κορυφῆς παρὰ τὰ ὦτα ἐν τοῖς
ἔμπροσθεν τοῦ τραχήλου ἑκατέρωθεν ἐς τὴν κοίλην
φλέβα καλεομένην φέρονται. ἡ δὲ κοίλη φλὲψ
περαίνεται μὲν ὡς ὁ οἰσοφάγος, πέφυκε δὲ μεταξὺ
τοῦ τε βρόγχου καὶ τοῦ οἰσοφάγου· φέρεται δὲ διὰ
τῶν φρενῶν καὶ διὰ τῆς καρδίης καὶ μεταξὺ τῶν
φρενῶν, καὶ σχίζεται ἐς τοὺς βουβῶνας καὶ ἐς τοὺς
μηροὺς ἐντός, καὶ τὰς διασφαγὰς ἐν τοῖσι μηροῖσι
ποιέεται, καὶ ἐς τὰς κνήμας φέρεται ἐντὸς παρὰ τὰ
σφυρά· αὗται καὶ ἄκαρπον ποιέουσι τὸν ἄνθρωπον
ὅταν ἀποτμηθῶσιν, αἱ καὶ ἐς τοὺς μεγάλους δακτύλους τελευτῶσιν. ἐκ δὲ τῆς κοίλης φλεβὸς ἀποπέφυκεν ἐς τὴν χεῖρα τὴν ἀριστερήν· φέρεται δ' ὑπο-
κάτω τοῦ σπληνὸς ἐς τὴν λαπάρην τὴν ἀριστερήν,
ὅθεν ὁ σπλὴν ἀποπέφυκε διὰ τοῦ ἐπιπλόου, καὶ τὴν
ἀποτελεύτησιν ἔχει ἐς τὸν κίθαρον· ἀποπέφυκε δὲ
κατὰ τὰς φρένας, καὶ ξυμβάλλει τῇ ὠμαίῃ κάτω
τοῦ ἄρθρου τοῦ ἀγκῶνος, καὶ τοῦ σπληνὸς[1] τάμνεται
αὕτη· καὶ ἄλλη ἐς τὴν δεξιὴν τὸν αὐτὸν τρόπον ἀποπέφυκεν ἀπὸ τῆς κοίλης.[2] κοινωνέουσι δὲ πᾶσαι αἱ
φλέβες καὶ διαρρέουσιν ἐς ἑωυτάς· αἱ μὲν γὰρ σφίσιν ἑωυταῖς ξυμβάλλουσιν, αἱ δὲ διὰ τῶν φλεβίων
τῶν διατεταμένων ἀπὸ τῶν φλεβῶν, αἳ τρέφουσι τὰς
σάρκας, ταύτῃ διαρρέουσι πρὸς ἑωυτάς.

4. Καὶ ἀπὸ τῶν φλεβῶν ὅ τι ἂν νόσημα γένηται,
ῥᾷόν ἐστιν ἢ ἀπὸ τῶν νεύρων· διαρρεῖ γὰρ σὺν τῷ

passes blood in his urine. Two other vessels run from the vertex of the head to the shoulders, and are in fact called "shoulder vessels". Two other vessels run from the vertex of the head, past the ears, through the front of the neck on each side, and into what is called the "hollow vessel" (*vena cava*). The hollow vessel takes the same course as the oesophagus, and is situated between the wind-pipe and the oesophagus. It passes over the surface of the diaphragm and the heart, between the halves of the diaphragm, and divides towards the groins and the insides of the thighs; it forms branches in the thighs and passes below the knee on the inside past the ankle; these vessels make a person sterile if they are cut, and they end at the large toes. From the hollow vessel there is a branch to the left arm: it passes beneath the spleen to meet the left flank at the point from which the spleen grows out along the omentum, and has its termination by entering the chest; it reappears on the other surface of the diaphragm, and finally joins the shoulder vessel below the elbow joint; this vessel is used for phlebotomy of the spleen. Another vessel branches to the right from the hollow vessel in the same way. All vessels communicate and flow into one another: for some join one another directly, while others join small vessels coming from the vessels that nourish the tissues and in this way flow into one another.

4. Any disease that arises from the vessels is easier than one that arises from the cords, for it flows away with

---

[1] V: ἀγκῶνος A.

[2] Foes in note 14, after Calvus' *de cava vena*: κοιλίης AV.

ὑγρῷ τῷ ἐνεόντι ἐν τῇσι φλεψὶ καὶ οὐκ ἀτρεμίζει·
καὶ ἡ φύσις τῇσι φλεψὶν ἐν ὑγρῷ ἐστιν ἐν τῇσι |
284   σαρκί. τὰ δὲ νεῦρα ξηρά τέ ἐστι καὶ ἀκοίλια, καὶ
πρὸς τῷ ὀστέῳ πεφύκασιν, καὶ τρέφονται δὲ τὸ
πλεῖστον ἀπὸ τοῦ ὀστέου, τρέφονται δὲ καὶ ἀπὸ τῆς
σαρκός, καὶ τὴν χροιὴν καὶ τὴν ἰσχὺν μεταξὺ τοῦ
ὀστέου καὶ τῆς σαρκὸς πεφύκασι, καὶ ὑγρότερα μέν
εἰσι τοῦ ὀστέου καὶ σαρκοειδέστερα, ξηρότερα δ'
εἰσὶν ἢ αἱ σάρκες καὶ ὀστοειδέστερα· νόσημα δ' ὅ τι
ἂν ἐς αὐτὰ ἔλθῃ, ῥώννυταί τε καὶ ἀτρεμίζει ἐν τῷ
αὐτῷ, καὶ χαλεπόν ἐστιν ἐξάγειν· μάλιστα δ' ἐσέρ-
χονται τέτανοί τε καὶ ἄλλα, ἀφ' ὧν τρόμος τὸ σῶμα
λαμβάνει καὶ τρέμειν ποιέει.

5.   Τὰ δὲ νεῦρα πιέζουσι τὰ ἄρθρα, παρατεταμένα
τέ εἰσι παρ' ὅλον τὸ σῶμα· ἰσχύουσι δὲ μάλιστα ἐν
ἐκείνοισι τοῦ σώματος καὶ αἰεὶ παχύτατά ἐστιν, ἐν
οἷσι τοῦ σώματος αἱ σάρκες ἐλάχισταί εἰσι. καὶ τὸ
μὲν σῶμα πᾶν ἔμπλεον νεύρων· περὶ δὲ τὸ πρόσ-
ωπον καὶ τὴν κεφαλὴν οὐκ ἔστι νεῦρα, ἀλλὰ ἶνες
παρόμοιαι νεύροις μεταξὺ τοῦ τε ὀστέου καὶ τῆς
σαρκὸς λεπτότεραι καὶ στερεώτεραι, αἱ δὲ νευρο-
κοίλιοι.

6.   Αἱ κεφαλαὶ ῥαφὰς ἔχουσιν, αἱ μὲν τρεῖς, αἱ δὲ
τέσσαρας· αἱ μὲν τέσσαρας ἔχουσαι, κατὰ τὰ ὦτα
ἑκατέρωθεν ῥαφή, ἄλλη ἔμπροσθεν, ἄλλη ἐξόπισθεν
τῆς κεφαλῆς, οὕτω μὲν ἡ τὰς τέσσαρας ἔχουσα· ἡ
δὲ τὰς τρεῖς, κατὰ τὰ ὦτα ἑκατέρωθεν, καὶ ἔμπροσ-
θεν· ὥσπερ δὲ ἡ τὰς τέσσαρας ἔχουσα, οὐ δια-

the moisture present in the vessels, and does not remain fixed. The home of vessels is in the moisture in the tissues. Cords, on the other hand, are dry and lack a lumen, and they grow next the bone; they receive most of their nourishment from the bone, and also some from the tissues; cords are in colour and strength midway between bone and tissue; they are moister than bone and more fleshy, but drier than tissue and more osseous. Any disease that enters the cords is severe and settles in one place, and it is difficult to drive out; the most frequent examples are tetanus and similar diseases, as the result of which trembling seizes the body and makes it shake.

5. The cords exert pressure on the joints, and they are present over the whole body. They have the greatest strength and are invariably thickest in the parts of the body where the fleshy parts are least. The whole body is quite full of cords; around the face and head, however, there are no cords, but instead fibers similar to cords, narrower and more rounded, midway between bone and muscle: they are the "hollow cords".

6. Skulls have sutures, some of them three, others four. In those with four, there is one suture in the region of the ears on each side, another at the front, and another behind: thus the head has four. The skull with three sutures has them by the ears and at the front. Here, as in the skull with four sutures, there is no joining suture.

πέφυκεν οὐδὲ ταύτῃ ῥαφή· ὑγιεινότεροι δ' εἰσὶ τὴν
κεφαλὴν οἱ τὰς πλείονας ῥαφὰς ἔχοντες. ἐν τῇσι
ὀφρύσι διπλόον[1] τὸ ὀστέον, καὶ ἡ σύγκλεισις τῶν
γενύων ἔν τε | τῷ γενείῳ μέσῳ καὶ ἄνω πρὸς τῇ
κεφαλῇ. σπονδύλους οἱ μὲν πλείονας, οἱ δὲ ἐλάσ-
σονας ἔχουσιν· καὶ οἱ μὲν πλείονας ἔχοντες, δυοῖν
δέοντας εἴκοσιν ἔχουσι, σὺν τοῖσι δ' ἐσχάτοισιν
εἴκοσίν εἰσιν, ὧν οἱ μὲν ἄνω πρὸς τῇ κεφαλῇ,[2] οἱ δὲ
κάτω πρὸς τῇ ἕδρῃ. πλευραὶ ἑπτά· ἄρθρα τῶν πλευρέων, τὰ μὲν ὄπισθεν τοῦ σώματος πρὸς τοὺς σπονδύλους, τὰ δ' ἔμπροσθεν ἐν τῷ στέρνῳ πρὸς ἑωυτάς.
κληῖδες ἄρθρα ἔχουσι, τὰ μὲν ἐν μέσῳ τοῦ στέρνου
κατὰ τὸν βρόγχον, κατὰ ταῦτα ἤρθρωνται· τὰ δὲ
πρὸς τοὺς ὤμους κεκλιμένα πρὸς τὰς πλάτας, αἱ ἐπὶ
τοῖς ὤμοις αἰεὶ[3] πεφύκασιν. αἱ δὲ πλάται πρὸς τὰ
γυῖα ἤρθρωνται,[4] ἐπιβάλλουσαι ἐπὶ τὸ ὀστέον τὸ ἐν
τῷ γυίῳ. παρὰ δὲ τὸ ὀστέον περόναι δύο παρήκουσιν, ἡ μὲν ἔνδοθεν, ἡ δὲ ἐκτός, αἱ πρὸς τὰς πλάτας
τῷ ὀστέῳ προσπεφυκυῖαι ἤρθρωνται.

Κάτω δ' ἐν τῷ ἀγκῶνι, κάτω μὲν περόνη ἤρθρωνται κάτω πεφυκυῖα, ἄνω δὲ[5] σμικρῷ τῆς περόνης ἐς
τὸν ἀγκῶνα, τό τε ὀστέον καὶ ἡ περόνη ἐς τὸ αὐτὸ
συμβάλλοντα ἄρθρον ἐν τῷ κυβίτῳ ποιέουσιν.
παρὰ τὸν πῆχυν δὲ περόναι παρήκουσι λεπταὶ πάνυ
τέσσαρες, αἱ μὲν δύο ἄνω, αἱ δὲ δύο κάτω· καὶ πρὸς

---

[1] A: διῆλθον V.   [2] Froben: τῇ ἕδρῃ τῆς κεφαλῆς AV.
[3] ὤμ. αἰεὶ V: ἄρθροις A.
[4] τὰ δὲ πρὸς τοὺς ὤμους add. V.

People have better health in the head who have the greater number of sutures. In the eyebrows the bone is double. The symphysis of the two halves of the jaw is in the middle of the chin, and this bone articulates above, next the head. Some individuals have a greater number of vertebrae, others less; those with the greater number have eighteen; with the ones at the ends, that makes twenty, of which one is at the top next the head, and one at the bottom in connection with the seat. Seven ribs: the ribs articulate at the back of the body with the vertebrae, at the front at the breast-bone with one another. The collar-bones have joints in the middle of the breast-bone by the windpipe, where they are fastened to one another, and in the shoulders, where they rest against the blades, which are always present on the shoulders. The shoulder-blades articulate with the limbs, lying next the bone in the limb. Alongside the bone (head of the humerus) are two processes—one on the inside (coracoid process) and one on the outside (acromion)—which, being joined to the shoulder-blades by bone, form the articulation.

Lower down in the bend of the elbow: on the lower side articulation is to a process growing on the lower side (olecranon), on the upper side, articulation occurs, a little above the process, into the elbow (olecranon fossa); by uniting, the bone and the process form a joint at the elbow. Alongside the forearm are four very slight processes, two on the upper side, two on the lower side: at

---

[5] κάτω πεφυκυῖα, ἄνω δὲ A: κατὰ τὸ πεφυκὸς κοιλανῶδες V.

μὲν τὸν ἀγκῶνα δύο περόναι πεφυκυῖαι ἄνω ἐκ τοῦ
ὀστέου πεφύκασιν, αὗται σὺν τῷ ὀστέῳ πεφυκυῖαι
παρὰ τὸ τοῦ ὀστέου ἄρθρον ἤρθρωνται ἐς τὸ κύβι-
τον· αἱ δὲ κάτω κείμεναι καὶ ἐντὸς κεκλιμέναι, αὗται
ἀμφότεραι ξυμβάλλουσαι πρὸς τὴν περόνην τὴν
288 ἄνωθεν ἀπὸ τοῦ γυίου φερομέ|νην, ἐντὸς τοῦ γυίου
ἤρθρωνται, καὶ περόνην καλευμένην ποιέουσιν,
αὗται ἑωυταῖς ξυμβάλλουσαι ἐν τῷ κυβίτῳ ἐντός.
κάτω δὲ πρὸς τὴν χεῖρα τὸ ὀστέον ἄρθρον ἔχει· αἱ
δὲ περόναι ταύτῃ ἁπαλαὶ ἐοῦσαι, αἱ μὲν δύο οὐκ
ἐξήκουσιν ἐς τὸ ἄρθρον, ἡ δ' ἄνω καὶ ἡ κάτω σὺν τῷ
ὀστέῳ ἤρθρωνται πρὸς τὴν χεῖρα. αἱ δὲ χεῖρες
ἄρθρα ἔχουσι πολλά· ὅσα γὰρ ὀστέα πρὸς ἑωυτὰ
συμβάλλουσι, πάντα ἄρθρα ποιέουσιν. δάκτυλοι
ἄρθρα ἔχουσι, ἕκαστος τρία, ἓν μὲν ὑπὸ τῷ ὄνυχι ἐν
μέσῳ τοῦ τε ὄνυχος καὶ τοῦ κονδύλου, ἄλλο ἐν τῷ
κονδύλῳ, ᾗ καὶ ξυγκάμπτουσι τοὺς δακτύλους, ἄλλο
τρίτον, ᾗ ὁ δάκτυλος ἀπὸ τῆς χειρὸς ἀποπέφυκεν.

Ἐν δὲ τοῖσιν ἰσχίοισιν ἄρθρα δύο εἰσὶν αἱ κοτύ-
λαι καλεύμεναι, καὶ οἱ μηροὶ ἐς ταῦτα ἐνήρθρωνται·
παρὰ δὲ τοὺς μηροὺς περόναι δύο παρήκουσιν, ἡ
μὲν ἐντός, ἡ δ' ἐκτός, καὶ ἐς τὸ ἄρθρον οὐδετέρη
ἐξήκει οὐδ' ἑτέρωθεν, ἀλλὰ πρὸς τῷ ὀστέῳ προσπε-
φύκασι πρὸς τῷ μηρῷ. ὁ δὲ μηρὸς ἄνωθεν μέν, ᾗ ἐς
τὴν κοτύλην ἐμβάλλει, δίκαιός ἐστι τοιῆδε δικαιό-
τητι· ἐπὶ μὲν τοῦ ἐντὸς κεκλιμένου τῶν δικραίων ἐπὶ
τοῦ ἄκρου ἐπιπέφυκεν στρογγύλον καὶ λεῖον, ὃ καὶ

the elbow two processes (lateral and medial epicondyles) grow above out of the bone; these are united with the bone (humerus) and articulate with the joint surface (trochlea and capitulum) into the elbow. The two processes lying on the lower side and inclining inward both (coronoid process with the trochlear notch and head of the radius) unite against the process from above, which comes from the limb (humerus), and articulate inside the limb; they form what is called the "process"; they unite with one another inside the elbow. Lower down the bone of the forearm articulates with the hand. The processes there are soft, and two of them are not at the joint (styloid processes?), while the upper one and the lower one articulate with the bone towards the hand. The hands have many joints; for all the bones that unite with one another form joints. The fingers each have three joints: one beneath the nail and the knuckle, another at the knuckle where people bend their fingers, and another third one where the finger grows out of the hand.

In the pelvis are two joints called "cups" (acetabula), and the thigh-bones are attached into these. Alongside the thigh-bones are two processes, one on the inside (lesser trochanter) and one on the outside (greater trochanter); neither of these is at the joint on either side, but are just attached against the thigh-bone. The upper end of the thigh-bone where it enters the cup is forked in the following way: on the end of the branch of the fork inclining inward grows a smooth sphere, which enters the

ἐς τὴν κοτύλην ἐμβάλλει, τὸ δ' ἕτερον τὸ ἔλασσον
τῶν δικραίων τὸ ἐκτὸς μᾶλλον ἔξω ἐξέχει, καὶ φαίνεται ἐν τῷ πυγαίῳ κάτω, καὶ ἰσχίον καλέεται. πρὸς
δὲ τὸ γόνυ τὸ ὀστέον τοῦ μηροῦ τοιόνδ' ἐστὶ
δίκραιον· τῷ δὲ δικραίῳ τούτῳ τὸ ὀστέον ἡ κνήμη
καλεομένη οἷον ἐν γιγγλύμῳ ἐνήρμοσται· ἄνωθεν δὲ
τοῦ ἐνηρμοσμένου ἡ μύλη ἐπίκειται, ἣ ἀποκωλύει ἐς
τὸ ἄρθρον ἀναπεπτάμενον ἐσβῆναι τὴν ὑγρότητα
τὴν ἀπὸ τῆς σαρκός. παρὰ δὲ τὴν κνήμην περόναι
δύο παρήκουσιν, αἱ κάτωθεν μὲν πρὸς τοῦ ποδὸς ἐς
τὰ σφυρὰ τελευτῶσιν, ἄνωθεν δὲ πρὸς τοῦ γούνατος
[οὐκ]¹ ἐξήκουσι πρὸς τὸ ἄρθρον. πρὸς δὲ τὸν πόδα
ἡ κνήμη κατὰ τὰ σφυρὰ ἄρθρον ἔχει, καὶ ἄλλο
290 κατώτερον | τῶν σφυρῶν· καὶ ἐν τοῖσι ποσὶν ἄρθρα
πολλά, ὥσπερ καὶ ἐν τῇσι χερσίν· ὅσα γὰρ τὰ
ὀστέα, τοσαῦτα καὶ τὰ ἄρθρα, καὶ ἐν τοῖσι δακτύλοισι τῶν ποδῶν τὸν ἀριθμὸν ἴσα κατὰ τὰ αὐτά.

Ἄρθρα δὲ πολλὰ ἐν τῷ σώματι σμικρά, οὐχ
ὁμοίως πᾶσιν, ἀλλὰ ἄλλα ἄλλοις· ταῦτα δὲ τὰ
γεγραμμένα πᾶσιν ὁμοίως εἰσίν, καὶ φλέβες αἱ
γεγραμμέναι πᾶσιν ὁμοίως εἰσίν, ἄλλα τε φλέβιά
εἰσιν ἄλλοις, ἀλλ' οὐκ ἄξια λόγου.

7. Μύξα πᾶσιν ἔνεστι φύσει,² καὶ ὅταν αὕτη
καθαρὴ ᾖ, ὑγιαίνουσι τὰ ἄρθρα, καὶ διὰ τοῦτο εὐκί-
νητά ἐστι, ὥστε ὀλισθαίνοντα πρὸς ἑωυτά. πόνος
δὲ καὶ ὀδύνη γίνεται, ὅταν ἀπὸ τῆς σαρκὸς ὑγρασίη

¹ Del. Potter.   ² A: Μύξαι πᾶσίν εἰσι φυσικαὶ V.

cup; the other lesser outward branch of the fork extends further outward and can be seen in the buttock below; it is called the hip. Towards the knee the thigh-bone is also forked, and in this fork the bone called the shin-bone is fixed, as if in a hinge. Over the joint lies the knee-cap, which prevents moisture out of the tissues from entering the joint when it is in the open position. Alongside the shin-bone are two processes, which, at the lower end towards the foot, end by meeting the ankle-bones (lateral and medial malleolus), and, at the upper end towards the knee, are at the joint (head of the fibula and medial condyle of the tibia). At the foot the shin-bone articulates with the ankle-bones, and again below the ankle-bones (with the calcaneum). In the feet there are many joints, just as in the hands; for there are as many joints as there are bones (in the fingers), and in the same way they are equal in number in the toes.

There are many small joints in the body, not the same in all, but different numbers in different individuals; the joints that have been described, however, are the same in all persons. So too the vessels that have been described are the same in all individuals; but there are small vessels that differ in different persons, but those are not worth discussing.

7. Fluid is present naturally in all the joints, and when it is clean the joints are healthy, and for this reason they move easily and their members slip smoothly over one another. Difficulty and pain arise when moisture flows

ρυῇ πονησάσης τι· πρῶτον μὲν πήγνυται τὸ ἄρθρον,
οὐ γὰρ ὀλισθηρὴ ἡ ὑγρότης ἡ ἐπερρυηκυῖα ἀπὸ τῆς
σαρκός· ἔπειτα, ὥστε πολλὴ λίην γενομένη, καὶ οὐκ
ἀρδομένη ἐκ τῆς σαρκὸς ἀεί, ξηραίνεται, καὶ ὥστε
πολλὴ ἐοῦσα καὶ οὐ χωρεῦντος τοῦ ἄρθρου ἐκρεῖ,
κακῶς τε πεπηγυῖα μετεωρίζει[1] τὰ νεῦρα, οἷσι τὸ
ἄρθρον συνδέδεται, καὶ ἄδετα ποιέει καὶ διαλελυ-
μένα. καὶ διὰ τοῦτο χωλοὶ γίνονται, καὶ ὅταν μὲν
τοῦτο μᾶλλον γίνηται, μᾶλλον, ὅταν δὲ ἧσσον,
ἧσσον.

8. Ἐς δὲ τὴν κοιλίην καὶ τὰ ἐσθιόμενα καὶ τὰ
πινόμενα χωρέουσιν, ἐκ δὲ τῆς κοιλίης ἶνες ἐς τὴν
κύστιν, ἡ διηθεῖ τὸ ὑγρόν, τεταμέναι εἰσίν.

9. Ῥόοι δὲ γίνονται καὶ διαψυχομένης τῆς σαρ-
κὸς λίην καὶ δια|θερμαινομένης καὶ ὑπερφλεγμαι-
νούσης.[2] ῥόοι δὲ διὰ μὲν τὸ ψῦχος γίνονται, ὁπόταν
τόδε γένηται, ὅταν ἡ σὰρξ ἡ ἐν τῇ κεφαλῇ καὶ αἱ
φλέβες τεταμέναι ἔωσιν· αὖται, φριξάσης τῆς σαρκὸς καὶ ἐς μικρὸν ἀφικνουμένης καὶ ἐκθλιψάσης,
ἐκθλίβουσι τὴν ὑγρότητα, καὶ αἱ σάρκες ἅμα αὖται
ἀντεκθλίβουσιν ἐς μικρὸν ἀφικνούμεναι. καὶ αἱ τρίχες ἄνω ὀρθαὶ γίνονται ὥστε πάντοθεν ἅμα ἰσχυρῶς
πιεζόμεναι· ἐντεῦθεν ὅ τι ἂν ἐκθλιβῇ, ῥεῖ ᾗ ἂν τύχῃ.
ῥεῖ δὲ καὶ διὰ τὴν θερμότητα, ὅταν αἱ σάρκες ἀραιαὶ
γινόμεναι διόδους ποιήσωσι, καὶ τὸ ὑγρὸν θερμαν-
θὲν λεπτότερον γένηται (πᾶν γὰρ τὸ ὑγρὸν θερμαινόμενον λεπτότερον γίνεται) καὶ πᾶν ἐς τὸ ὑπεῖκον
ῥεῖ.

out of tissue that has been damaged in some way. First, the joint becomes fixed, since the moisture flowing into it from the tissue is not slippery. Then, as the amount of moisture is too great, and it is not continually renewed out of the tissue, it dries out, and as it is great in amount and the joint does not have room for it, it flows out of the joint, and, congealing badly, raises up the cords by which the joint is held together, and so unbinds and dissolves the joint. This is why persons become lame; when the process has taken place to a greater degree, more lame, when to a lesser degree, less lame.

8. Into the cavity passes what is eaten and drunk. From the cavity bands extend to the bladder, which filters the liquid.

9. Fluxes occur when the tissue is over-chilled, when it is over-heated, or when there is an excess of phlegm. Fluxes that arise from excessive cold occur when, any time there is over-chilling, the tissue in the head and the vessels are stretched tight: as the tissue shivers, contracts, and squeezes out what is in it, the vessels too squeeze out their moisture, and these tissues, contracting, simultaneously squeeze back. Since there is great pressure on them from all sides at once, the hairs stand on end; the fluid squeezed out of this area flows wherever it chances to go. Fluxes also arise from excessive heat, when the tissues, on becoming rarefied, develop passages, and the moisture, being heated, becomes thinner (for every fluid becomes thinner on being heated) and all flows in the direction of least resistance.

---

[1] τὸ ἄρθρον add. A.
[2] V: ὑποφλ- A.

Μάλιστα δ' ὅταν λίην ὑπερφλεγμήνῃ, διὰ τόδε ῥεῖ· αἱ σάρκες λίην[1] ἔμπλεαι γενόμεναι ὅ τι ἂν μὴ δύνωνται χωρέειν, ῥεῖ τὸ ὑγρὸν τὸ μὴ δυνάμενον χωρέεσθαι, ῥεῖ δὲ ᾗ ἂν τύχῃ· ἐπὴν δὲ ἅπαξ εὔροοι αἱ ῥοαὶ γένωνται, ῥεῖ κατὰ[2] χωρίον ᾗ ἂν τύχῃ, ἔστ' ἂν συμπιεχθῶσιν αἱ δίοδοι τοῦ ῥόου. δι' ἰσχνότητα ὅταν τὸ σῶμα ξηρανθῇ· ὥστε γὰρ τὸ σῶμα κοινωνέον αὐτὸ ἑωυτῷ διαλαμβάνει καὶ ἄγει, καθ' ὅ τι ἂν ὑγρὸν ἐπιτύχῃ, ἐς αὐτὸ[3] τὸ ξηρόν· ἄγειν δὲ οὐ χαλεπὸν αὐτό ἐστιν, ὥστε τοῦ σώματος κενοῦ τε καὶ οὐ συνοιδέοντος ὑπὸ ἰσχνότητος. ὅταν δὲ τὰ κάτω ξηρὰ γένωνται, τὰ δ' ἄνω ὑγρά (μᾶλλον δὲ τὰ ἄνω ὑγρά ἐστι τεύχεα, αἱ γὰρ φλέβες ἄνω πλείονές εἰσιν ἢ κάτω, καὶ αἱ σάρκες ἐλάττονος ὑγρότητος δεόμεναι αἱ ἐν τῇ κεφαλῇ), ἄγει δὴ τὸ ξηρὸν τοῦ σώματος τὸ ἐκ τῆς κεφαλῆς ὑγρόν· καὶ ἅμα καὶ δίοδοί εἰσι τῷ ἄγοντι μᾶλλον ἢ τῷ ἀγομένῳ· καὶ γὰρ αὗται κερδαίνουσιν ὥστε ξηραὶ ἐοῦσαι, καὶ ἅμα καὶ τὰ ὑγρὰ πέφυκε κάτω χωρέειν, καὶ ἢν βραχέη τις ἀνάγκη γένηται.

294   10.   Ῥόοι δὲ ἀπὸ τῆς κεφαλῆς ἑπτά· ὁ μὲν κατὰ τὰς ῥῖνας, ὁ δὲ κατὰ τὰ ὦτα, ὁ δὲ κατὰ τοὺς ὀφθαλμούς· οὗτοι οἱ ῥόοι καταφανέες ἐκ τῆς κεφαλῆς τοῖσιν ὀφθαλμοῖσιν. ἐπὴν δ' ἐς τὸν κίθαρον ῥυῇ ὑπὸ ψύχεος, χολὴ γίνεται, μάλιστα δὲ ῥεῖ ἐς τὸν κίθαρον

[1] λίην A: αἱ ἢν μὲν V.
[2] A. Anastassiou, *Gnomon* 52 (1980), p. 314: καὶ τὸ A.

Most especially do fluxes arise on account of an excess of phlegm. The tissues becoming too full are not able to make room for it all, and whatever fluid they are not able to make room for flows wherever it chances to go. When the paths of the flux have once been well opened, it flows to any place it chances, until the passages are pinched in. Fluxes arising as the result of withering, when the body becomes dry: since the body communicates with itself, it divides and draws moisture, wherever it chances upon it, to the dryness. To draw the moisture is not difficult, as the body is hollow, and, on account of its withering, the tissue is not expanded so as to be tight. When the lower parts become dry, but the upper ones are moist (in the upper parts there is more moisture, for the vessels above are greater than those below, and the tissues of the head require less moisture), then obviously the dry part of the body draws moisture out of the head; besides, the passages favour the area that is drawing over the one from which is being drawn, since they, being dry, gain in the exchange; furthermore, moisture naturally moves downwards, even if it is subjected to this compulsion only briefly.

10. Seven fluxes from the head:[a] A. through the nostrils; B. through the ears; C. through the eyes. These fluxes are, as may be clearly seen, from the head. D. When there is a flux to the chest on account of cold, bile is present. Usually such a flux occurs because there is

[a] See *Glands* 11 for a similar series of seven fluxes from the head.

---

[3] Joly: ἑωυτῶ (-τὸ) AV.

ὑπὸ ψύχους διὰ τόδε, ὅτι εὔροον γίνεται ἐς τὸν
βρόγχον, ὥστε οὐδὲ ξυγκεκαλυμμένον· ὑπὸ δὲ τοῦ
ψύχους καὶ κόπος ἔχει διὰ τοῦτο τοὺς ὑπὸ τῆς χολῆς
ἐχομένους, ὅτι αἱ σάρκες, ὅταν χειμὴ ᾖ, οὐκ ἀτρε-
μίζουσιν, ἀλλὰ σείονται, καὶ σειόμεναι μοχθέουσι,
καὶ κοπιῶσιν, ὥστε σειόμεναι ὥσπερ ἐν τῇσιν ὁδοι-
πορίῃσιν· καὶ ἔμπυοι γίνονται, ὅταν ἐς τὸν κίθαρον
ῥέῃ, καὶ φθισιῶντες.

Ὅταν δ' ἐς τὸν μυελὸν ῥόος γένηται, φθίσις
ἀλαΐα[1] γίνεται. ὅταν δ' ὄπισθεν ἐς τοὺς σπονδύλους
καὶ ἐς τὰς σάρκας ῥυῇ, ὕδρωψ γίνεται, καὶ τῷδ' ἐστὶ
γιγνώσκειν, ξηρὰ τὰ ἔμπροσθεν, ἡ κεφαλὴ καὶ
αἱ ῥῖνες καὶ οἱ ὀφθαλμοί· καὶ τοῖσιν ὀφθαλμοῖσι
γίνεται ἀμβλυώσσειν, καὶ χλωροὶ γίνονται καὶ τὸ
ἄλλο σῶμα, καὶ οὐκ ἀποπτύει οὐδέν, οὐδ' ἢν πολὺ
ῥέῃ· ὅδε γὰρ ὁ ῥόος, διὰ τῆς σαρκὸς μέσης ῥέων
τῆς ὄπισθεν καὶ τῆς ἔμπροσθεν ἀπεστραμμένος,
ξηρὰ τὰ ἔμπροσθεν ποιέει, τὴν δ' ὄπισθεν ἄρδει
σάρκα, καὶ τὴν ἐντὸς μᾶλλον πρὸς τὴν κοιλίην, ἢ
ἐκτὸς πρὸς τὴν ῥινόν· διὰ τοῦτο δὲ ἐκτὸς μᾶλλον ἢ
ἐντὸς στερεώτερον τὸ σῶμα, καὶ στενωτέρας διατρήσιας ἔχει· ὥστε δὲ λεπταὶ ἐοῦσαι ξυμπιλέονται, καὶ
ἀκέουσιν αὗται σφίσιν ἑωυταῖς, καὶ ῥόος οὐ δύναται
ταύτῃ ἰέναι οὐδείς· εὐρύτεραι δέ[2] εἰσιν αἱ ἐντὸς καὶ
296 λεπτότερα τὰ μεταξὺ ἔχουσαι. | ὁ δὲ ῥόος, ὥστε ἀφ'
ὑψηλοτέρων, καὶ λεπτὰ τὰ ἀντικωλύοντα ἔχων, ῥεῖ
καὶ πίμπλησιν ὑγρότητος τὰς σάρκας· τὰς ἀπὸ τῶν
σιτίων ἐς τὸ αὐτὸ χωρέουσα ἡ ὑγρότης διέφθαρται·

free flow into the windpipe, since it is not closed on the top. From the cold, weariness too is felt in these patients, who suffer from bile, because their tissues, when it is wintery, do not remain at rest, but shake, and in shaking labour and become weary, because they shake just as in walking. Internal suppurations occur when the flux is to the chest, and also consumptions.

E. When a flux to the marrow occurs, an undetected consumption ensues. F. When there is a flux posteriorly to the vertebrae and the tissues, dropsy arises. This condition is to be recognized by the following: the front regions are dry, the head, nostrils and eyes; dullness of vision comes over the eyes, they become yellow-green, and so too does the rest of the body; the patient expectorates nothing, not even if the flux is great. For this flux, flowing through the middle of the tissue of the back, and turning away from the tissue in the front, leaves the front dry, but waters the tissue in the back, and more so what is inside next the cavity than what is outside towards the skin. Because of this the body is firmer outside than it is inside, and outside it has narrower channels; being fine, these channels are squeezed together and adhere to themselves, so that no flux can pass through them. Inside, on the other hand, the channels are wider and what is between them is less firm; the flux, then, coming from a higher level and encountering no firm resistance, flows and fills the tissues with moisture. As for the moistures from foods, the moisture as it arrives at the same place is

---

[1] Foes in note 35, after Cornarius' *tabes occulta ac inconspicua*: ἄλλη AV.   [2] Schubring: αἵδ' εὑρ. τε AV.

διεφθαρμένη δ' αὐτὴ ὑπὸ τῆς συμμίξιος, καὶ τὸ ῥέον
σὺν αὐτῇ ἀπὸ τῆς κεφαλῆς, τρέφουσι τὸ σῶμα· λίην
δὲ πολλῷ ὑγρῷ αἱ σάρκες τρεφόμεναι καὶ νοσηλῷ
θάλλουσαι[1] ὕδρωπος ἔμπλεαί εἰσιν.

Ἢν δ' ὀλίγον ῥεύσῃ, ἰσχιάδα καὶ κέδματα ἐποίησεν, ἐπὴν ῥέον παύσηται· ὥστε γὰρ ὀλίγον ἐρρυη-
κὸς καὶ πάντοθεν ὠθεύμενον ὑπὸ παντὸς κρέσσονος
ἐόντος ὥστε ὀλίγον ἐόν, καὶ οὐκ ἔχον ἐπιρροὴν
[καὶ][2] ὥστε πάντοθεν ὠθεύμενον, ἐς τὰ ἄρθρα ἀπο-
φυγὴν ποιέεται. γίνεται δὲ κέδματα καὶ ἰσχιάδες
καὶ ἀπὸ τοιούτων νοσημάτων ὑγιῶν γινομένων· ὅταν
τὸ μὲν νόσημα ποιέον ὑγιὲς γένηται, καταλειφθῇ δέ
τι ἐν τῇ σαρκὶ καὶ μὴ ᾖ αὐτῷ ἔξοδος, μήτ' αὖ ἔσω
μήτε[3] ἐς τὸ δέρμα φῦμα ποιήσῃ ἐξιόν, φεύγει ἐς τὸ
ὑπεῖκον, ἐς τὰ ἄρθρα, καὶ ἢ κέδματα ἢ ἰσχιάδα
ἐποίησεν.

11. Ἢν δὲ συνοιδήσωσιν αἱ ῥῖνες, καὶ φλέγματος ἔμπλεαι ἔωσιν συμπεπηγότος, τοῦτο χρὴ τὸ
φλέγμα τὸ συμπεπηγὸς λεπτύνειν ἢ πυρίησιν, ἢ
φαρμάκῳ, καὶ μὴ ἀποτρέπειν· ἢν γὰρ ἀποτρεφθὲν
ἄλλῃ πῃ ῥεύσῃ, ταύτῃ[4] τὸ ῥέον μέζονα νόσον ποιέοι.

12. Ὁπόταν δ' ἐς τὰ ὦτα ῥέῃ, τὸ πρῶτον ὀδύνην
παρέχει, βίῃ γὰρ χωρέει· πόνον δὲ παρέχει, ἔστ' ἂν
ἀποσυριγγωθῇ· ἐπὴν δὲ μάθῃ ῥεῖν, οὐκέτι πόνον
ποιέει. τῷ ὑπὸ τῆς ὀδύνης ἐχομένῳ φάρμακον θερμαῖνον φύσει χλιαρὸν ποιήσαντα, διέντα νετώπῳ,

[1] τε λίην add. A.   [2] Del. Littré.

corrupted by being mixed together with the flux from the head, and these nourish the body—in fact, the tissues, over-nourished by the excessive moisture and swollen by its morbid part, are flooded with dropsy.

G. If the flux is small in amount, it produces sciatica and disease in the joints (*kedmata*) after it has stopped; for inasmuch as the flow was slight and was forced back on all sides by everything being more powerful than it, the flux, being small and not profiting from any additional flux since it is forced back on all sides, makes an escape for itself into the joints. Disease of the joints and sciatica also arise from disorders like these after they have recovered. For when the active disease has gone away, but something is left behind in the tissue and has no way out, and it neither escapes back inside nor comes out and forms a tubercle in the skin, it flees in the direction of least resistance, i.e. to the joints, and it gives rise to either disease of the joints or sciatica.

11. If the nostrils swell shut and become filled with congealed phlegm, you must thin this congealed phlegm with vapour-baths or a medication, and take care not to turn it away; for if, on being turned away, the flux goes somewhere else, it produces a greater disease there.

12. When there is a flux to the ears, first it produces pain, for the flow is violent; this distress continues until the flux gains a pipe through which to escape; when flow is once established, there is no longer distress. For the patient afflicted with this pain warm a naturally warming medication, dissolve it in oil of bitter almonds, and infuse

---

[3] ἔξω add. V.   [4] Potter: παντὶ (-τῃ) AV.

298

ἐγχεῖν, καὶ ὄπισθεν σικύην προσβάλλειν, ἢν τὸ ἀρι-
στερὸν ἀλγέῃ, ἐς τὸ δεξιόν, καὶ | <ἢν τὸ δεξιόν, ἐς>[1]
τὸ σκαιόν· μὴ κατακρούειν δέ, ἀλλ᾽ ὡς ἂν ἕλκῃ μοῦ-
νον· ἢν δὲ μὴ πρὸς ταῦτα παύηται, ψύχοντα ἐγχεῖν
φύσει ψυχρά, καὶ φάρμακον πῖσαι ὅ τι ἂν κάτω
ὑποχώρησιν ποιέῃ, ἄνω δὲ μή, ὥσπερ οὐδ᾽ ἀρήγει
ἐμέειν, καὶ τὰ ἄλλα ψύχειν. καὶ αἰεὶ δὲ ἐκ τοῦ ὑγιὲς
μὴ ποιέοντος τρόπου μεταλλάσσειν· καὶ ἢν μὲν
κάκιον ποιέῃ, ἔρχεο[2] ἐς τὸ ὑπεναντίον· ἢν δὲ ῥέπῃ ἐς
τὸ ὑγιές, τὸ πάμπαν μὴ ἀφελεῖν τι τῶν προσφερομένων, μηδ᾽ ἀποζευγῆσαι, μηδὲ προσθεῖναι ἄλλο τι.
ἢν δὲ σεσυριγγωμένον ἤδη ᾖ, καὶ πεπυωμένος ῥέῃ
ἰχὼρ πολὺς καὶ κακὸν ὀζόμενος, τοῦτο ὧδε ποιέειν·
σπογγία δεύων ξηραίνων τῷ[3] φαρμάκῳ ξηρῷ πρὸς
τὴν ἀκοὴν ὡς πελαστάτω προσθεῖναι, καὶ πρὸς τὰς
ῥῖνας καθαρτήριον, ὅπως, τοῦ ἐς τὰ ὦτα ῥέοντος,
πρόσθεν ἐς τὰς ῥῖνας φέρηται,[4] καὶ μὴ ἐς τὴν κεφαλὴν πάλιν ἀποχωρέῃ, νοσηλὸν ἐόν.

13. Ὅταν δ᾽ ἐς ὀφθαλμοὺς ῥεῦμα ἴῃ, φλεγμαί-
νουσιν οἱ ὀφθαλμοὶ καὶ οἰδέουσιν· τοῦτον χρὴ τῷ
φαρμάκῳ ἢ τῷ ὑγρῷ ἢ τῷ ξηρῷ ἐν παστῷ ἰᾶσθαι·
ἢν δ᾽ εὐθέως φλεγμήνωσι, μὴ ἔγχριε μηδέν, ἀλλ᾽ ἢ
κλύσαι[5] κάτω τῷ ἰσχυροτάτῳ, ἢ ἄλλῳ τινὶ ἀπισχνῆ-
ναι ὑποχωρητικῷ φαρμάκῳ ἀνακῶς,[6] μὴ ἔμετον
ποιήσῃς· ἢν δὲ οἷον λίθοι ὑποτρέχωσιν, φάρμακον

---

[1] Aldina. [2] Mack after Cornarius' *converte*: ἔχου AV.
[3] Joly: ξηραίνοντι (τινὶ) AV.
[4] V: ἀποτρέπηται A.

44

it into the ear. Also apply a cupping-instrument behind the ear; if the pain is in the left ear, on the right side, and <if in the right ear, then on> the left. Do not make incisions, but let the cup alone draw. If with this treatment the pain does not stop, cool some naturally cool substance and infuse it into the ear; let the patient drink a medication that will stimulate a downward movement, but not an upward one, as it is not good for him to vomit; for the rest, cool. (Always change from a course of management that is not leading to recovery; if some measure is making things worse, move to its opposite; if the patient is inclining towards health, remove nothing whatsoever of what is being administered, nor discontinue anything, nor add anything new.) If a pipe has already formed, and purulent fluid is flowing out copious and ill-smelling, do the following. Moisten a sponge, and, drying it with a dry medication, apply it as tightly as possible to the ear; then apply a cleaning medication to the nostrils, in order that what is flowing into the ears will be carried forward into the nose and not move back into the head, for it is peccant.

13. When there is a flux to the eyes, they fill with phlegm and swell up. You must treat this patient with a medication, either moistening or drying, in the form of a powder. If the eyes fill suddenly with phlegm, do not apply ointment, but either administer a good strong enema or reduce the patient's swelling with some other downwards-acting medication carefully, so as not to provoke vomiting. If what seem to be little stones run down

---

[5] Mack after Foes' note 40: καῦσαι AV.

[6] Schubring: φυλασσόμενος AV.

ἐγχρίειν ὅ τι πλεῖστον δάκρυον ἄγειν μέλλει, καὶ τὸ
ἄλλο σῶμα ὑγραίνοντα καὶ φλεγμαίνειν ποιέοντα,
ὡς ὑγρότεροι οἱ ὀφθαλμοὶ γένωνται καὶ ἐκκεκλυσμέ-
νοι, ὡς τὸ δάκρυον συμπεπηγὸς ὑποτρέχειν ποιέῃς.
300 ὅταν δ' ἐς τοὺς ὀφθαλμοὺς κατὰ σμικρὸν ῥέῃ, καὶ
κνιπότητα παρέχῃ, τοῦτον ἐγχρίειν μαλθακώδει, ὅ τι
μέλλει ξηραίνειν ἅμα καὶ δάκρυον ὀλίγον ἄγειν, καὶ
πρὸς τὰς ῥῖνας φάρμακον προσφέρειν ἢ ἑκάστης
ἡμέρης, ἢ διὰ τρίτης, γνώμῃ τῇ αὐτῇ χρώμενος·
τοιόνδε ἔστω τὸ φάρμακον, ὅ τι μὴ πλεῖον ἢ ἐμβά-
φιον ἀπάγειν μέλλει κατὰ τὰς ῥῖνας, ἀπάγειν δὲ
κατὰ σμικρόν, τὸ δὲ κατὰ τοὺς ὀφθαλμοὺς ἀποξηραίνειν, ὡς ὅ τι ἂν τὸ τῶν ὀφθαλμῶν φάρμακον
ἀποξηρήνῃ καὶ ἀποφράξῃ, κατὰ τὰς ῥῖνας ἀποτρά-
πηται. τὰ δὲ φάρμακα τὰ τῆς κεφαλῆς καθαρτήρια,
ἃ μὲν αὐτῶν ἰσχυρά ἐστιν, ἀπὸ τῆς κεφαλῆς ἄγου-
σιν ὅλης· ἅσσα δὲ ἀσθενέα, ἀπὸ τῶν ὀφθαλμῶν, καὶ
αὐτόθεν ἀπὸ τῶν πέλας τῆς ῥινός.

Ἢν δ' ἀπὸ τῆς σαρκὸς καὶ τοῦ ὀστέου, μύξης
ὑποστάσης μεταξὺ τοῦ ὀστέου καὶ τῆς σαρκός,
ῥεῦμα ἐς τοὺς ὀφθαλμοὺς γένηται, τῷδε δῆλόν ἐστιν,
ὅτι ἐντεῦθεν ῥεῖ· τὸ δέρμα τὸ ἐπὶ τῇ κεφαλῇ φλιβό-
μενον ὑπείκει, καὶ ἕλκεα ἐς τὴν κεφαλὴν ἐκθύουσι,
καὶ κατὰ τοὺς ὀφθαλμοὺς δακρύουσι, καὶ οὐχ ἑλ-
κοῦνται τὰ βλέφαρα, οὐδὲ δάκνει, οὐδ' ἀμβλυώσσειν
ποιέει, ἀλλ' ὀξὺ ὁρῶν γίνεται· τὸ γὰρ ῥεῦμα οὐχ
ἁλμυρόν ἐστιν, ὡς οὐκ ἀπὸ τοῦ ἐγκεφάλου, ἀλλὰ
μυξῶδες μᾶλλον. τοῦτον ὧδε χρὴ ἰᾶσθαι· φαρμάκῳ

out of the eyes, smear on whichever medication will best draw tears, and, moistening the rest of the body and making it full of phlegm in order that the eyes will become moister and be washed out, cause the congealed tears to run down. When moisture flows to the eyes a little at a time and produces irritation, anoint this patient with some emollient that will at the same time both dry the eyes and draw a few tears; to the nostrils apply every day or every other day a medication that has the same purpose: let the medication be such as not to draw off through the nostrils more than an oxybaphon, this a little at a time, while at the same time having a drying effect on the eyes, so that anything the medication dries up and turns away from the eyes comes out through the nostrils. Cleaning medications of the head: those that are powerful draw from the whole head, those that are weak, from the eyes and just from the area near the nose.

If—when fluid has collected between the bone and the flesh—a flux occurs from the flesh and the bone to the eyes, the flux's origin is revealed by the following: the skin on the head gives way when pressed and ulcers break out on the head; patients weep from their eyes, but the eyelids do not ulcerate, nor does the flux sting or cause dullness of vision, but the patient sees keenly. The flux is not salty, not being from the brain, but rather serous. You must treat this patient as follows. Clean his head with a

καθαίρειν χρὴ τὴν κεφαλὴν μὴ ἰσχυρῷ, καὶ τὸ
σῶμα ἰσχναίνειν καὶ σιτίοισι καὶ φαρμάκοισι κάτω
ὑπάγοντα, ὡς ἀποξηρανθῇ ἰσχναινομένου τοῦ
σώματος, ἢ ἐκτρεφθῇ τῷ κατὰ τὰς ῥῖνας προστιθε-
μένῳ φαρμάκῳ· πρὸς δὲ τοὺς ὀφθαλμοὺς οὐδὲν δεῖ
φάρμακον προσφέρειν. ἢν δὲ δὴ μηδ' οὕτως ὑγιὴς
γίνηται, τὴν κεφαλὴν κατατάμνειν ἔστ' ἂν πρὸς τὸ
ὀστέον ἴῃς, μὴ μετεώρους μηδ' ἐπικαρσίους τὰς
τομὰς ποιέειν· τάμνειν δὲ ἄχρι τούτου, ἄχρις ἂν τοῦ
ὀστέου θίγῃς· τάμνειν δὲ πυκνά, ὡς ἂν τὸ συνεστηκὸς ἐξέλθῃ θᾶσσον διὰ τῶν ἑλκέων ἀπορρέον, ἅμα
302   δὲ αἱ τομαὶ πυκναὶ ἐοῦσαι πρόστασιν ποιῶσι | τῇ
σαρκὶ πρὸς τὸ ὀστέον. οὕτως ἰᾶσθαι δεῖ· τούτῳ
τοιάδε ἡ ἀποτελεύτησις γίνεται, ἢν μή τις ἐντρε-
πίσῃ· οὐκ ἐκκέκλυσται, ὥστ' ἐκκλυζόμενον ὀξὺ ὁρᾶν
ποιέειν, αἰεὶ τῷ ἐφισταμένῳ μαρμαρυγώδης μᾶλλον
γίνεται, καὶ τὸ ὀξὺ ὁρῶν τοῦ ἀνθρώπου ἀποσβέν-
νυται.

Ἢν δ' ἐς τὴν ὄψιν ἐς τὸ ὑγρὸν καθαρὸν αἱματῶ-
δές τι ἐσέλθῃ ὑγρόν, τούτῳ ἡ ὄψις ἔνδον ἐμφαίνεται
τοῦ ὀφθαλμοῦ οὐ στρογγύλον ἐὸν διὰ τόδε· ἐν ᾧ ἂν
τὸ αἱματῶδες ἐνῇ, τοῦτο οὐκ ἐμφαίνεται, τούτῳ δὴ
ἐλλείπει τὸ φαινόμενον περιφερὲς εἶναι, καὶ προκινέ-
εσθαι αὐτῷ δοκέει πρὸ τῶν ὀφθαλμῶν, καὶ οὐδὲν
κατ' ἀλήθειαν ὁρᾷ. τούτου χρὴ τὰς φλέβας ἀποκαί-
ειν τὰς πιεζούσας τὰς ὄψιας, αἳ σφύζουσιν αἰεὶ καὶ
μεταξὺ τοῦ τε ὠτὸς καὶ τοῦ κροτάφου πεφύκασιν·
καὶ ἐπειδὰν ταύτας ἀποφράξῃς, πρὸς τοὺς ὀφθαλ-

mild medication, and reduce his body's swelling with foods and medications that act downwards, so that the flux will be dried out as the body's swelling goes down, or turned aside by a medication applied to the nostrils. Do not administer any medication to the eyes themselves. If, indeed, the patient fails to recover with this treatment, make an incision into his head right down to the bone; do not make the incisions superficial or at an angle—cut until you touch the bone. Make the cuts close together in order that whatever collects in them will escape more quickly by flowing off through the wounds, and at the same time the cuts, being close together, will make the flesh adhere against the bone. This is the way you must treat. Unless someone treats, the case ends as follows: since the eyes are not being washed out—if the eye is washed out it enables the person to see clearly—the continual addition of fluid causes the person more and more to see sparks and his keenness of vision is extinguished.

If anything bloody gets into the clear moisture in the pupil, the person's pupil forms an image inside the eye that is no longer circular, because wherever the bloody material happens to be the image is not reflected, so that what is seen lacks being spherical, something appears to move before the eyes, and the person sees nothing correctly. You must burn the blood out of the vessels that are pressing on the patient's eyes, i.e. the ones between the ear and the temple that continually throb. When you have turned the flow in these aside, apply to the eyes

μοὺς φάρμακα, ὅσα ὑγραίνει, πρόσφερε, καὶ
δάκρυον ἄπαγε ὡς πλεῖστον, ὅπως τὸ συνεστηκὸς ἐν
τοῖσιν ὀφθαλμοῖσιν ἐκκλυσθῇ τὸ τὴν νοῦσον παρέχον. ἢν δὲ ὁ ὀφθαλμὸς ῥαγῇ, μαλθακοῖσι φαρμάκοισι χρῆσθαι καὶ στρυφνοῖσιν, ὡς στυφόμενον τὸ
ἕλκος ἐς σμικρὸν συνίῃ, καὶ ἡ οὐλὴ λεπτὴ ᾖ. ὅταν
δ᾿ ἄργεμον ᾖ, δακρύειν τῷ ὀφθαλμῷ ἀρήγει.

14. Ὁπόταν δὲ εἰς τὸν κίθαρον ῥέῃ καὶ χολὴ ᾖ,
τῷδε δῆλόν ἐστιν· ὀδύνη ἔχει ἐς τὴν λαπάρην καὶ ἐς
τὴν κληῖδα τὴν ἐς τὴν λαπάρην, καὶ πυρετός, καὶ ἡ
γλῶσσα τὰ ἄνω χλωρὴ γίνεται, καὶ ἀποχρέμπτεται
ξυμπεπηγότα· ταύτης τῆς νούσου ἑβδομαίῳ ὁ κίνδυνός ἐστιν ἢ ἐναταίῳ. [ἄλλη χολή·][1] ὁκόταν ἀμφότερα τὰ πλευρὰ ἀλγέῃ, τὰ δ᾿ ἄλλα ὅμοια ᾖ τῇ ἑτέρῃ,
αὕτη μὲν περιπλευμονίη ἐστίν, ἡ δ᾿ ἑτέρη πλευρῖτις.

Αὗται δὲ γίνονται διὰ τόδε· ὅταν ἐς τὸν πλεύμονα
ῥεύσῃ ἐκ τῆς κεφαλῆς διὰ τοῦ βρόγχου καὶ τῶν
ἀρτηριῶν, ὁ πλεύμων, ἅτε ψαφαρὸς ἐὼν καὶ ξηρὸς
φύσει, ἕλκει ἐφ᾿ ἑωυτὸν τὸ ὑγρὸν ὅ τι ἂν δύνηται·
καὶ ἐπὴν εἰρύσῃ, μέζων γίνεται, καὶ ὅταν μὲν ἐς
ὅλον ῥεύσῃ, μέζων ὁ λοβὸς γενόμενος ἀμφοτέρων
ἔψαυσε τῶν πλευρέων, καὶ περιπλευμονίην ἐποίησεν·
ὅταν δὲ τῆς ἑτέρης μοῦνον, πλευρῖτιν. ἡ περιπλευμονίη πολὺ ἐπικινδυνοτέρη ἐστί, καὶ ὀδύναι πολὺ
ἰσχυρότεραί εἰσιν αἱ ἐς τὰς λαπάρας καὶ ἐς τὰς
κληῖδας, καὶ ἡ γλῶσσα πολὺ ὠχροτέρη, καὶ τὴν
φάρυγγα ἀλγέει ὑπὸ τοῦ ῥεύματος, καὶ κόπος ἔχει

304

medications that moisten, and provoke tears as energetically as possible, in order that what is collected in the eyes and producing the disease will be washed away. If the eye has suffered a rupture, apply mild astringent medications in order that, being drawn together, the lesion will contract and the scar will be small. When an argema is present, it is good for the eye to weep.

14. When there is a flux to the chest and bile is present, this is revealed by the following. Pain is present in the flank and in the corresponding collar-bone; there is fever, the tongue becomes yellow-green on top, and clotted sputum is coughed up. This disease is dangerous on the seventh day, or on the ninth. [Another bile.] It is called pleurisy; if the symptoms are identical except that there is pain on both sides, it is pneumonia.

These arise in the following way. When there is a flux from the head through the windpipe and the bronchial tubes to the lung, the lung, being of loose texture and by nature dry, attracts any moisture that it can. When it has drawn this in, it swells, and when the flux is a complete one, the swollen lobe touches both sides and gives rise to pneumonia; when the lobe touches only one side, it produces pleurisy. Pneumonia is much the more dangerous; the pains to the flanks and the collar-bones are much more violent, and the tongue much deeper yellow-green. There is pain in the throat as a result of the flux, great

---

[1] Del. recentiores.

ἰσχυρός, καὶ πνεῦμα ἑκαταῖον ἢ ἑβδομαῖον λάζεται.
τοῦτον ἢν μὴ ἑβδομαῖον ὁ πυρετὸς ἀφῇ, ἀποθνήσκει, ἢ ἀποπυΐσκεται,[1] ἢ ἀμφότερον· ἢν δ' ἐναταῖον
δύο ἡμέρας διαλιπὼν λάζηται, ὡς τὰ πολλὰ καὶ
οὗτος ἢ ἀποθνήσκει, ἢ ἔμπυος διαφεύγει· ἢν δὲ δω-
δεκαταῖον ἔμπυος γίνεται· ἢν δὲ τεσσαρεσκαιδεκα-
ταῖον, ὑγιὴς γίνεται. καὶ ἔμπυοι ὅσοι ὑπὸ περιπλευμονίης ἢ πλευρίτιδος γίνονται, οὐκ ἀποθνήσκουσιν,
ἀλλ' ὑγιέες γίνονται.

Ὡς τὰ πολλὰ ἔμπυοι γίνονται, ὅταν ῥεῦμα ἐς τὸ
αὐτὸ ὥσπερ ἐπὶ τῇσι χολῇσι γένηται· ἀλλὰ τῇσι
μὲν χολῇσι πολὺ ἀπορρεῖ, καὶ ἀπορρεῦσαν παύεται·
τοῖσι δ' ἐμπύοισιν ἔλασσόν τε ῥεῖ καὶ οὐ παύεται,
καὶ ἔμπυοι γίνονται, ὅταν ἔλασσον ἀποχρέμπτωνται
ἢ ἐπιρρεῖ ἐς τὸν πλεύμονα. τοῦτο γάρ, τὸ ἐν τῷ
πλεύμονι συνιστάμενόν τε καὶ ἐπιρρέον, πῦον γίνε-
ται· τὸ δὲ πῦον συνιστάμενον ἐν τῷ πλεύμονι καὶ ἐν
τῷ κιθάρῳ ἑλκοῖ καὶ σήπει· καὶ ἐπὴν ἑλκωθῇ, ἀπὸ
τοῦ ἡλκωμένου ἐπιρρεῖ· καὶ ἐπαναχρεμπτομένου ἅμα
μὲν ἡ κεφαλὴ μᾶλλον ῥεῖ σειομένη, ἅμα δὲ ἐκ τοῦ
306   ἡλκωμένου ἐν τῷ κιθάρῳ καὶ | τῷ πλεύμονι μᾶλλον
ῥεῖ, καὶ τὰ ἕλκεα κινεύμενα ἐπαναρρήγνυται, ὥστε
καὶ εἰ παύσαιτο τὸ ἀπὸ τῆς κεφαλῆς ῥέον, τὸ ἀπ'
αὐτῶν τῶν ἑλκέων ἱκανὸν ἔσται νοῦσον παρασχεῖν.
γίνεται δὲ καὶ ἀπὸ ἕλκους ἔμπυος, καὶ ῥάων αὕτη ἡ
νοῦσος· γίνεται δὲ καὶ ἐκτὸς τοῦ πλεύμονος μάλιστα
μὲν ἀπὸ ῥήγματος, καὶ ὅταν ἡ σὰρξ φλασθῇ· κατὰ

[1] Littré: -πνίγεται AV.

weariness is felt, and difficult breathing seizes the patient on the sixth or seventh day. If fever does not leave this patient on the seventh day, he dies, or suppurates, or both. If, after intermitting for two days, the fever attacks again on the ninth day, usually this patient, too, dies or, if he escapes with his life, has an internal suppuration. If there is fever on the twelfth day, the patient suppurates internally, if on the fourteenth day, he recovers. Those who suppurate internally as the result of pneumonia or pleurisy do not die, but recover.

Generally internal suppuration arises when a flux occurs to the same spot (i.e. the lung) as in the case of bilious fluxes; but whereas with bilious fluxes much flows away, and when it has flowed away the disease stops, in patients with internal suppuration less flows away and the disease does not stop, and the patients suppurate internally when they cough up less fluid than is flowing to the lung. For what flows together and collects in the lung turns to pus, and this pus collected in the lung and the chest putrefies and causes ulceration; when such an ulceration takes place, new fluid is added from the area that is ulcerated. As this is coughed up, the head, being shaken, increases its flow, and at the same time the flow from the ulcerated area in the chest and the lung increases, and the ulcers, being moved, are torn open again, so that even if the flux from the head were to stop, that from the ulcers alone would be sufficient to maintain the disease. A person can also suppurate internally as the result of an ulcer, and this disease is easier. Suppuration also occurs outside the lung, especially from a tear or when the tissue is

τοῦτο γὰρ πῦον ξυνίσταται, καὶ ξυνιστάμενον, εἴ τις
σείοι τὸ σῶμα, κλυδάζεται, καὶ ψόφον παρέχει, καὶ
καίονται ταύτῃ.[1]

Φθίσις δὲ γίνεται, ὅταν ἐς τὸ αὐτό, ὥσπερ τῷ
ἐμπύῳ, ὁ ῥόος γένηται διὰ τοῦ βρόγχου καὶ τῶν
ἀορτρέων, αἳ ξυνέχουσι τὸν πλεύμονα καὶ τὸν
βρόγχον· ἐς δὲ τὸν πλεύμονα ῥέῃ θαμινὰ κατ' ὀλίγον, καὶ ὑγρότητα ἐν τῷ πλεύμονι μὴ ποιέῃ πολλήν·
ξηραινόμενον γὰρ τὸ ἐπιρρέον ἐν τῷ βρόγχῳ πεπηγός, ὥστε οὐκ ἐκκλυζόμενον, ἀλλὰ κατ' ὀλίγον ἐπιρρέον καὶ ἐνεχόμενον βῆχα ποιέει· ἔν τε τῇσιν ἀορ-
τρῇσιν ἐνεχόμενον τὸ ῥέον, ὥστε στενὰς διατρήσιας
ἐχούσας τὰς ἀορτράς, στενοχωρίην τῷ πνεύματι
παρέχει, καὶ τοῦτο ποιέει πνεῦμα ἔχειν· ὥστε γὰρ
αἰεὶ λειπόμενον αἰεὶ ἐπιθυμέει ἀναπνεῖν· καὶ ἐν τῷ
πλεύμονι, ὥστε οὐκ ἰσχυρῶς ὑγρῷ ἐόντι, ξυσμὸς
ἐγγίνεται· ὅταν δὲ πολὺ ἀπορρυῇ τῆς κεφαλῆς, οὔτ'
ἐν τῷ πλεύμονι ξυσμὸς γίνεται· πολὺ γὰρ αὐτῷ τὸ
ἐπιρρέον ἐστί. καὶ ἔμπυοι ἐκ τῶν φθισίων τούτων
γίνονται, ὅταν ὑγρότερον τὸ σῶμα γένηται· καὶ ὅταν
ξηρότερον γένηται, ἐκ τῶν ἐμπύων φθισιῶντες.
ἔμπυοι τῷδε δῆλοι γίνονται· τὴν λαπάρην ἀρχομέ-
νων πόνος ἔχει· ἐπὴν δὲ πῦον ἤδη ξυνεστήκῃ, ὅ τε
πόνος ὁμοίως ἔχει, βήξ τε γίνεται, καὶ ἐπαναχρέμ-
πτεται πῦον, καὶ πνεῦμα | ἔχει. ἢν δὲ μήπω ἐρρώγῃ,
ἐν τῇ λαπάρῃ σείεται καὶ ψοφέει οἷον ἐν ἀσκῷ· ἢν δὲ
τούτων μηδὲν προσημήνῃ, ἔμπυος δὲ ᾖ, τοισίδε χρὴ

308

[1] Ermerins: ταῦτα AV.

crushed, for pus collects at the spot; when there is a collection of pus, if someone shakes the patient's body, the pus splashes and makes a sound, and that is where they are cauterised.

Consumption arises when, just as in internal suppuration, a flux to the same spot (i.e. the lung) has occurred through the windpipe and the bronchial tubes, which connect the lung and the windpipe, and there is a continual, small flux which does not produce any great amount of moisture in the lung; for what is flowing down gets dried up in the windpipe and congeals, so that it does not splash, but, flowing down a little at a time, is caught and provokes coughing. Now since the bronchial tubes have narrow channels, this flux, being caught in them, produces narrowing of the airway, and this provokes difficult breathing, so that the patient, his breath continually failing, continually desires to inspire. And since there is no great amount of moisture in the lung, irritation is felt; when, on the other hand, the flux from the head is great, there is no irritation in the lung, since the afflux to the lung is copious. Patients suppurate internally after these consumptions when the body becomes more moist; they become consumptive after internal suppuration when it becomes more dry. Internal suppuration is revealed by the following: at the beginning there is pain in the flank; then, after pus has already collected, with the pain continuing the same, coughing begins, pus is expectorated, and breathing becomes difficult. If the pus has not yet broken out of its cavity, if the patient is shaken it makes a sound in the flank as if in a wineskin. If none of these signs is present, but the patient is definitely suppurating inter-

τεκμαίρεσθαι· πνεῦμα πολὺ ἔχει, φθέγγεταί τε ὑπο-
βραγχότερον, καὶ οἱ πόδες οἰδέουσι καὶ τὰ γούνατα,
μᾶλλον δὲ κατὰ τὴν λαπάρην, ἐν ᾗ τὸ πῦον ἔνεστι·
καὶ ὁ κίθαρος συγκεκαμμένος ἐστί, καὶ λυσιγυῖα
γίνεται, καὶ ἱδρὼς περιχεῖται ὅλον τὸ σῶμα, καὶ τοτὲ
μὲν δοκέει θερμὸς αὐτὸς ἑωυτῷ εἶναι, τοτὲ δὲ
ψυχρός· καὶ οἱ ὄνυχες περιτεταμένοι εἰσὶ καὶ ἡ κοιλίη θερμὴ γίνεται· τούτοισι χρὴ γινώσκειν τοὺς
ἐμπύους.

15. Ὅταν δ᾽ ὄπισθεν ῥεύσῃ ἐς τὴν ῥάχιν, φθίσις
γίνεται τούτῳ τοιάδε· τὴν ὀσφῦν ἀλγέει, καὶ τὰ
ἔμπροσθεν τῆς κεφαλῆς κενὰ δοκέουσιν αὐτῷ εἶναι.

16. Χολῇ δὲ τάδ᾽ ἐστὶν ἐπικίνδυνα, ἴκτερος ἢν
ἐπιγένηται, ἐν τοῖσιν ὀφθαλμοῖσι καὶ ἐν τοῖσιν
ὄνυξι πελιδνὰ ὅταν γένωνται, καὶ ἐς τὸ σῶμα ὅταν
ἔχῃ ἕλκεα καὶ τὰ περὶ τὰ ἕλκεα πελιδνὰ ᾖ, καὶ ὁ
ἱδρὼς ὁκόταν μὴ καθ᾽ ὅλον τὸ σῶμα ἐκθύῃ, ἀλλὰ
καθ᾽ ἓν μέρος τοῦ σώματος, καὶ ὅταν τοῦ πυρετοῦ
ἔτι ὄντος ἐπαναχρέμπτηται χλωρόν, ἤ, ἐόντος ἐντὸς
ἐν τῷ πλεύμονι ἔτι τοῦ χλωροῦ, ἡ ἐπανάχρεμψις
παύσηται· τοῦτο δεῖ γινώσκειν ὅταν ἐνῇ καὶ ὅταν μὴ
ἐνῇ· ὅταν ἐνῇ, ἐμψοφεῖ ἐν τῇ φάρυγγι ἀναπνέοντος,
καὶ πνεῦμα ἐπικίνδυνον, καὶ λύγξ, καὶ ὁ πυρετὸς ἔτι
ὤν, ἀποχρέμματος ἔτι ἐν τῷ πλεύμονι ἐνεόντος, καὶ
ἡ κοιλίη ἀσθενέος ἤδη ἐόντος ὑποχωρέουσα· ταῦτα
πλευρίτιδος καὶ περιπλευμονίης ἐπικίνδυνα.

17. Πλευρῖτιν ὧδε χρὴ ἰᾶσθαι· τὸν πυρετὸν μὴ
παύειν ἑπτὰ ἡμερέων, πότῳ χρῆσθαι ἢ ὀξυμελι-

nally, you must form your judgement from the following: there is great difficulty of breathing, the patient's voice is somewhat hoarser than normal, and his feet and knees swell, especially on the side where the pus is located; the chest is compressed, there is slackness of the limbs, sweat pours out over his whole body, and sometimes he feels warm, sometimes cold; the digits become clubbed, and his cavity is hot. These are the signs by which you must recognize that patients are suppurating internally.

15. When there is a flux posteriorly into the back, it produces a consumption of the following nature: the patient suffers pain in the loins, and he has the sensation that the front of his head is empty.

16. As far as bile goes, the following disorders are dangerous, if jaundice comes on: when there is lividness in the eyes and under the nails; when there are ulcers on the body, and the areas around the ulcers are livid; when sweat breaks out not over the whole body but in one part of it; when, with fever still present, yellow-green material is coughed up, or, with yellow-green material still present in the lung, expectoration stops. You must recognize when such material is present and when it is not: when it is present, a noise is heard in the throat when the patient breathes in, breathing is dangerously difficult, the patient hiccups, his fever persists, material that should be coughed up remains in his lung, and the cavity of the already weakened patient has a downward motion. These are the dangerous signs of pleurisy and pneumonia.

17. You must treat pleurisy thus. Do not check the fever for seven days. As drink give either oxymelicrat or

κρήτῳ, ἢ ὄξει καὶ ὕδατι· ταῦτα δὲ χρὴ προσφέρειν
ὡς πλεῖστα, ὡς ἐπίτεγξις γένηται καὶ γινομένη |
310 ἐπανάχρεμψιν ποιέῃ· καὶ τὴν ὀδύνην παύειν θερμαντηρίοισι φαρμάκοισι, καὶ καταρροφεῖν διδόναι ὅ τι
ἐπανάχρεμψιν ποιήσει, καὶ λουτροῖσι χρῆσθαι
τεταρταίοισιν· τῇ δὲ πέμπτῃ καὶ τῇ ἕκτῃ χρίειν
ἐλαίῳ· τῇ δὲ ἑβδόμῃ λούειν, ἢν μὴ ὁ πυρετὸς μέλλῃ
ἀφιέναι, ὡς ὑπὸ τοῦ λουτροῦ ὁ ἱδρὼς ἐγγένηται· καὶ
ἔτι τῇ πέμπτῃ καὶ τῇ ἕκτῃ ἰσχυροτάτοισι χρῆσθαι
τοῖσιν ἐπαναχρεμπτηρίοισι φαρμάκοις, ὡς τὴν
ἑβδόμην ὅτι ῥήϊστα ἡμέρην ἀγάγῃ· ἢν δὲ μηδὲ τῇ
ἑβδόμῃ ἡμέρῃ παύσηται, τῇ ἐνάτῃ παύσεται, ἢν μή
τι ἄλλο τῶν ἐπικινδύνων γένηται· ἐπὴν δὲ ὁ πυρετὸς
ἀφῇ, ῥοφήματα ὡς ἀσθενέστατα ποιέων προσφέ-
ρειν· ἢν δὲ ἰνηθμὸς[1] ἐγγένηται, ἢν μὲν ἔτι θερμοῦ[2]
τοῦ σώματος ἐόντος, τῶν ποτῶν ἀφαιρέειν· ἢν δὲ τοῦ
πυρετοῦ ἀφεικότος, πυρίνοις τοῖσι ῥοφήμασι χρῶ.
καὶ τὴν περιπλευμονίην τὸν αὐτὸν τρόπον ἰῶ.

18. Τοὺς ἐμπύους καθαίρειν τὴν κεφαλὴν μὴ
ἰσχυροῖσι φαρμάκοισιν, ἀλλὰ κατὰ μικρὸν ἀποτρέ-
πειν ἐς τὰς ῥῖνας, καὶ ἅμα διαχωρητικοῖσι σιτίοισι
χρῆσθαι· καὶ ἐπὴν ἡ ἀρχὴ τοῦ νοσήματος μηκέτι ᾖ,
ἀλλ’ ἐκτρέπηται ὁ ῥόος, ἐπανάχρεμψιν ποιέεσθαι,
καὶ βῆχα ποιέειν, καὶ ἐγχύτοισι φαρμάκοισι χρῆσθαι καὶ σιτίοισιν ἅμα· ὁπόταν δὲ δέῃ ἀπόχρεμψιν
ποιέεσθαι, καὶ πλέοσι σιτίοισι καὶ ἀλυκοῖσι χρῆσθαι καὶ λιπαροῖσι, καὶ οἴνῳ αὐστηρῷ, καὶ βῆχα
ποιέειν ὅταν ὧδε ἔχῃ.

vinegar and water; these you should administer in very generous amounts, in order that moistening occurs and leads to expectoration. Check the pain with warming medications. As gruel give what will promote expectoration. On the fourth day give baths, on the fifth and sixth days anoint with olive oil, and on the seventh day wash—unless the fever is about to remit—in order that through bathing sweating will be induced. Begin even on the fifth and sixth days to give powerful expectorant medications, in order that on the seventh day the patient will bring up his sputum as easily as possible. If the fever does not cease on the seventh day, it will do so on the ninth, unless some other danger intervenes. When the fever has remitted, prepare and give a very mild gruel. If purging occurs, should the body still be warm withhold drinks, but if fever is no longer present, give wheaten gruels. Also treat pneumonia in the same way.

18. In patients that suppurate internally, clean the head—not with powerful medications, but by turning away the flux a little at a time to the nostrils—and at the same time prescribe laxative foods. When the beginning of the disease has passed and the flux has turned outward, provoke expectoration and coughing, and at the same time infuse medications and give food. When you must provoke expectoration, give copious salty rich foods and dry wine, and stimulate coughing, when the case is as described.

---

[1] Froben: ἰνθμὸς A: νυγμὸς V.

[2] Littré: νεαροῦ AV.

19. Καὶ τοὺς ὑπὸ τῆς φθίσιος τὸν αὐτὸν τρόπον τἆλλα, πλὴν τὰ σιτία μὴ πολλὰ ἅμα, καὶ τὰ ὄψα μὴ πλέονα ἢ τὰ σιτία, καὶ τῷ | οἴνῳ ὑδαρεῖ χρῶ ἐπὶ τῷ σιτίῳ, ὡς μὴ θερμαίνῃ, καὶ τῷ σώματι ἀσθενεῖ ἐόντι θερμότητα παρέχῃ, καὶ ἅμα ἀμφότερα θερμαίνωσιν ἐν τῷ αὐτῷ χρόνῳ καὶ θερμωλὴν ποιέωσιν.

20. Ῥεῦμα πολὺ ὁπόταν διὰ τοῦ οἰσοφάγου ἐς τὴν κοιλίην ῥεύσῃ, ἴνησις γίνεται κάτω, ἔστι δ' ὅτε καὶ ἄνω· τούτῳ ἢν μὲν ὀδύνη ἐνῇ ἐν τῇ γαστρί, ὑπεξάγειν πρῶτον φαρμάκῳ ἢ χυλῷ, ἔπειτα φαρμάκῳ ἰσχητηρίῳ χρῆσθαι, τοῖσι δὲ σιτίοισι διαχωρητικοῖσιν ἕως ἂν ἡ ὀδύνη ἔχῃ· ἐπὴν δὲ παύσηται ἡ ὀδύνη, καὶ τοῖσι σιτίοισιν ἰσχητηρίοισι χρῆσθαι· τὸν αὐτὸν δὲ τρόπον καὶ ἐπὴν πολλὰς ἡμέρας ἴνησις ἔχῃ, ἰᾶσθαι· ἢν δὲ ἀσθενὴς ᾖ καὶ μὴ δύνηται προσφέρεσθαι ὑπὸ ἀσθενείης, κλύζειν πρῶτον μὲν χρὴ χυλῷ πτισάνης, ἔπειτα ἐπὴν τούτῳ καθήρῃς, τῶν στυφόντων τινί [ὕστερον ἐπὴν τοῦτο κλύσῃς].[1]

21. Ὁπόταν δ' ἐς τὴν σάρκα ὄπισθεν παρὰ τοὺς σπονδύλους ῥεῦσαν ὕδρωπα ποιήσῃ, ὧδε χρὴ ἰᾶσθαι· καίειν τὴν σάρκα τὴν ἐν τῷ τραχήλῳ μεταξὺ τῶν φλεβῶν ἐσχάρας τρεῖς, καὶ ἐπὴν καύσῃς, ξυνάγειν καὶ ποιέειν ὡς ἰσχνοτάτας οὐλάς· καὶ ἐπὴν ἀποφράξῃς, πρὸς τὰς ῥῖνας φάρμακον πρόσφερε, ὡς ἐκτρέπηται, καὶ ἀσθενὲς αὖθις καὶ αὖθις, ἕως ἂν ἀποτρεφθῇ· καὶ τὰ μὲν ἔμπροσθεν τῆς κεφαλῆς θέρ-

[1] Del. Littré.

19. Treat patients with consumption generally the same, except do not give large amounts of cereals at one time; let the quantity of main dishes not exceed that of the cereals; after the meal give wine well-diluted with water in order that it will not be warm and provoke heating in the weakened body, and both of them together (i.e. the wine and the body) become warm at the same time and cause feverish heat.

20. When a massive flux passes through the oesophagus into the cavity, there is an evacuation downwards, and sometimes also upwards. If there is pain in the person's belly, first clean downwards with a medication or a juice, and then give an astringent medication and laxative foods for as long as the pain is present. When the pain stops, prescribe astringent foods as well. Treat in this same way even when the evacuation continues for many days; if the patient is weak and cannot take anything because of his weakness, you must first give him an enema of barley-water, and then, when you have purged with that, some astringent medication.

21. When a flux into the tissue at the back along the vertebrae produces dropsy, treat as follows. Burn three eschars in the tissue of the neck between its vessels, and after you have cauterized, draw the edges of the wound together and make them as flat as possible. Having thus blocked the flux's passage, apply a medication to the nostrils—a weak one administered repeatedly—in order that the flux will be turned in that direction, and continue until it turns aside; warm the front of the head and cool

μαινε, τὰ δ' ὄπισθεν ψύχε· καὶ ἐπήν σοι ἐκτεθερμασμένος ᾖ τὰ ἔμπροσθεν τῆς κεφαλῆς, σιτία ἐσθίειν τὰ φλεγματωδέστατα καὶ ἥκιστα | διαχωρητικά, ὡς ὅτι μάλιστα διευρυνθῶσιν αἱ ῥοαὶ αἱ ἔμπροσθεν τῆς κεφαλῆς· ἔπειτα δ' ἐπὴν ἀποφράξῃς καὶ ἀποτρέψῃς τὴν ἐπίρρυσιν, ἤν τι πρὶν ἢ εὐτρεπίζειν τὸ ῥεῦμα ἐς τὸ σῶμα ἔλθῃ, ὧδε χρὴ ἰᾶσθαι· ἢν μὲν μᾶλλον πρὸς τὸ δέρμα ἐκκεχωρηκὸς ᾖ, τὰ ἔξωθεν πυριῶντα· ἢν δ' ἔνδον πρὸς τὴν κοιλίην, ἔξωθεν δὲ μὴ δῆλον ᾖ, φάρμακον πιπίσκοντα· ἢν δ' ἐπ' ἀμφότερα, ἀμφοτέρων ἀφαιρέειν· ἐπιτηδεύειν δὲ χρὴ ἐγγυτάτην ἔξοδον ποιέειν, ἤν τε κάτω, ἤν τε ἄνω, ἤν τε ἄλλη ὅπη τοῦ σώματος ἔξοδοί εἰσιν.

22. Ὁπόταν ἰσχιὰς ἀπὸ ῥόου γένηται, σικύην χρὴ προσβάλλειν, καὶ ἕλκειν ἔξω, καὶ μὴ κατακρούειν, καὶ ἔνδοθεν θερμαντήρια φάρμακα πιπίσκοντα διαθερμαίνειν, ὅπως ἔξοδος ᾖ καὶ ἔξω ἐς τὸ δέρμα ὑπὸ τῆς ἑλκύσιος τῆς σικύης, καὶ ἐντὸς πρὸς τὴν κοιλίην ὑπὸ τῆς θερμασίης· ὁπόταν γὰρ ἀποφραχθῇ καὶ μὴ ἔχῃ ὅπη ὁδοιπορέῃ, ὁδοιπορέουσα ἐς τὰ ἄρθρα ῥέει ἐς τὸ ὑπεῖκον, καὶ ἰσχιάδα ποιέει.

23. Ἡ ὄπισθεν φθίσις· τούτῳ τὴν κεφαλὴν καθαρτέον ἀσθενεῖ φαρμάκῳ, ἕως ἂν ἀποτρεφθῇ ὁ ῥόος, καὶ τῇ διαίτῃ ὥσπερ ἔμπροσθεν χρῶ, φάρμακον δὲ πίσον ἐλατήριον, καὶ κάτω γάλακτι κλύσον, τὰ δ' ἄλλα πυρίῃσιν ἰῶ.

24. Ὕδωρ ἐς τὸ ἐπίπλοον· ἐπὴν ὁ σπλὴν ὑπὸ πυρετοῦ μέγας γένηται, γίνεται δὲ ὅταν τὸ σῶμα

the back of it. When you have succeeded in warming the front of the head, let the patient eat foods that are the most productive of phlegm and the least laxative, in order that the channels at the front of the head will dilate as much as possible. Then, after you have blocked the flux and turned it aside, if before being treated some of the flux had already gone to the body, you must treat as follows. If the flux moved mainly outwards to the skin, apply vapour-baths externally; if it moved inside to the cavity, and is not evident from the outside, have the patient drink a medication. If the flux moved in both directions, remove it from both. You must take care to establish the exit as close to the source as possible, no matter whether you are employing the downward pathway, or the upward, or any of the other exits of the body.

22. When sciatica arises from a flux, you must apply a cupping instrument and draw off fluid without making any punctures. Warm the patient internally by having him drink warming medications, in order that there will be both an exit externally into the skin by the drawing of the cupping instrument, and an exit internally into the cavity brought about by heating. For when fluid is blocked up and has nowhere else to go, it goes into the joints, flowing in the direction of least resistance, and produces sciatica.

23. Consumption in the back. In this patient, the head is to be cleaned out with a mild medication continued until the flux is turned aside. Employ the same regimen as in the case before, have the patient drink a diuretic medication, clean him downwards with an enema of milk, and for the rest treat him with vapour-baths.

24. Water entering the omentum. When the spleen is enlarged as the result of fever—this happens simultane-

λεπτυνθῇ, τοῖσι γὰρ αὐτοῖσιν ὅ τε σπλὴν θάλλει καὶ
τὸ σῶμα φθίνει· ὅταν δὲ τὸ σῶμα λεπτὸν ᾖ καὶ ὁ
σπλὴν θάλλῃ καὶ τὸ ἐπίπλοον ἅμα τῷ σώματι λεπτυνθῇ, ἡ πιμελὴ ἡ ἐν τῷ ἐπιπλόῳ ἐστί, τήκεται· ἐπὴν
δὲ ταῦτά τε κενὰ πιμελῆς γένηται καὶ ἀπὸ τοῦ
σπληνὸς θάλλοντος ἀπορρέῃ ἐς τὸ ἐπίπλοον, ὡς
ἐγγύτατα ἐὸν τὸ ἐπίπλοον, ὥς τε τεύχεα ἔχον καὶ
ταῦτα κενεά, ἐσδέχεται· καὶ ἐπὴν τὸ | νόσημα ἅπαξ
ἐν τῷ σώματι γένηται, ἐς τὸ νοσέον τρέπεται, ἢν μή
τις εὐτρεπίζῃ, ὡς καὶ τὸ εὐτρεπιζόμενον ἐπικίνδυνον.
τοῦτον ὧδε ἰᾶσθαι· φάρμακα πιπίσκειν ὑφ' ὧν ὕδωρ
καθαίρεται, καὶ σιτία τὰ φλεγματωδέστατα διδόναι·
ἢν δὲ μηδ' οὕτω ῥᾴων γένηται, καίειν ὡς λεπτότατα
καὶ ὡς ἐπιπολαιότατα, ὅπως τὸ ὕδωρ ἴσχειν δύνῃ,
πέριξ τοῦ ὀμφαλοῦ κύκλον, καὶ ἐς τὸν ὀμφαλὸν μή,[1]
καὶ ἀφιέναι ἑκάστης ἡμέρης. τῶν νοσημάτων ὅ τι
ἂν ἐπικίνδυνον παραλάβῃς,[2] ἐν τούτοισι παρακινδυ-
νεύειν χρή· ἐπιτυχὼν μὲν γὰρ ὑγιᾶ ποιήσεις, ἀτυχή-
σας δέ, ὅπερ καὶ ὡς ἔμελλε γίνεσθαι, τοῦτ' ἔπαθεν.

25. Παιδίῳ δὲ χρὴ ὕδρωπα ὧδε ἰᾶσθαι· τὰ οἰδέ-
οντα καὶ ὕδατος ἔμπλεα ἐξοίγειν μαχαιρίῳ πυκνὰ
καὶ σμικρὰ ἐξοίγοντα, ἐξοίγειν δ' ἐν μέρει ἑκάστῳ
τοῦ σώματος, καὶ πυρίῃσι χρῆσθαι, καὶ αἰεὶ τὸ ἐξοι-
γόμενον χρίειν θερμαντηρίῳ φαρμάκῳ.

26. Πλευρῖτις ξηρή ἄνευ ῥόου γίνεται ὅταν ὁ
πλεύμων λίην ξηρανθῇ ὑπὸ δίψης ἀναγκαίης· ὁ γὰρ
πλεύμων, ἅτε ξηρὸς ἐών, ἐπήν τι μᾶλλον ξηρανθῇ

[1] Linden after Foes in note 76: μίαν AV.

ously with the body becoming lean, since the same process makes both the spleen swell and the body waste—when the body is lean and the spleen swells and the omentum wastes along with the body, then the fat that is in the omentum melts. When these parts have been emptied of fat, and there is a flux from the swollen spleen into the omentum, its neighbouring organ, the omentum, containing cavities and these being empty, accepts the flux. When the disease has once developed in the body, everything goes to the ailing part, unless someone takes up the treatment, for even cases that are treated are dangerous. Treat this patient as follows. Have him drink medications that will clean out water, and give foods that are the most productive of phlegm. If he does not improve even with this treatment, cauterize very lightly and superficially in order that the eschar will be able to hold the water—about the umbilicus in a circle, but not in the umbilicus itself—and draw this off each day. In any of the dangerous diseases you take on, you must accept some degree of risk: for if you are lucky, you will make the patient well, but if you fail, he only suffers what was likely to have happened anyway.

25. In a child you must treat dropsy as follows. Open the areas that are swollen and full of water by making numerous small incisions with a scalpel; make these openings in every part of the body. Administer vapourbaths, and anoint all the openings with a warming medication.

26. Dry pleurisy without any flux arises when the lung dries out too much, as the result of an involuntary thirst. For when the lung, which is a dry part, becomes even

---

[2] A: ἐπικινδυνότατόν ἐστιν V.

τῆς φύσιος, ἰσχνὸς γίνεται, καὶ ἀκρατὴς γενόμενος,
κλιθεὶς ἐς τὸ πλευρὸν ὑπ' ἀκρασίης, ψαύει τοῦ πλεύ-
ρου· καὶ ἐπὴν θίγῃ ὑγροῦ ἐόντος, ἅπτεται, καὶ πλευρῖτιν ποιέει· τότε δὲ καὶ ὀδύνη γίνεται ἐς τὸ πλευρὸν
καὶ ἐς τὴν κληῖδα, καὶ πυρετός, καὶ ἐπαναχρέμπτε-
ται λευκόν. τοῦτον χρὴ πολλοῖσι πότοισιν ἰᾶσθαι,
καὶ λούειν, καὶ τῆς ὀδύνης φάρμακον διδόναι καὶ
τἆλλα τὰ ἀνάχρεμψιν ποιεῦντα· | οὗτος ἐν ἑπτὰ ἡμέ-
ρῃσιν ὑγιὴς γίνεται, καὶ τὸ νόσημα ἀκίνδυνόν ἐστι,
καὶ σιτία οὐ χρὴ διδόναι.

27. Πυρετοὶ διὰ τόδε γίνονται, ὅταν τοῦ σώματος
ὑπερφλεγμήναντος αἱ σάρκες ἀνοιδήσωσιν, καὶ τὸ
φλέγμα καὶ ἡ χολὴ κατακλεισθέντα ἀτρεμίζωσι, καὶ
μὴ ἀναψύχηται μηδὲν μήτ' ἐξιὸν μήτε κινεύμενον,
μήτ' ἄλλου ὑπιόντος. ὁπόταν κόπος ἔχῃ καὶ πυρετὸς καὶ πλησμονή, λούειν χρὴ πολλῷ, καὶ χρίειν
ὑγρῷ, καὶ θερμαίνειν ὡς μάλιστα, ὡς ἡ θερμωλή,
ἀνοιχθέντος τοῦ σώματος, ὑπὸ τοῦ ἱδρῶτος ἐξέλθῃ·
ἑξῆς δὲ ταῦτα ποιέειν καὶ τρεῖς καὶ τέσσαρες ἡμέ-
ρας· καὶ ἢν μὴ παύηται, φάρμακον πῖσαι χοληγόν,
καὶ <μὴ>[1] ψύχειν τὸν πυρετόν, πρὶν ἢ τεταρταῖος ᾖ,
μηδ' ἕως ἂν τὸ σῶμα θάλλῃ, πιπίσκειν φάρμακον·
οὐ γὰρ ἰνῶνται[2] εἰ μὴ σμικρόν, ὥστε συνοιδέοντος
τοῦ σώματος· ἐπὴν δὲ ἰσχνὸς ᾖ, πιπίσκειν, καὶ ἰνή-
σεται. πυρετῷ σιτίον μὴ προσφέρειν, μηδὲ ῥοφήμα-
σιν ὑπεξάγειν, καὶ ποτὸν ὕδωρ θερμὸν καὶ μελίκρη-
τον καὶ ὄξος σὺν ὕδατι, ταῦτα δὲ πιπίσκειν ὡς
πλεῖστα· ἢν γὰρ ψυχρὸν μὴ ἐσίῃ, τὸ ποτὸν θερμὸν

drier than it naturally is, it withers, becomes weak, and, because of its weakness, leans against the side and comes into contact with it. When it touches the side, a moist part, it fastens itself to it and this produces pleurisy. Then pain invades the side and the collar-bone, there is fever, and white sputum is expectorated. Treat this patient with frequent drinks, wash him, give him a medication for the pain, and do the other things that promote expectoration. He recovers in seven days, and the disease is not dangerous; you need not give foods.

27. Fevers arise in the following way: when from an excess of phlegm in the body the tissues swell up, when phlegm and bile are closed in and become immobile, and when nothing is cooled by going out, or by moving, or by anything else passing off. When there are weariness, fever and fullness, you must wash with copious water and anoint with oil, and warm as much as possible in order that, as the body is opened, the fever heat will make its exit with the sweat. Do this successively for three or four days. If the fever does not stop, have the patient drink a medication that draws bile. Do not cool the fever until the fourth day, nor give the patient a purgative medication to drink as long as the body is swollen, for the evacuations will not be effective, except perhaps faintly, since the body is swollen up tight. After the swelling has gone down, give a medication to drink, and he will be evacuated. In a fever do not give cereal, nor evacuate downwards with gruels; as drink give warm water, melicrat and vinegar with water, and have the patient drink generous amounts; for if no cold enters the body, the drink, being

---

[1] Littré. [2] A: κρίνονται V.

ἐὸν καὶ μένον ἐκ τοῦ σώματος τοῦ νοσέοντος ἀφαι-
ρέει, ἤν τε διουρήσῃ, ἤν τε διϊδρώσῃ· πάντῃ δὲ
ἀνοιγόμενόν τε καὶ ἀναπνέον καὶ κινεύμενον τὸ
σῶμα συμφέρον πρήσσει. ἐπὴν δὲ ἰσχνὸν ὄντα
καίῃ, δῆλον ὅτι οὐ διὰ τὸ φλεγμαίνειν ὁ πυρετὸς
ἔχει· καὶ ἢν μὴ παύηται, τρέφειν, καὶ φλεγμαίνειν
ποιέειν· καὶ ἢν μηδ' οὕτω ξυμφέρῃ, δῆλον ὅτι οὐκ ἡ
ἰσχνίη[1] τὸν πυρετὸν παρέχει· τοῦτον χρὴ φάρμακον
πῖσαι, ὡς ὑπεξάγῃ, ὅπῃ ἂν μᾶλλον ὁ πυρετὸς ἔχῃ,
ἤν τ' ἄνω, ἤν τε κάτω, ἢν | μὲν ἄνω, ἄνω, ἢν δὲ
κάτω, κάτω. οὐδὲν δ' ἧσσον δεῖ τοὺς ἀσθενέας τῶν
ἰσχυρῶν φαρμάκων πιπίσκειν, ἀλλ' ὁμοίως ἢ μοῦνον
οὕτω, τοῖσι μὲν ἰσχυροῖς ἰσχυρόν, τοῖσι δ' ἀσθενέ-
σιν ἀσθενές. τὰς δὲ πυρώσιας ποτοῖσι καὶ ῥοφήμα-
σιν, ὥστε τὸν πυρετὸν ψυκτηρίῳ φαρμάκῳ ἐκλύειν,
καμμάρῳ ἢ ἄλλῳ τινὶ τοιούτῳ· καὶ ἐπὴν ⟨μὴ⟩
λύσῃς[2] τῷ ψυκτηρίῳ, θερμαντηρίοισι χρῶ ἑξῆς·
ἐπὴν δὲ μὴ παύηται, ψυκτηρίοισι πάλιν χρῆσθαι.

28. Ἴκτερον ὧδε χρὴ ἰῆσθαι· ἐπὴν παραλάβῃς,
τρέφε, καὶ λουτροῖσι καὶ πιαντηρίοισι καὶ ποτοῖσι
καὶ σιτίοισι καθυγραίνειν ἢ τρεῖς ἢ τέσσαρας ἡμέ-
ρας· ἐπὴν δ' ὑγρανθῇ τὸ σῶμα, καθαίρειν καὶ
ξηραίνειν τὸ σῶμα, λιπαρά τε ἐξαίφνης ἐξαρύσαι,
πάντῃ προσφέρων φάρμακον ᾗ δυνατὸν ὑγρότητα
ἐξάγειν· πρὸς δὲ τὴν κεφαλὴν καθαρτηρίῳ ἀσθενεῖ·
καὶ οὐρητικὰ πιπίσκειν· καὶ πρὸ τῶν σιτίων τοῦτον

---

[1] ἡ ἰσχνίη Potter: ἐχρῆν AV.
[2] Littré: λύσῃς A: ναυτιῶτο V.

warm and staying in place, will take something away from the ailing body, whether it passes off in the urine or as sweat, and the body, being opened, taking in breath and moving on every side, fares better. When the body is feverish without being swollen, clearly the fever is not being caused by swelling. If such a fever does not go away, feed the body and make it swell up with phlegm. If this does not help either, it is clear that thinness is not producing the fever. Have this patient drink a medication that will evacuate from the region that the fever is mainly occupying, whether that be the upper or the lower; i.e. if the fever is mainly in the upper region, give a medication to evacuate upwards, if mainly in the lower region, a medication to evacuate downwards. Weak patients need to drink a medication just as much as do strong ones, only whereas you give strong patients a strong medication, give weak patients a weak one. Relieve feverish conditions with drinks and gruels just as you relieve a fever with a cooling medication, employing aconite or something else of the same kind. When you do not succeed in relieving it with the cooling medication, use in turn heating ones. When that does not bring a stop, revert to cooling ones.

28. You must treat jaundice as follows. When you first take on the patient, build him up and moisten him thoroughly for three or four days with baths, with substances that fatten, with drinks, and with foods. When the body has been moistened, clean and dry it, and all at once draw off its fullness by applying medication everywhere that moisture can be drawn out; for the head use a mild cleaning medication, and have the patient drink diuretics. At

τὸν χρόνον, ὃν¹ καθαίρεις τὴν τεταραγμένην ὑγρότητα, κατάποτον δίδου, ὡς μὴ τρέφηται ἀπὸ τούτου
322 τοῦ ὑγροῦ² τὸ σῶμα· ὅταν δὲ | ἰσχναίνηται, καὶ λου-
τροῖσι κάθαιρε· τοῦ δὲ σικύου τοῦ ἀγρίου τὴν ῥίζαν
κόψας, ἐς ὕδωρ ἐμβαλών, ἀπὸ τούτου λοῦε· χοληγὰ
δὲ φάρμακα μὴ πίπισκε, ὡς μὴ ταράσσῃ μᾶλλον τὸ
σῶμα τοῦτον· ἐπὴν δὲ ξηρὸν ᾖ τὸ τεταραγμένον,
τρέφε, μηδενὶ ὑποχωρητικῷ φαρμάκῳ, μηδὲ διουρη-
τικῷ, ἀλλ' οἴνῳ οἰνώδει καὶ ἄσσα ἐρυθρότερον ποιέει
τὸν ἄνθρωπον, τούτοισιν· ἢν δὲ χλωρὸς ᾖ, πάλιν
ἐξαρύσαι, ξηραίνειν δὲ μηδαμά, ὡς μὴ παγῇ χλωρὸς
ἐών.

29. Θηρίον ἐπέρχεται ἐπὶ τὸ σῶμα διὰ τόδε· ἐπὴν
φλεγμήνῃ ἡ σὰρξ ἡ πέριξ, καὶ οἱ κρημνοὶ μεγάλοι
ἔωσι τοῦ ἕλκεος, καὶ τὸ ἕλκος ὑγρόν, καὶ ἐπὶ τοῦ
ἕλκεος ἐξηρασμένος ἐπῇ ἰχώρ, ᾖ τὸ ἕλκος συμπεπηγὸς ᾖ ᾖ ξυνσεσηπός, ὁ ἰχὼρ ὁ ἀπὸ τοῦ ἕλκεος
ἀπορρέων κωλύεται ἔξω χωρέειν ὑπὸ τοῦ ἐπιπεπηγό-
τος ἐπὶ τοῦ ἕλκεος πρὸς τὴν σάρκα· ἡ δὲ σὰρξ ὑπο-
δέχεται, ὥστε μετέωρός γ' ἐοῦσα αὐτὴ ὑπὸ φλεγμασίης, καὶ ὅταν ἀφίκηται ὁ ἰχὼρ ὑπορρέων, σήπει
καὶ μετεωρίζει. τοῦτον φαρμάκοισιν ὑγραίνουσιν³
αὐτὸ τὸ ἕλκος χρίειν, ὡς ὑγραινομένου ἔξω τὸ ῥεῦμα
ῥέῃ ἐκ τοῦ ἕλκεος, καὶ μὴ ὑπὸ τὴν σάρκα, καὶ τὰ
κατάρροα τοῦ ἕλκεος ψύχουσι φαρμάκοισιν, ὡς χει-
μιοῦσα συμπιλῆται ἡ σὰρξ καὶ μὴ διαρραγεῖσα

---

¹ Littré: om. A: ἢν V.   ² Potter: χρόνου AV.
³ Joly: -νοντας AV.

this period, when you are cleaning out the disordered moisture, give a pill before meals in order that the body will not take up nourishment from the moist part. When the moisture has been removed, clean with baths, too: pound root of squirting cucumber, put it in water, and wash with this. Do not have the person drink medications that drive off bile, lest these disturb his body even more. When what was stirred up by the moisture has become dry, build the person up without employing either a laxative medication or diuretic, but give him strong wine and substances that promote ruddiness. If he appears yellow-green, draw off fluid again, but be careful not to dry him even a little, lest he become fixed in the yellow-green state.

29. A malignant ulcer assails the body because of the following: when the tissue around the ulcer swells up and its edges are enlarged, and the ulcer is moist, and on the ulcer there is serum that has dried up, or the ulcer is clotted together or putrefied into a mass, the serum flowing out of the ulcer is prevented from moving outwards by the material clotted on top of the ulcer against the tissue. The tissue, being raised up in its phlegmasia, receives the serum, and when it flows under the tissue, the tissue suppurates and becomes raised. In this patient anoint the ulcer itself with moistening medications in order that, the ulcer being moistened, the flux will move outward from the ulcer and not under the tissue, and anoint the suffusion of the ulcer with cooling medications in order that, being chilled, the tissue will felt together rather than be torn apart and provoke a counter-fluxion. Generally,

ΠΕΡΙ ΤΟΠΩΝ ΤΩΝ ΚΑΤΑ ΑΝΘΡΩΠΟΝ

ἀντεπιρρέῃ· καὶ τἆλλα δὲ ἕλκεα ψύχουσι περιχρίειν,
καὶ ἐπ' αὐτὰ τὰ ὑγραίνοντα ἐπιτιθέναι.

30. Κύναγχος ἀπὸ αἵματος γίνεται, ὅταν τὸ αἷμα
παγῇ τὸ ἐν τῇσι φλεψὶ τῇσιν ἐν τῷ τραχήλῳ· τούτῳ[1]
ἀπὸ τῶν ἐν τοῖσι γυίοισι φλεβῶν αἷμα ἀφαιρέειν,
καὶ ἅμα κάτω ὑπεξάγειν, ὡς τὸ τὴν νοῦσον παρέχον
τοῦτο κατασπασθῇ· καὶ γλῶσσαν, ὁπόταν ἕλκεα
ἔχῃ μεγάλα, ὡσαύτως εὐτρεπιστέον.

324   31. Τὰ νοσήματα χρὴ ἀπ' ἀρχῆς ἰᾶσθαι· ὅσα
μὲν ἀπὸ τῶν ῥόων γίνεται, τοὺς ῥόους παύειν πρῶ-
τον· ὅσα δ' ἀπ' ἄλλου, παύειν τὴν ἀρχὴν τοῦ νοσήματος, καὶ εὐτρεπίζειν· ἔπειτα τὸ συνερρυηκός, ἢν
μὲν πολὺ ᾖ, ἐξάγειν· ἢν δὲ ὀλίγον, διαιτῶν καθι-
στάναι.

32. Κεφαλῆς κατάγματα· ἢν μὲν[2] τὸ ὀστέον
καταγῇ καὶ ξυντριβῇ, ἀκίνδυνον· καὶ ἰᾶσθαι χρὴ
τοῦτον ὑγραίνουσι φαρμάκοισιν· ἢν δὲ ῥαγῇ καὶ
ῥωγμὴ ἐγγένηται, ἐπικίνδυνον· τοῦτον πρίειν, ὡς
μὴ κατὰ τὴν ῥωγμὴν τοῦ ὀστέου ἰχὼρ ῥέων τὴν
μήνιγγα σήπῃ· ὥστε γὰρ κατὰ στενὸν ἐσιὼν μέν,
ἐξιὼν δὲ οὔ, λυπέει καὶ μαίνεσθαι ποιέει τὸν ἄνθρω-
πον· τοῦτον χρὴ πρίειν, ὡς ἔξοδος ᾖ τῷ ἰχῶρι, μὴ
μοῦνον ἔσοδος, εὐρέως διαπρισθέντος, καὶ φαρμάκοισι χρῆσθαι, ἅσσα ἐφ' ἑωυτὰ τὸ ὑγρὸν ἕλκουσι,
καὶ λούειν.

[1] Linden: τούτων AV.
[2] V: μὴ A.

72

you must anoint the area around ulcers with cooling agents and on the ulcers themselves apply moistening substances.

30. Angina arises from blood, when the blood in the vessels of the neck congeals. In this patient let blood from the vessels of the limbs; at the same time, purge below in order that what is producing the disease will be drawn downwards. The tongue, too, when it has large ulcers, is to be treated likewise.

31. Diseases must be treated from their origin.[a] In those that arise from fluxes, first stop the fluxes; in those that arise from something else, stop the source of the disease and turn it to the better. Then, if what has flowed together is copious, evacuate it, if it is little, restore the patient by means of regimen.

32. Fractures in the head: if the bone is fractured and shatters, this is not dangerous:[b] you must heal such a patient with moistening medications. But if the bone fractures and a cleft arises, this is dangerous. Trephine the patient in order that the serum flowing down through the cleft in the bone does not make the membrane suppurate, for inasmuch as it enters through a narrow space and does not come out again, it produces pain and makes the patient delirious. You must trephine this patient, in order that there will be a way out for the serum—not just a way in—by sawing open a wide space. Apply medications that attract moisture, and bathe.

[a] See chapter 1 above.
[b] See *Wounds in the Head* 17 for the same view: "where the bones are broken in with many and rather wide fractures they are still less dangerous, and are more readily removed."

33. Πυρεταίνοντι κεφαλὴν μὴ κάθαιρε, ὡς μὴ
μαίνηται· θερμαίνουσι γὰρ τὰ τὴν κεφαλὴν καθαί-
ροντα φάρμακα· πρὸς δὴ τὸ ἀπὸ τοῦ πυρετοῦ θερμὸν
τὸ ἀπὸ τοῦ φαρμάκου προσελθὸν μανίην ποιέει.
θανάσιμα τρώματα· ἐφ' ᾧ ἄν τινι κακῶς ἔχοντι
χολὴν μέλαιναν ἀπεμέσῃ, ἀποθνήσκει ὁ τὸ τρῶμα
ἔχων. καὶ ὑπὸ ἰνηθμοῦ ὃς ἂν ἐχόμενος καὶ ἔχων
ἀσθενέως καὶ λεπτὸς ἐὼν ἐξαπίνης ξηρὸν καθίζῃ,
ἀποθνήσκει. ἐπὴν ὑπὸ θερμωλῆς ἐχομένῳ ἑλκύδρια
ἐκθύωσιν ἀσθενεῖ ἐόντι πέριξ πελιδνά, ἀποθνήσκει.
ἐπὴν ὑπό τινος νοσήματος ἐχομένῳ ἀσθενεῖ ἤδη
ἐόντι πελιδνὰ ἐκθύῃ, θανάσιμον. ἐπὴν φάρμακόν τις
πιὼν ὑπέρινος ᾖ[1] καὶ κάτω καὶ ἄνω ὑπεκχωρέῃ,
οἶνον καταρροφεῖν τὸ μὲν πρῶτον κεκρημένον,
ἔπειτα ἄκρητον θαμινὰ διιδόναι, καὶ παύεται. φάρμακον δὲ μήτ' ἰνηθμῷ μήτ' ἐμετήριον,[2] χολὴ δ',
ἐπὴν αὐτομάτη ῥαγῇ ἢ κάτω ἢ ἄνω, χαλεπωτέρα
παύειν· ἡ γὰρ αὐτομάτη ὑπὸ βίης γινομένης τῷ
σώματι βίαται· ἢν δ' ὑπὸ φαρμάκου ῥέῃ, οὐχ ὑπὸ
συγγενέος βίαται. ἐπὴν παραλάβῃς ἰνώμενόν τε καὶ
ἐμεῦντα, μὴ παύειν τὸν ἔμετον· ὁ γὰρ ἔμετος τὸν
ἰνηθμὸν παύει· ῥάων δὲ ὁ ἔμετος[3] ὕστερον παύσαιτο
ἄν· ἢν δ' ἀσθενὴς ᾖ ὁ ταῦτα πάσχων, ὕπνου φάρμακον ἐμετηρίσας διδόναι. τὸ μὲν αἷμα ὁπόταν νοῦσον
ποιέῃ, ὀδύνην παρέχει, τὸ δὲ φλέγμα βάρος, ὡς τὰ
πολλά.

34. Τῶν νοσημάτων ὧν μὴ ἐπίστηταί τις, φάρ-

---

[1] A: ὑπερνοσῇ V.

33. In a patient with fever do not clean the head, lest he become delirious, for medications that clean the head warm, and when the heat coming from the medication is added to that coming from the fever, it produces delirium. Mortal wounds: if it is any kind of a severe one and besides the wounded person vomits dark bile, he dies. Anyone who suffers an evacuation, is weak and thin, and then suddenly becomes dry, dies. When in a patient with fever heat small ulcers livid all around break out when he is in a weakened state, he dies. When, in a patient with some disease who is already weak, livid spots break out, this is a fatal sign. When, on drinking a medication, someone is cleaned violently and evacuates both downwards and upwards, first let him drink wine mixed with water, and then regularly give him undiluted wine; the evacuation will stop. When bile breaks out spontaneously either downwards or upwards without a purgative or emetic medication, it is more difficult to stop; for what is spontaneous is driven by a force originating in the body, whereas what flows as the result of a medication is not driven by an inherent force. When you take on a person evacuating downwards and also vomiting, do not stop his vomiting, for vomiting stops downward evacuation, and the vomiting will stop more easily later. If the person suffering these things is weak, give him a sleep-producing medication after the emesis. Blood, when it provokes disease, produces pain, whereas phlegm produces heaviness in most cases.

34. In diseases that you do not know, have the patient

---

[2] AV. Littré comments: "La construction est embarrassée; mais le sens est clair." Joly adds πιόντος.   [3] V: πυρετὸς A.

μακον πῖσαι μὴ ἰσχυρόν· ἢν δὲ ῥᾴων γένηται,
δέδεικται ὁδός, εὐτρεπιστέον ἐστὶν ἰσχνήναντα· ἢν
δὲ μὴ ῥᾴων ᾖ, ἀλλὰ χαλεπώτερον ἔχῃ, τἀναντία. ἢν
μὴ ἰσχναίνειν ξυμφέρῃ, φλεγμαίνειν συνοίσει καὶ
θαμινὰ μεταλλάσσειν, ταύτῃ τῇ γνώμῃ χρώμενος.
τῶν δὲ νοσημάτων ἤν τι,[1] ἰσχύοντος μὲν τοῦ ἀλγέ-
οντος, παραλάζηται, τοῦ δὲ νοσήματος ἀσθενέος,
ἐνταῦθα μὲν ἰσχυροτέρῳ θαρσεῦντα τῷ φαρμάκῳ
τοῦ νοσήματος χρῆσθαι, ὥστε καὶ ἤν τι τοῦ ὑγιαί-
νοντος ᾖ ἀπάγειν σὺν τῷ ἀσθενέοντι, οὐδεμία
βλάβη ἐστίν· ἐπὴν δὲ τὸ μὲν νόσημα ἰσχυρότερον,
τὸν δὲ νοσέοντα ἀσθενέα λάβῃς, ἀσθενέσι τοῖσι
φαρμάκοισιν εὐτρεπίζειν, ἅσσα αὐτοῦ τοῦ νοσήματος περιέσονται καὶ ἀπάξουσιν, ἀσθενέστερον δὲ
μηδὲν ποιήσουσι τὸν ἀλγέοντα.

35. Γυμναστικὴ δὲ καὶ ἰητρικὴ ὑπεναντία πέφυκεν, ἡ μὲν γὰρ γυμναστικὴ οὐ δεῖται μεταλλαγὰς
ποιέειν, ἀλλ' ἡ ἰητρική· τῷ μὲν | γὰρ ὑγιαίνοντι οὐκ
ἀρήγει ἐκ τοῦ παρεόντος μεταλλάσσειν, τῷ δὲ
ἀλγέοντι.

36. Τῶν δὲ νοσημάτων ἄσσα μὲν ἕλκεα ἐόντα
ὑπερέχοντα τοῦ ἄλλου σώματός εἰσιν, ἅμα τοῖσι
φαρμάκοισι καὶ λιμῷ χρὴ ἰῆσθαι.

37. Ῥόου ξυμφέρον ἐκ κεφαλῆς ῥέοντος, ἔμετος.

38. Τὰ παλαιὰ νοσήματα χαλεπώτερον ἰᾶσθαι
τῶν νέων· ἀλλὰ νοσήματα τὰ παλαιὰ νέα πρῶτον
ποιέειν· ἕλκος πεπωρωμένον, ἐκβάλλοντα τὸ σκληρὸν σηπτηρίῳ φαρμάκῳ, ἔπειτα συνάγειν. τῶν

328

drink a purgative medication that is not too strong. If he improves, the path is revealed: treatment is to be the removal of moisture. If the patient does not improve, but gets worse, treatment is to be the opposite: if the removal of moisture does not benefit, it will be of help to promote phlegm, and to vary the treatment frequently, with the same purpose. If some disease attacks a person, and the person is strong but the disease is weak, you may confidently employ a purgative medication stronger than the disease, since if anything of what is healthy happens to get carried off with what is diseased, no damage will result. However, when you take on a case where the disease is stronger but the patient is weak, you must treat with mild medications that will overcome the disease and get rid of it, but in no way weaken the patient.

35. Gymnastics and medicine are by nature opposites, for gymnastics is not intended to bring about any changes, whereas medicine must, since the healthy person is not benefited by changes from his present state, but the ill one is.

36. Diseases involving ulcers that cover the rest of the body you must treat simultaneously with purgative medications and a fasting regimen.

37. Vomiting is beneficial for a flux flowing from the head.

38. Old diseases are more difficult to treat than new ones; therefore you must first make old diseases into new ones. When a wound has become hardened, remove the hard part with a putrefactive medication, and then draw it

---

[1] Joly: τις AV.

φαρμάκων ὅσα φλεγμαίνειν ποιέει μάλιστα, ταῦτα
συνάγουσι τὰ καθαρά· τὰ δ' ἰσχναίνοντα, ταῦτα δὲ
καθαίρουσιν. ἢν δέ τις συνάγῃ τὰ μήπω ὡραῖα
ἐόντα, τὸ νοσέον τρέφει σῶμα ὃ ἂν ἕλκος ἔχῃ· καὶ
ἢν μὲν συνάγειν δέῃ τὸ ἕλκος καὶ ἐμπλῆσαι, φλεγμαίνειν ἀρήγει, καὶ ἢν ἐν κεφαλῇ σάρκα βούλῃ·
ἐπαναφερομένη[1] γὰρ ἡ σὰρξ ὑπὸ τῶν σιτίων ὠθέει
τὴν ὑπὸ τοῦ φαρμάκου σηπομένην καὶ ξυμμαχεῖ· ἢν
δὲ μετέωρον ᾖ λίην, ἰσχναίνειν τοῖσι σιτίοισι.

39. Τοὺς ἀνιωμένους καὶ νοσέοντας καὶ ἀπάγχε-
σθαι βουλομένους, μανδραγόρου ῥίζαν πρωῒ πιπίσκειν ἔλασσον ἢ ὡς μαίνεσθαι. σπασμὸν ὧδε χρὴ
ἰᾶσθαι· πῦρ παρακαίειν ἑκατέρωθεν τῆς κλίνης, καὶ
μανδραγόρου ῥίζαν πιπίσκειν ἔλασσον ἢ ὡς μαίνεσθαι, καὶ πρὸς τοὺς τένοντας τοὺς ὀπισθίους σακκία
προστιθέναι θερμά. ἀπὸ σπασμοῦ πυρετὸς ἢν ἐπι-
λάβῃ, παύεται αὐθημερὸν ἢ τῇ ὑστεραίῃ ἢ τῇ τρίτῃ
ἡμέρῃ. ἀπὸ ῥήγματος πυρετὸς οὐ λάζεται πλεῖον
ἢ τρεῖς ἢ τέσσαρας ἡμέρας· ἢν δὲ λάζηται, οἰόμενος
ἀπὸ ῥήγματος ἔχειν, ἀπ' ἄλλου τινὸς λάζοιτο ἄν,
καὶ οὐ χρὴ ὡς ἂν ἀπὸ ῥήγματος εὐτρεπίζειν. ὁπόταν ἄνθρωπος συντεταμένος ᾖ τοὺς πόδας καὶ τὰς
χεῖρας, μανίην ἑωυτῷ ποιέει.

40. Φλέβα δὲ ὧδε χρὴ καίειν ἐπιτήδειαν,[2] ὥστε
τὸ νόσημα ὃ ἂν καὶ ᾖ νοσέων <. . .>[3] ἢν κεκαυμέ-
νος ᾖ ὤνθρωπος, ῥέῃ δέ τι τοῦ αἵματος, ὡς μὴ ἐπι-

330

---

[1] A: ἐπανατρεφομ- V.

78

together. Medications that provoke the greatest swelling draw together what is clean; those that remove moisture, clean. If you draw together what has not yet become mature, you nourish the ailing body that has a wound. If it is necessary to draw the wound together and to fill it full, it helps to provoke swelling, even if you wish to promote tissue in the head. For the tissue, being restored by the food, thrusts off what has been putrefied by the medication, and acts as an ally. If the wound is excessively thick, remove its moisture by means of foods.

39. To those who are troubled and ill and want to hang themselves give mandrake root to drink early in the morning, an amount less than would make them delirious. A convulsion must be treated as follows. Keep fires lighted on each side of the bed, and have the patient drink mandrake root to an amount less than would make him delirious; to the posterior tendons apply warm poultices. If a fever comes on in consequence of a convulsion, the convulsion stops on the same day, on the next day, or on the third day. Fever arising from a tear does not last for more than three or four days; if a fever does last, although it may be held to be from a tear, in fact it is from something else, and you must not treat it as if it were from a tear. When a person is suffering spasms in his legs and arms, he becomes delirious.

40. You must cauterize the appropriate vessel as follows, so that the disease, wherever it is present ... If, when a person has been cauterized, some haemorrhaging

---

[2] Potter: -δειον AV.

[3] Ermerins conj. lacun.

κίνδυνον ἦ τοῦτο αὐτῷ, ἀμφότερα ταῦτ' ἐστὶ ποιέειν·
ἢν διακαύσῃς ταύτῃ ἐν τῷ πόνῳ οὗ εἵνεκα ἐκαίετο,
οὐ ξυμφύεται, ὠφέλησε δὲ τῷ ῥόῳ· ἢν γὰρ διακαῇ,
οὐ ῥέει· ἐπὴν γὰρ διακαῇ, τὸ ἄκρον ἑκάτερον ἀνα-
τρέχει τῆς φλεβὸς ἢ διεκάη, καὶ συναυαίνεται· ἢν δέ
<τι>[1] καταλελειμμένον ἦ, ὑπὸ τοῦ καταλελειμμένου,
διαρρέοντος τοῦ αἵματος,[2] ὑγραίνεται· ἢν δὲ αἷμα
ῥέῃ ἐκ φλεβός, διακαίειν ἐπικαρσίην· ἢν δὲ μὴ
παύηται πρὸς ταῦτα, ἄνω καὶ κάτω ἑκατέρωθεν δια-
τάμνειν, ὡς ἀποτρεφθῇ τὸ αἷμα ῥέον· διαλελαμμέ-
νον γὰρ φαρμάκῳ ῥᾷον παύειν ἢ τὸ ἀθρόον. ὀδύνης
ἐν κεφαλῇ τοῦ αἵματος ἀφαιρέειν ἀπὸ τῶν φλεβῶν·
ἢν δὲ μὴ παύηται, ἀλλὰ πολυχρόνιον ἦ, διάκαιε τὰς
φλέβας, καὶ ὑγιὴς γίνεται· ἢν δὲ τὴν κεφαλὴν καθήρῃς, μᾶλλον πονέεις.

41. Ἰητρικὴν οὐ δυνατόν ἐστι ταχὺ μαθεῖν διὰ
τόδε, ὅτι ἀδύνατόν ἐστι καθεστηκός τι ἐν αὐτῇ
σόφισμα γενέσθαι, οἷον ὁ τὸ γράφειν ἕνα τρόπον
μαθὼν ὃν διδάσκουσι, πάντα ἐπίσταται· καὶ οἱ ἐπιστάμενοι πάντες ὁμοίως διὰ τόδε, ὅτι τὸ αὐτὸ καὶ
ὁμοίως ποιεύμενον νῦν τε καὶ οὐ νῦν οὐκ ἂν τὸ ὑπεν-
αντίον γένοιτο, ἀλλ' αἰεὶ ἐνδυκέως | ὅμοιόν ἐστι,
καὶ οὐ δεῖ καιροῦ. ἡ δὲ ἰητρικὴ νῦν τε καὶ αὐτίκα οὐ
τὸ αὐτὸ ποιέει, καὶ πρὸς τὸν αὐτὸν ὑπεναντία ποιέει,
καὶ ταῦτα ὑπεναντία σφίσιν ἑωυτοῖσιν· πρῶτον ὑπεκ-
χώρησιν κοιλίης τὰ ὑπεχωρητικὰ οὐκ αἰεὶ ποιέ-

332

---

[1] Joly.   [2] A: ῥεύματος V.

occurs, in order that this will not be dangerous for him, both these things are to be done. (If you burn a vessel through at the point of the pain for which you have cauterized, it does not grow together, but you have helped the flux; for if the vessel is burnt through, there is no haemorrhage, since when that is done each end of the vessel that was burnt through retracts and dries together. But if anything is left behind,[a] the wound is moistened by blood flowing through this.) If blood flows out of a vessel, burn it through at right angles. If the flow does not stop with this treatment, cut the vessel through above and below on both sides, in order that the blood as it flows will be turned away; for blood whose flow is divided is easier to stop with a medication than what flows in a mass. For pain in the head, draw off blood from the vessels; if the pain does not stop, but lasts for a long time, burn the vessels through, and the patient will recover. If you clean the head, you make the pain worse.

41. Medicine cannot be learned quickly because it is impossible to create any established principle in it, the way that a person who learns writing according to one system that people teach understands everything; for all who understand writing in the same way do so because the same symbol does not sometimes become opposite, but is always steadfastly the same and not subject to chance. Medicine, on the other hand, does not do the same thing at this moment and the next, and it does opposite things to the same person, and at that things that are self-contradictory. First, laxatives do not always provoke

[a] I.e. if the vessel is not burnt completely through.

ουσι, καὶ τὰ ὑπεκχωρητικὰ ἀμφότερα ποιέουσι,
τάχα δὲ οὐδ' οὕτως ἔχουσι τὰ ὑπεκχωρητικὰ τοῖσι
στασίμοις ὡς ὑπεναντία. ἐπιστάσης τῆς κοιλίης,
διὰ τὴν λίην στάσιν φλεγμῆναν τὸ σῶμα, ἐς τὴν
κοιλίην φλέγματος ἀφικομένου, οὕτως ἡ στάσις
ὑπεχώρησιν ἐποιήσεν· ἐπὴν γὰρ τὸ φλέγμα ἐς τὴν
κοιλίην ἐσέλθῃ, ἰνηθμὸς γίνεται· ἐν τούτῳ δὲ τὰ
ὑπεκχωρητικὰ φύσει στάσιν ποιέουσιν ἐν τῇ κοιλίῃ.
ἢν μὲν μὴ ὑπεκχωρητικὰ προσφέρῃς, ἐκκλύζεται[1] δὲ
τὸ νοσεῖν ποιέον καὶ ὑγραίνηται· ἐπὴν ἐκκλυσθῇ,
γίνεται ὑγιής. καὶ οὕτω τά τε στάσιμα τοῖσιν ὑπεκ-
χωρητικοῖσι ταὐτὸ[2] ποιέουσι τῆς κοιλίης, καὶ τοῖσι
στασίμοισι τὰ ὑπεκχωρητικά.

Τὸν αὐτὸν δὲ τρόπον [καὶ τοὺς ἐρυθροὺς καὶ τοὺς
χλωρούς·][3] καὶ τὰ φλεγματώδεα χλωροὺς ποιέει καὶ
ἀχρόους, καὶ τὰ ἰσχναίνοντα εὐχρόους· ἑκατέρου δ'
ἐστὶ φάρμακον τὰ ὑπεναντία τῷ ὑπεναντίῳ· αὐτίκα
ὅταν φλεγμαίνῃ χλωρὸς[4] ἐών, ἐκλύεται, ἢν [μή][5] τι
ἰσχναίνῃ φάρμακον προσενεχθέν· ἐνθάδε τῷ φλεγ-
μαίνοντι τὸ ἰσχναῖνον ὠφέλησεν· τούτων δὲ τό ποτε
ὠφελεύμενον τῷ ὠφελεῦντι νῦν ὠφελεῖ ἐνταῦθα, ὁπόταν ὑπὸ ἰσχνότητος ἄχροος καὶ χλωρὸς ᾖ· ἢν γάρ
τις φλεγματώδες προσφέρῃ, παύεται τὸ χλωρόν.

---

[1] Ermerins: ἐκλύεται (-ηται) AV.
[2] Littré: στάσιν AV.
[3] Del. Ermerins.
[4] V: χολὸς A.
[5] Del. Foes in note 108 after Cornarius.

evacuation of the cavity, they may provoke evacuation both upwards and downwards, and it is possible for laxatives not even to act in the opposite way from things that promote stasis. For example, with the cavity stopped by a medication, the excessive stasis has often filled the body with phlegm, and as the phlegm came to the cavity, in this way the stasis has brought about evacuation; for when phlegm enters the cavity, evacuation occurs; in this patient, then, naturally laxative substances bring about a stasis in the cavity.[a] (If you do not administer laxatives, let what is making the patient ill be washed out with an enema, and moisten him; when it has been washed out, he recovers.) In this way, agents that promote stasis do the same thing for the cavity as do laxatives, and laxatives the same as agents that promote stasis.

In the same way [in both ruddy and jaundiced persons], agents that promote phlegm make patients both yellow-green and pallid, and those that remove moisture make them ruddy; in each case there is a medication— opposite agents to the opposite condition. Immediately when a jaundiced person swells up, the swelling is resolved, if some medication that is administered dries; thence, what dries has helped the patient who was swollen. But what was helped at one time by something helping it, now helps in turn, e.g. when a person is pallid and yellow-green because of a drying agent: if someone gives him a medication that promotes phlegm, the yellow-greenness goes away.

[a] I.e. as a step in their mode of action.

ΠΕΡΙ ΤΟΠΩΝ ΤΩΝ ΚΑΤΑ ΑΝΘΡΩΠΟΝ

334   42. Ὀδύνη τε γίνεται καὶ διὰ τὸ ψυχρὸν καὶ διὰ
τὸ θερμόν, καὶ διὰ τὸ πλέον καὶ διὰ τὸ ἔλασσον· καὶ
ἐν μὲν τοῖσιν ἐψυγμένοισι φύσει ἐκ τοῦ σώματος
πρὸς τὸ δέρμα διὰ τὸ θερμαῖνον λίην ὀδύνη γίνε-
ται·¹ ἐν δὲ τοῖσι θερμοῖσι φύσει διὰ τὸ ψυχρόν· καὶ
ἐν μὲν τοῖσι ξηροῖσι φύσει ὑγραινομένοις, ἐν δὲ τοῖς
ὑγροῖσι φύσει ξηραινομένοισι· τὴν γὰρ φύσιν διαλ-
λασσομένοις ἑκάστοισι καὶ διαφθειρομένοις αἱ ὀδύναι γίνονται· ὑγιαίνονταί τε αἱ ὀδύναι τοῖσιν ὑπε-
ναντίοισιν· ἴδιον ἑκάστῳ νοσήματί <τί>² ἐστι· τοῖσι
θερμοῖσι φύσει, διὰ δὲ τὸ ψυχρὸν νοσέουσι, θερμαῖ-
νόν τε καὶ τἆλλα τούτων κατὰ λόγον.

Ἄλλος ὅδε τρόπος· διὰ τὰ ὅμοια νόσος γίνεται,
καὶ διὰ τὰ ὅμοια προσφερόμενα ἐκ νοσεύντων ὑγιαί-
νονται· οἷον στραγγουρίην τὸ αὐτὸ ποιέει οὐκ ἐοῦσαν, καὶ ἐοῦσαν τὸ αὐτὸ παύει· καὶ βὴξ κατὰ τὸ
αὐτό, ὥσπερ καὶ στραγγουρίη, ὑπὸ τῶν αὐτῶν γίνεται καὶ παύεται. ἄλλος ὅδε τρόπος· πυρετὸς ὁ διὰ
φλεγμασίην γινόμενος, τοτὲ μὲν ὑπὸ τῶν αὐτῶν
γίνεται καὶ παύεται, τοτὲ δὲ τοῖσιν ὑπεναντίοις ἢ
ἐγένετο· τοῦτον μὲν γὰρ εἴ τις βούλεται λούειν ὕδατι
θερμῷ καὶ ποτὰ πολλὰ διδόναι, ὑγιὴς γίνεται διὰ
τοῦτο τὸ φλεγμαῖνον· τοῖσι φλεγμαίνειν ποιέουσι
προσφερομένοις ὁ ὢν πυρετὸς γίνεται ὑγιής· καὶ εἴ
τις βούλεται φάρμακον πῖσαι ὑποχωρητικὸν καὶ
336   ἐμετικόν· τὸν αὐτὸν τρόπον τοῖσί | τε ποιέουσι παύε-
ται, καὶ τοῖσι παύουσι γίνεται. τοῦτο μὲν γὰρ εἴ τις
ἐμέοντι ἀνθρώπῳ βούλεται ὕδωρ δοῦναι πιεῖν πολύ,

84

42. Pain arises both from cold and from heat, and both from excessively great amounts and from too little. In persons that are cooled by nature out of their body towards the skin, pain arises from excessive heating, in those by nature hot, from cold, in those by nature dry, when they are moistened, and in those by nature moist, when they are dried. For in each thing that is altered with respect to its nature, and destroyed, pains arise. Pains are cured by opposites, and there is a specific thing for each disease: in persons by nature hot, and who are ill because of cooling, it is what heats, and so on according to this principle.

Another principle is the following: a disease arises because of similars, and, by being treated with similars, patients recover from such diseases. For example, the same thing produces strangury when it is not present, and stops it when it is present; cough, in the same way as strangury, is engendered and is halted by the same things. Another principle is the following: fever that has arisen due to phlegmasia sometimes arises and is stopped by the same things, and sometimes is stopped by things opposite to those from which it arose. For if someone washes this patient with hot water and gives copious drinks, the patient recovers as a result of this swelling; when things that promote swelling are administered, the fever that is present becomes well; also if someone decides to make this patient drink a downward evacuant and an emetic. Thus, in the same way it is stopped by the things that produce it and produced by those that stop it. That is to say, if to a patient that is vomiting someone gives much water

---

[1] AV. See Schubring p. 66.   [2] Joly.

ἐκκλυσθήσεται δι' ἃ ἐμέει σὺν τῷ ἐμέτῳ· οὕτω μὲν
διὰ τὸ ἐμέειν ὁ ἔμετος παύεται· ὧδε <δὲ>[1] διὰ τὸ
παύειν, ὅτι κάτω ποιήσει αὐτῷ ἐλθεῖν ἐκεῖνο,[2] ὃ
ἐνεὸν ἔμετον ποιέει, ἀμφοτέροισι τοῖσιν ὑπεναντί-
οισι τρόποις ὑγιὴς γίνεται. καὶ εἰ μὲν οὕτως εἶχε
πᾶσι, καθεστήκει ἄν, οὕτω τὰ μὲν τοῖς ὑπεναντίοισιν
εὐτρεπίζεσθαι οἷά τέ ἐστι καὶ ἀφ' ὅτου ἐγένετο, τὰ
δὲ τοῖσιν ὁμοίοισιν οἷά τέ ἐστι καὶ ἀφ' ὅτου ἐγένετο.

43. Τούτου δ' αἴτιόν ἐστιν ἡ τοῦ σώματος ἀσθε-
νείη· τὸ γὰρ σῶμα ὑπὸ μὲν τῶν σιτίων ἴσων ἴσως
τρέφεται, ὑπὸ δὲ τοῦ σώματος τὰ σιτία κρατέεται·
ἐπὴν δὲ μάσσον[3] προσενέγκηται ἢ ἀλλοίως μεταλ-
λάξας κρατέηται, κρατέει δὴ[4] τὰ σιτία· καὶ ὁπόταν
κρατέηται τὸ σῶμα ὑπὸ τῶν προσοισμάτων, <ἃ>[5]
θάλλειν ποιέει ταὐτὰ καὶ κρατέει ἅμα τοῦ σώματος
τά τε ὑπεναντία ποιέουσιν. αὐτίκα τὸ λοῦσθαι
θερμῷ, ἕως μὲν ἂν τὸ σῶμα κρατέῃ τοῦ προσοίσματος, θάλλει· ἐπὴν δὲ κρατηθῇ, ἰσχνὸν ποιέει τὸ
σῶμα· καὶ τὸ εὐωχέεσθαι ὁμοίως τῷ λοῦσθαι ποιέει·
ταῦτα μὲν γὰρ ἕως μὲν ἂν κρατέωνται, θάλλειν ποι-
έουσιν· ἐπὴν δὲ κρατέωσιν, ὑπεκχωρήσεις τε ποιέουσιν καὶ ἀλλοίας κακίας· ὁπότε δὲ ὃ προσφέρεται
αὐτὸ[6] μεταλλάσσεται, | ἀνάγκη καὶ ᾧ προσφέρεται[7]

---

[1] Joly.
[2] Joly: ἐξ ἐκείνου AV.
[3] Joly: πλέον ἢ ἔλασσον AV.
[4] Ermerins: κρατέεται δὲ καὶ AV.

338

86

to drink, the patient will, by vomiting, be washed clean by what he vomits; thus, through vomiting, the vomiting stops; and in this way through its stopping—because it will make the patient evacuate downwards that which, when present in the body, provokes vomiting—the patient recovers in two contrary ways. And if this were so in all cases, the principle would be established, that sometimes conditions can be treated by things opposite to those from which they arose, and sometimes by things like to those from which they arose.

43. The cause of this is the body's weakness. For from the correct measure of foods the body is nourished equally and the foods are mastered by the body, but when more foods are administered than can, in changing to something different, be mastered, the foods take mastery, and when the body is mastered by the things administered to it, the same things that otherwise make it thrive prevail over the body and produce the opposite effect. As long as the body has mastery over what has been administered to it, being washed at once with warm water makes it thrive, but when the body has been mastered by what was administered to it, being bathed makes it lean. Being well fed has the same effects as being washed: for as long as foods are under the mastery of the body, they make the body thrive, but when they themselves have the mastery, they provoke downward evacuations and other sorts of ills. When what is administered is changed itself, then it follows necessarily that the person to whom it is adminis-

---

[5] Littré.　　[6] Potter: τοῦτο AV.

[7] ᾧ πρ. Linden after Cornarius: τὸ προσφερόμενον AV.

μετατρέπεσθαι· τὸ γὰρ σῶμα μετατρεπόμενον καὶ
ὀλιγοεργὲς ὂν καὶ ὑπὸ παντὸς νικώμενον τὰς παλιγκο-
τίας παρέχει. τοῦτο δὲ ποιέει καὶ τὰ ὑποχωρητικά,
καὶ τὰ θάλλειν ποιέοντα, ἰσχναίνοντά τε· ταῦτα τὸ
σῶμα ποιέουσι, καὶ τἆλλα πάντα τὰ ὑπεναντία τούτοισι πάσχοντα.

44. Ἡ δὲ ἰητρικὴ ὀλιγόκαιρός[1] ἐστιν· καὶ ὃς
τοῦτο ἐπίσταται, ἐκείνῳ καθέστηκε, καὶ ἐπίσταται
[τὰ εἴδεα καὶ τὰ μὴ εἴδεα, ἅ ἐστιν ἐν ἰητρικῇ ὁ καιρὸς γνῶναι·][2] ὅτι τὰ ὑποχωρητικὰ οὐχ ὑποχωρητικὰ
γίνεται, καὶ τἆλλα ὅτι ὑπεναντία ἐστί, καὶ τὰ ὑπε-
ναντιώτατα οὐχ ὑπεναντιώτατά ἐστιν. ὁ δὲ καιρὸς
ὅδ' ἐστί· τὰ σιτία προσφέρειν, ὅσων μέλλει τὸ σῶμα
προσφερόμενον τὸ πλῆθος κρατέειν, ὥστ' ἢν μὲν
οὕτω ποιέῃ, πᾶσα ἀνάγκη τὸ ὑποχωρητικὸν σιτίον
προσφερόμενον ὑποχωρητικὸν εἶναι, καὶ τὸ φλεγματῶδες φλεγματῶδες. ἢν κρατέῃ μὲν γὰρ[3] τὸ σῶμα
τῶν σιτίων, οὔτε νοῦσος οὔτε ὑπεναντίωσις γίνεται
προσφερομένων, καὶ οὗτος ὁ καιρός ἐστιν ὃν δεῖ τὸν
ἰητρὸν εἰδέναι· [ἐπὴν δὲ τὸν καιρὸν ὑπερβάλλῃ, τὸ
ὑπεναντίον γίνεται, καὶ οἳ πρὶν ὑπερπέσσειν οἴονται
ἔχειν, καὶ τὸ θερμαίνεσθαι·][4] ἕως μὲν γὰρ ἂν τὸ
σῶμα τοῦ προσοίσματος κρατέῃ, τρέφεται τὸ σῶμα·
ἐπὴν δὲ τοῦτον τὸν καιρὸν ὑπερβάλλῃ, τὸ ὑπεναν-

---

[1] V: -χρόνιος A.
[2] Del. Potter as an intruded marginal gloss.
[3] μὲν γὰρ Potter: γὰρ ἐν A: om. V.
[4] Del. Potter. καὶ οἳ—θερμαίνεσθαι del. Ermerins.

tered also be changed; so the body, being changed, of little strength, and overwhelmed by everything, makes things get worse; this can result also from evacuations, from medications that provoke growth, and from those that reduce moisture. The body is affected by these things, and by all the other things that are their opposites.

44. In medicine the correct measure is narrow, and the person who understands this has a fixed principle, and he understands [the forms in which correct measure in medicine is and is not to be recognized] that laxatives become non-laxative, that other things turn to their opposites, and that things that were the most opposite are no longer the most opposite. Correct measure is the following: to administer as much food as, being administered, will be mastered by the body, so that, if a person does this, there is every necessity that the laxative food you administer have a laxative effect, and the phlegm-promoting a phlegm-promoting effect. Now if the body gains a mastery over the food, neither will any disease nor anything else contrary to the nature of what is administered result; this is the correct measure the physician must recognize. [But when he goes beyond the correct measure, the opposite happens; . . . .][a] For, as long as the body gains the mastery over what is administered, the body is nourished; but when one goes beyond this correct measure,

[a] The latter part of the deleted passage is unintelligible.

τίον γίνεται, ἰσχναίνεται γάρ· καὶ τἆλλα πάντα δὲ
τὰ φλεγμαίνειν ποιεῦντα, ἕως μὲν ἂν κρατέῃ τὸ
σῶμα, ἔτι τὰ πρὸς τὸν καιρὸν καὶ τὰ κατὰ φύσιν
340 ποιέουσιν ἕκαστον, | τὰ φλεγματώδεα φλεγμαίνειν·
ἐπὴν δὲ ὑπερβάλλῃ τὸν καιρόν, τὰ ὑπεναντία γίνεται.[1]

45. Πάντα φάρμακά εἰσι τὰ μετακινέοντα τὸ
παρεόν· πάντα δὲ τὰ ἰσχυρότερα μετακινέουσιν·
ἔξεστι δέ, ἢν μὲν βούλῃ, φαρμάκῳ μετακινέειν· ἢν
δὲ μὴ βούλῃ, σιτίῳ· ἅπαντα δὲ νοσέοντι μετακινέειν
ἐκ τοῦ παρεόντος ἀρήγει· ἢν γὰρ μὴ μετακινήσῃς τὸ
νοσέον, αὔξεται. φάρμακα οὐ χρὴ τὰ ἰσχυρὰ φύσει
ἐπὶ τῶν ἀσθενέων νοσημάτων διδόναι, ὀλιγότητι τοῦ
φαρμάκου ἀσθενὲς ποιεῦντα· ἀλλὰ τοῖσι μὲν ἰσχυροῖσι φύσει φαρμάκοις [ἐπὶ][2] ἰσχυροῖσι χρῆσθαι,
τοῖς δ' ἀσθενέσι φαρμάκοις μὴ ἰσχυροῖσι, μηδὲ
μεταποιεῦντα τὸ φάρμακον. ἀλλὰ κατὰ φύσιν
ἑκάστοισιν· τοῖσι μὲν ἀσθενέσι ἀσθενῆ φάρμακα
φύσει, τοῖσι δὲ ἰσχυροῖσι νοσήμασιν ἰσχυρὰ φύσει
τὰ φάρμακα. τὰ δὲ νοσήματα ᾗ πελαστάτω πέφυκεν, ἐξάγειν, ταύτῃ δ' ἐξάγειν ᾗ ἑκάστῳ ἔξοδος
ἐγγυτάτω.

Τὰ ὑποχωρητικὰ τοιάδε ἐστίν, ὅσα ὀλισθηρὰ καὶ
τμηματώδεα, καὶ ὅσα ἐν τοῖσι θερμοῖσι λεπτύνον-
ται· ἡ γὰρ κοιλίη θερμή ἐστι· καὶ τἆλλα τὰ ἁλμυρά,
καὶ ὅσα τῶν τοιούτων πλεῖστον ἔχουσιν. τὰ δ' οὐ
διαχωρητικά, ἀλλὰ στάσιμα, ὅσα φῦσαν παρέχου-
σιν· τὰ γὰρ ὑγρὰ ξηραινόμενα φῦσαν ποιέουσι, καὶ

the opposite happens, and it becomes lean. All the other
things, too, which promote phlegm each do what is opportune and according to their nature, as long as the body has
the mastery; e.g. the phlegm-promoting promote phlegm.
But when the physician goes beyond the correct measure,
the opposite happens.

45. All substances that change the state of the patient
are medications, and all the more forceful substances
change in this way. It is possible, if you wish, to bring
about change by means of a medication, or, if you do not
wish to use a medication, you can bring about change by
means of foods. Everything that changes from the exist-
ing state benefits what is ill, for if you do not change what
is ill, it increases. Medications that are by nature strong
you must not give in weak diseases, attempting to make
the medication weak by giving it in a small amount; rather
you should administer strong medications in diseases that
are themselves by nature strong, and weak medications in
diseases that are not strong, and not alter the medication.
Give to each disease according to its nature: in weak diseases give medications by nature weak, in strong diseases,
medications by nature strong. Draw off diseases at the
point that is nearest; draw off where the exit is nearest for
each.

Laxatives are these kinds of things: things that are slippery and separable, and that are thinned with warming
(for the cavity is warm) and also salty things, and substances that are the most like these. The non-laxative—
i.e. substances promoting stasis—are things that produce
flatulence (for when what is moist becomes dry it pro-

---

[1] The text in A ceases here, with the remark τέλος περὶ
τόπων τῶν κατὰ ἄνθρωπον.   [2] Del. Aldina.

τὰ στύφοντα καὶ τὰ ὑπὸ θερμοῦ πηγνύμενα καὶ τὰ
ψαθυρὰ καὶ τὰ ξηρά. πάντα δὲ [τὰ]¹ ἐντὸς φλεγμαίνειν ποιέουσι προσφερόμενα, ἄσσα τὰ ἐκτὸς ἰσχναί-
νουσιν· ταῦτα δὲ καὶ ἰσχητήριά ἐστι καὶ φλεγματώδεα. καὶ τὰ ὑποχωρητικὰ ἰσχναίνοντα θερμαίνουσι·
τὰ τοιαῦτα ἔτι δὲ τὰ ὀξέα καὶ φλεγματώδεα. πάντα
δὲ τὰ ψύχοντα τὰ ἐν τῇ κοιλίῃ· τὰ δὲ τοιαῦτα ὑποχωρητικά ἐστι· καὶ τὰ ψυχρὰ καὶ τὰ ὑγρά· ὁπόταν
δὲ μὴ ὑποχωρητικὰ ἔωσι, θερμαίνουσιν. ψύχουσι δὲ
καὶ τὰ θερμὰ ἐς τὴν κοιλίην προσφερόμενα καὶ ταχὺ
342 <διαχώρησιν ποιεύμενα,>² διαχώ|ρησιν δὲ μὴ ποι-
εύμενα θερμά ἐστιν ἐν τῇ κοιλίῃ. τούτων ὅσα πλησ-
μονὴν ποιέει, μάλιστα φλεγματώδεά ἐστιν· ἃ δὲ
πλεῖστα προσφερόμενα, οὐ ποιέει πλησμονήν, διαχωρητικά.

46. Ἰητρικὴ δή μοι δοκέει ἤδη ἀνευρῆσθαι ὅλη,
ἥτις οὕτως ἔχει, ἥτις διδάσκει ἕκαστα καὶ τὰ ἔθεα
καὶ τοὺς καιρούς. ὃς γὰρ οὕτως ἰητρικὴν ἐπίσταται,
ἐλάχιστα τὴν τύχην ἐπιμένει, ἀλλὰ καὶ ἄνευ τύχης
καὶ ξὺν τύχῃ εὖ ποιηθείη ἄν. βέβηκε γὰρ ἰητρικὴ
πᾶσα, καὶ φαίνεται τῶν σοφισμάτων τὰ κάλλιστα ἐν
αὐτῇ συγκείμενα ἐλάχιστα τύχης δεῖσθαι· ἡ γὰρ
τύχη αὐτοκρατὴς καὶ οὐκ ἄρχεται, οὐδ' ἐπ' εὐχῇ
ἐστιν αὐτὴν ἐλθεῖν· ἡ δ' ἐπιστήμη ἄρχεταί τε καὶ
εὐτυχής ἐστιν, ὁπόταν βούληται ὁ ἐπιστάμενος χρῆσθαι. ἔπειτα τί καὶ δεῖται ἰητρικὴ τύχης; εἰ μὲν γὰρ
ἔστι τῶν νοσημάτων φάρμακα σαφῆ, οἶμαι, οὐκ
ἐπιμένει τὴν τύχην τὰ φάρμακα ὑγιᾶ ποιῆσαι τὰ

duces flatulence), the astringent, things that congeal when they are heated, the friable, and the dry. All substances administered in order to remove moisture externally promote phlegm within; thus, they both remove moisture and promote phlegm. Laxatives that remove moisture warm; it is the same with substances that are sharp and promote phlegm. All things that cool what is in the cavity, such are laxative; i.e. both the cold and the moist; when they are not laxative, they warm. Warm things that enter the cavity and quickly bring about an evacuation cool, but if they do not stimulate an evacuation, they are warm in the cavity. Substances that produce fullness are usually phlegm-promoting, whereas those that—even when administered in very great amounts—do not lead to fullness, are laxative.

46. Medicine in its present state is, it seems to me, by now completely discovered, insofar as it teaches in each instance the particular details and the correct measures. For anyone who has an understanding of medicine in this way depends very little upon good luck, but is able to do good with or without luck. For the whole of medicine has been established, and the excellent principles discovered in it clearly have very little need of good luck. Good luck is arbitrary and cannot be commanded, and even prayer cannot make it come; but understanding can be commanded, and of itself represents good luck, whenever the person who has knowledge employs it. Then, in what sense does medicine need luck? For if there are obvious medications indicated in diseases, I hold that these medications do not depend upon good luck to make the

---

[1] Del. Ermerins.   [2] Littré; Foes ($\pi o \iota \acute{e} o \upsilon \sigma \iota$) in note 124 after Cornarius' *secessum faciunt.*

νοσήματα, εἴ πέρ ἐστι [τὰ]¹ φάρμακα· εἰ δὲ σὺν τῇ
τύχῃ διδόναι ὠφελέει, οὐδὲν μᾶλλον τὰ φάρμακα ἢ
καὶ τὰ μὴ φάρμακα σύν γε τῇ τύχῃ ὑγιᾶ ποιέουσι
προσφερόμενα τοῖσι νοσήμασιν. ὅστις δὲ τὴν
τύχην ἐξ ἰητρικῆς ἢ ἐξ ἄλλου τινὸς ἐξελάσει, φάμε-
νος οὐ τοὺς καλῶς τι πρῆγμα ἐπισταμένους χρῆσθαι τύχῃ, τὸ ὑπεναντίον δοκέει μοι γινώσκειν· ἐμοὶ
γὰρ δοκέουσι μοῦνοι καὶ ἐπιτυγχάνειν καὶ ἀτυχεῖν
οἱ καλῶς τι καὶ κακῶς πρῆξαι ἐπιστάμενοι· ἐπιτυγχάνειν τε γὰρ τοῦτ' ἐστὶ τὸ καλῶς ποιέειν, τοῦτο
δὲ οἱ ἐπιστάμενοι ποιέουσιν· ἀτυχεῖν δὲ τοῦτ' ἐστίν,
ὃ ἄν τις μὴ ἐπίστηται, τοῦτο μὴ καλῶς ποιέειν· ἀμα-
θὴς δὲ ἐών, πῶς ἂν ἐπιτύχοι; εἰ γάρ τι καὶ ἐπιτύχοι,
344   οὐκ ἂν ἀξίως λό|γου τὴν ἐπιτυχίην ποιήσαιτο· ὁ γὰρ
μὴ καλῶς ποιέων οὐκ ἂν ἐπιτύχοι τἆλλα τὰ εἰκότα
μὴ πράσσων.

47. Τὰ γυναικεῖα νοσεύματα καλεύμενα· αἱ ὑστέ-
ραι πάντων τῶν νοσημάτων αἴτιαί εἰσιν· αὗται γὰρ
ὅπῃ ἂν ἐκ τῆς φύσεως μετακινηθέωσι, νούσους
παρέχουσιν, ἤν τε προέλθωσιν, ἤν τε παραχωρήσω-
σιν. καὶ ὅταν μὲν μὴ βάλλουσαι τὸ στόμα αἱ
μῆτραι καὶ μὴ ψαύουσαι τῶν κρημνῶν μετακεκινη-
μέναι ἔωσιν ἔξω, σμικροτάτη νοῦσός ἐστιν· ἐπὴν δὲ
προκινηθέωσιν ἐς τὸ ἔμπροσθεν καὶ ἐμβάλλωσι τὸ
στόμα ἐς τὸν κρημνόν, πρῶτον μὲν ψαύσασα πόνον
παρέσχεν, εἶτα ἀποφραχθεῖσα ἡ μήτρη καὶ ἐπιπω-
μασθεῖσα ὑπὸ τῆς ἐμβλήσεως τῆς ἐς τὸν κρημνόν,
οὐ γίνεται ῥόος τὰ καταμήνια· τοῦτο δὲ συνιστάμε-

diseases better—if they really are medications: for if these medications administered in diseases only help when they have good luck, then they are no more making the patients well than non-medications would with good luck. On the other hand, anyone who excludes "good luck" from medicine or any other art, saying that those who understand a given matter thoroughly do not employ "good luck", seems to me to know himself that this is incorrect. For I hold that those alone have good and bad luck respectively, who understand well and poorly how to do something. For to have good luck is to do the correct thing, and those with understanding do this. But to have bad luck is this: to do badly what one does not understand. If someone is ignorant how could he ever have good luck? For even if he should have good luck in some particular thing, he would not make this good luck into anything worth mentioning: by not acting correctly he would not have good luck, since he did not do the other appropriate things.

47. Diseases of women, as they are called. The uterus is the cause of all these diseases; for however it changes from its normal position—whether it moves forward, or whether it withdraws—it produces diseases. When the uterus does not drop its os and does not move so that it is outside and touching the labia, the disease is very minor. But when it moves ahead towards the front and inserts its os against the labium, first this produces pain because of the contact, and also the menstrual flow fails to take place because the uterus is obstructed and capped by its impaction against the labia, and when this flow is held

---

[1] Del. Ermerins.

νον οἶδός τε καὶ ὀδύνην παρέχει. καὶ ἢν μὲν κάτω
κατελθοῦσα καὶ ἀποστραφεῖσα ἐμβάλῃ ἐς τὸν βουβῶνα, ὀδύνην παρέξει· ἢν δὲ ἄνω ἐπαναχωρήσασα
ἀποστραφῇ καὶ ἀποφραχθῇ, καὶ οὕτω διὰ τὴν
ἀραιότητα νοῦσον παρέχει· ὁπόταν δὲ διὰ τοῦτο
νοσέῃ, ἐς τὰ ἰσχία καὶ τὴν κεφαλὴν ὀδύνην ποιέει.
ὁπόταν δὲ αἱ μῆτραι πλησθεῖσαι συνοιδήσωσιν, οὐ
ῥεῖ οὐδὲν καὶ πλέαι γίνονται· ἐπὴν δὲ πλέαι γίνων-
ται, ψαύουσι τῶν ἰσχίων· ἐπὴν δὲ πλησθεῖσαι αἱ
μῆτραι ὑγρότητος καὶ διευρυνόμεναι οὐ χωρέωνται,
ψαύωσι δὲ τῶν ἰσχίων, ὀδύνας παρέχουσι καὶ ἐς τὰ
ἰσχία καὶ <ἐς>[1] τὸν βουβῶνα, καὶ οἷον σφαῖραι ἐν
τῇ γαστρὶ ὑποτρέχουσι, καὶ τὴν κεφαλὴν πονέουσι,
τοτὲ μὲν ἐς τὸ ἕτερον μέρος, τοτὲ δὲ ὅλην, οἵη γίνεται καὶ ἡ νοῦσος.

Ὧδε δὴ ταῦτα εὐτρεπιστέον· ἢν μὲν προέλθῃ
346   μοῦνον καὶ ᾗ διαχρίειν, χρῶ ᾧ | τινι βούλει τῶν
κακόδμων, ἢ κέδρῳ, ἢ μυσσωτῷ, ἢ ἄλλῳ τινὶ τῶν
βαρύτερον καὶ κακὸν ὀζομένων, καὶ κάπνιζε, καὶ μὴ
πυρία, μηδὲ σιτίῳ μηδὲ ποτῷ οὐρητικῷ χρῶ τούτου
τοῦ χρόνου, μηδὲ λοῦε θερμῷ. ἢν δὲ ἀνακεχωρήκῃ
καὶ μὴ ἀπεστραμμένη ᾖ, τοῖσιν εὐόδμοισι προσθετοῖσι χρῶ, ὅσα ἀναθερμαίνουσιν ἅμα· ταῦτα δὲ
τοιάδε εἰσί· σμύρνῃ, ἢ μύρῳ, εἴθ᾽ ἑνὶ <εἴτ᾽ ἐν>[2]
ἄλλῳ εὐόδμῳ καὶ θερμαίνοντι ἅμα· τοιούτοισι προσθετοῖσι χρῶ· καὶ πυριᾶν οἴνῳ κάτωθεν, καὶ θερμῷ
ὕδατι λούειν, καὶ διουρητικοῖσι χρῶ. τὸ δὲ δῆλόν
ἐστιν, ἢν μὴ ἀποστραφῇ ἀνακεχωρηκυῖα, ῥεῦμα

back, it produces swelling and pain. If the uterus descends downwards and turns aside to fall against the groin, it will produce pain; and if it ascends upwards, turns aside and becomes obstructed, in this way too it produces a disease, on account of its porousness; when the uterus is diseased in this way, it provokes pain in the hips and in the head. When the uterus becomes filled and swells shut, nothing flows and it fills up; when it is full, it touches the hip-joints. When the uterus has become filled with fluid, dilated and immobile, and when it touches the hip-joints, it produces pains both in the hip-joints and in the groin, and something like spheres pass by in the belly, and the patient has pain in her head, sometimes in one half, sometimes in the whole head; such is the disease that arises.

These conditions are to be treated as follows. If the uterus only protrudes and it can be anointed, apply any of the evil-smelling substances you wish, cedar, *mussotos*,[a] or any other of the heavy- and evil-smelling ones, fumigate, and neither administer a vapour-bath nor give food or a diuretic potion at this time, nor wash with warm water. If the uterus has moved upwards but not turned aside, employ pleasant-smelling applications to the uterus, that are at the same time warming: e.g. myrrh or sweet oil, either alone or together with another sweet-smelling and warming substance. Make these applications, apply a vapour-bath of wine from below, wash with warm water, and use diuretics. This is a clear indication: if the uterus has withdrawn upwards but not turned aside,

---

[a] "A savoury dish of cheese, honey, garlic, etc." (LSJ).

---

[1] Linden.   [2] Potter.

γίνεται· ἢν δὲ ἀπεστραμμένη ᾖ, οὐ γίνεται ῥόος τὰ
καταμήνια καλεύμενα· τοῦτο τὸ νόσημα πυρίη πρῶτον τοιῇδε χρὴ ἰᾶσθαι, ἐς οἶνον ἐρινεὰ ἐμβάλλοντα,
θερμαίνοντα τοῦτον, περιθέντα σικυωνίην περὶ τὸ
στόμα τοῦ τείχεος, ἐν ᾧ ἂν θερμαίνηται, ὧδε ποιῆ-
σαι· σικυωνίην μέσην διαταμών, ἐκκενώσας, τὸ
ἄκρον ἀποταμὼν σμικρόν, ὡς ἐπ᾽ ἀσκίων τοῦτο
περιπωμάσαι, ὅπως ἂν ἡ ὀδμὴ διὰ τοῦ στενοῦ ἰοῦσα
πρὸς τὴν μήτρην ἀφίκηται· καὶ θερμῷ ὕδατι αἰονᾶν,
καὶ φαρμάκοισι θερμαίνουσι χρῆσθαι προσθετοῖσι.
θερμαίνοντα δ᾽ ἐστὶ τὰ ἄγοντα τῶν πρόσθεν, τὰ δὲ
τοιάδε, βόλβιτον, χολὴ βοός, σμύρνα, στυπτηρίη,
χαλβάνη, καὶ ἄλλο ὃ τοιοῦτόν ἐστι, τούτων ὡς πλείστοισι, καὶ ὑπεξάγειν ἐλατηρίοισι φαρμάκοισι κάτω
ὅσα ἔμετον <οὐ>[1] ποιέουσιν, ἀσθενέουσιν, ὅπως μὴ
ἰνηθμὸς γένηται ἐκ τῆς ὑπερινήσιος.

Τὰ δὲ προσθετὰ ὧδε χρὴ ποιέειν, ἢν βούλῃ
ἰσχυρὰ ποιέειν· μέλι ἡμίεφθον ποιέων, ἔμβαλε τῶν
γεγραμμένων προσθετῶν τῶν ἄγειν ποιούντων, καὶ
ἐπὴν ἐμβάλλῃς, ποίησον ὥσπερ τὰς βαλάνους τὰς
πρὸς τὴν ἕδρην προστιθεμένας, μακρὰς δὲ ποίει καὶ
λεπτὰς ταύτας· τὴν δὲ γυναῖκα ὑπτίην κατακλίνας,
348   ἄνω | τοὺς πόδας ποιήσας τῆς κλίνης τοὺς πρὸς
ποδῶν, ἔπειτα πρόσθες [καὶ θερμαίνει][2] ἐν ῥάκει
δέων ἢ[3] ἄλλῳ τινί, ἕως ἂν κατακατῇ· ἢν δὲ ἀσθενέστερον βούλῃ τὸ προσθετὸν προστιθέναι, ἐς ὀθόνιον

---

[1] Linden.   [2] Del. Ermerins.

a flux occurs, but if it has turned aside, the flux called the menses does not occur. This latter condition you must treat first with the following vapour-bath: add wild figs to some wine, and heat; place a gourd over the mouth of the vessel in which the mixture is being heated as follows: cut a gourd open in the middle, empty it out, and cut off a little at the tip, so that the gourd will cover the skin vessel tightly as a lid, in order that the odour will pass through the narrow part of the gourd and arrive at the uterus. Also foment with hot water, and employ direct applications of warming medications—warming are those among the above that draw the menses, and also the following: cow's dung, bull's gall, myrrh, alum, all-heal juice, and anything else that is similar—apply a great amount of these, and evacuate downwards with laxative medications that do not provoke vomiting and are mild, in order that purging does not become excessive.

You must make uterine applications as follows, if you wish them to be forceful: prepare semi-boiled honey, add some of the emmenagogues named, and when you are ready to insert them form the mixture into suppositories like those administered per anum, long and narrow. Lie the woman down on her back, raise the legs of the bed at the foot end, then insert the pessary, wrapping it in a cloth or something else of the kind, until it melts.[a] If you wish to make a milder application, bind the pessary

---

[a] I agree with Fuchs that, although Littré's text is palaeographically preferable, its sense seems questionable: "et on fera chauffer la partie soit sur un pot de chambre, soit . . ."

[3] ῥάκει δέων ἢ Foes in note 136 after Cornarius' *panniculo intecto aut*: ἀκηδίη V: ἀμίδι ἢ Littré.

ἐνδέων. καὶ ἢν ὑγρότητος ἔμπλεαι οὖσαι αἱ μῆτραι
τὸ στόμα συνοιδήσωσιν καὶ ἀρροίην παράσχωσι,
ῥόον χρὴ ποιέοντα ἰᾶσθαι προσθετοῖσι φαρμάκοισι,
καὶ πυριῶντα ὡς γέγραπται, οὕτω ποιεῦντα, ὥσπερ
κατὰ τὴν πρόσθεν ἀρροίην· καὶ εἰ ἐς τὸ πρόσθεν
προσχωρέουσα ἀποστραφῇ, ῥόον χρὴ ποιέειν
ὥσπερ ἐπὶ τῆς πρότερον ἀρροίης. ὅταν δὲ ῥόος ᾖ
λίην, οὔτε θερμαίνειν χρὴ θερμῷ ὕδατι οὔτ' ἄλλῳ
οὐδενὶ οὔτε οὐρητικοῖσι χρῆσθαι οὔτε σιτίοισι δια-
χωρητικοῖσι· τῆς γε κλίνης τὰ πρὸς ποδῶν ὑψηλό-
τερα εἶναι, ὡς μὴ ἡ κατάκλισις εὔροος ᾖ· καὶ προσ-
θετοῖσιν ἅμα τοῖς στύφουσι χρῶ. οἱ δὲ ῥόοι, ὁπόταν μὲν εὐθέως ἵκηται ἡ κάθαρσις, εὐθέως ὑφαίμο-
νες γίνονται, ὁπόταν δ' ἧσσον ἴῃ, πυώδεες· καὶ τῇσι
νεωτέρῃσιν ὕφαιμα μᾶλλον, αἱ δὲ πρεσβύτεραι
μυξώδεα μᾶλλον ἔχουσι τὰ καταμήνια καλεύμενα.

in fine linen. If the uterus becomes full of moisture, and its mouth swells shut, so that it cuts off the outflow, you must restore the flow by treating with medicinal pessaries and vapour-baths as described, i.e. by doing the same things as in the amenorrhea above. Also if the uterus moves forward and turns aside, you must provoke its flow as in the amenorrhea above. When the menstrual flow is excessive, it is imperative that you neither warm with hot water or anything else, nor employ diuretics or laxative foods. Rather the part of the bed towards the feet must be higher than the rest, in order that the downward slant will not favour flow, and at the same time you should use astringent pessaries. The flow, when the cleaning takes place at once, becomes bloody at once, but when the cleaning occurs less quickly, it is purulent. In younger women it is more bloody, while older women have menses, as they are called, that are more mucous.

GLANDS

INTRODUCTION

In chapter 11 of *Joints*, where the writer is discussing cautery in the axilla, we read:

> All men have glands, smaller or larger, in the axilla and many other parts of the body. But the general nature of glands (περὶ ἀδένων οὐλομελίης) will be described in another treatise, both what they are, and their signification and action in the parts they occupy.[1]

Galen remarks on this passage in his *Commentary to Joints*:

> The complete system of the nature of glands, which he promised to expound elsewhere, he called "οὐλομελίη"; however such a book of Hippocrates about the general nature of glands is not extant. But one of the more recent Hippocrateans wrote a pamphlet to which he affixed the inscription "Hippocrates on the general nature of glands"; this work falls far short of the genuine Hippocratic writings in both language and thought, nor has any of the earlier physicians made mention of it, nor do those who made lists (sc. of Hippocratic works) know the book.[2]

[1] Loeb *Hippocrates* vol. 3, 226f.　　[2] Galen vol. 18A, 379.

Erotian, although not including the title in the list of
Hippocratic works in his preface, does include several
words present in *Glands* in his glossary, and one that must
derive from it: Λ24 λύματα (*Glands* ch. 12), otherwise
unknown in the Corpus, and cited in the correct number
and case.[3]

Down to Littré and Petrequin, Galen's judgement was
accepted as sufficient in itself, and final. Ermerins, however, argued that *Glands* is in fact the treatise promised
by the author of *Joints*. He defends this view by pointing
out Erotian's knowledge of the work, and referring to the
fact that its title and contents correspond exactly to the
notice in *Joints*. Furthermore, he records a number of
rare words that occur only in these two books. Generally,
subsequent scholars have rejected Ermerins' conclusion,
seeking to buttress with linguistic evidence the impossi-
bility of a unity of authorship.[4]

A number of points should be borne in mind. First,
the *prima facie* case favours unity of authorship; onus of
proof is on the side of denial. Second, Galen's opinion in
the matter is worthless as evidence: his absolute inability
to separate the historical Hippocrates from the instrument of his own medical propaganda is now patent.[5]
Third, the evidence put forth by Galen and his modern

[3] See Nachmanson p. 450 and note 1.

[4] See Joly pp. 104–10.

[5] See W. D. Smith, *The Hippocratic Tradition*, Ithaca, 1979,
pp. 123–76, and several of the articles in J. Kollesch and
D. Nickel, *Galen und das hellenistische Erbe*, Stuttgart, 1993:
e.g. P. Potter, "Apollonius and Galen 'On Joints'", pp. 117–23;
G. E. R. Lloyd, "Galen on Hellenistics and Hippocrateans: Contemporary Battles and Past Authorities", pp. 125–43.

followers ranges greatly in value: none of it seems absolutely irresistible, at least to me. The same, however, can be said of Ermerins' efforts. All of which leaves me a weak unitarian, leaning towards agnosticism.

Very systematic in its structure, the treatise progresses from a histological definition of glands (ch. 1), via brief general remarks on their pathology (ch. 2), and an account of their function (ch. 3), location, and relationship to hairs (chs. 4–6), to a discussion of the various particular glands and their diseases: tonsils (ch. 7), axillary nodes (ch. 8), lymph glands of the intestine (ch. 9), brain (chs. 10–15), and breasts (chs. 16–17).

Glands is represented in all editions of the collected Hippocratic works, including Zwinger's. A new edition by Robert Joly has recently appeared in the Budé series; it is upon Joly's edition that mine mainly depends.

ΠΕΡΙ ΑΔΕΝΩΝ ΟΥΛΟΜΕΛΙΗΣ

VIII 556
Littré

1. Περὶ δὲ ἀδένων οὐλομελίης ὧδε ἔχει. φύσις μὲν αὐτῆσι σπογγώδης, ἀραιαὶ μὲν καὶ πίονες, καὶ ἔστιν οὔτε σαρκία ἴκελα τῷ ἄλλῳ σώματι, οὔτε ἄλλο τι ὅμοιον τῷ σώματι, ἀλλὰ ψαφαρὰ καὶ φλέβας ἔχει συχνάς· εἰ δὲ διατάμοις, αἱμορραγέει λάβρως· τὸ εἶδος λευκαὶ καὶ οἷον φλέγμα, ἐπαφωμένῳ δὲ οἷον εἴρια· κἢν ὀργάσῃς[1] τοῖς δακτύλοις ἐπὶ πολὺ βιησάμενος, ἡ ἀδὴν ὑγρὸν ἀφίησιν ἐλαιῶδες, καὶ αὐτὴ θρύπτεται πολλὰ καὶ ἐξαπόλλυται.

2. Πονέουσι δὲ οὐ κάρτα, ἀλλὰ τῷ ἄλλῳ σώματι ἐπὴν πονέωσι διδοῦσιν[2] ἰδίην νοῦσον· παῦρα δὲ καὶ τῷ σώματι ξυμπονέουσιν. αἱ νοῦσοι φύματα γίνονται, καὶ χοιράδες ἀναπηδῶσι, καὶ πῦρ ἔχει τὸ σῶμα· πάσχουσι δὲ ταῦτα, ἐπὴν ὑγρασίης πληρωθῶσι τῆς ἀπὸ[3] τοῦ ἄλλου σώματος ἐπιρρεούσης εἰς αὐτάς· ἐπιρρέει δὲ ἐκ τοῦ ἄλλου σώματος διὰ τῶν φλεβῶν, αἱ δι' αὐτῶν τέτανται πολλαὶ καὶ κοῖλαι, ὥστε ἀκολουθεῖν τὸ ὑγρὸν ὅ τι ἂν ἕλκωσιν εὐπετέως ἐς αὐτάς· κἢν πολὺ ἔῃ καὶ νοσῶδες ἡ ῥοή, ξυντείνουσιν αἱ ἀδένες ἐπὶ σφᾶς τὸ ἄλλο σῶμα· οὕτω πυρετὸς ἐξά-

[1] Ermerins: ἐργ- V.

GLANDS

1. The general nature of glands is as follows. Their substance is spongy and they are rarefied and fatty; their tissues neither look like the rest of the body, nor resemble it in any other way, being instead friable and possessing many vessels: if you cut through a gland, it bleeds violently. In appearance glands are white and phlegmy, to the touch they are like wool. If you knead one in your fingers, pressing it firmly, the gland discharges an oily fluid and itself breaks almost completely into small pieces, and disappears.

2. Glands rarely become ill, but when they do, they give their disease to the rest of the body; seldom, though, do they ail in sympathy with the body. Their diseases: tubercles form, scrofulas erupt, fever seizes the body. Glands suffer these things when they become filled with moisture flowing to them from the rest of the body. This moisture flows to them out of the rest of the body along the vessels that run through them, so numerous and hollow that whatever moisture they attract easily reaches them. If the flux is copious and diseased, the glands contract the rest of the body upon themselves: in this way

---

[2] Joly: δὲ ἢ δι' V.
[3] Ermerins: ἐπὶ V.

πτεται, καὶ ἀείρονται καὶ φλογιῶσιν[1] αἱ ἀδένες.

3. Ἀδένες δὲ ὕπεισιν ἐν τῷ σώματι πλείους καὶ[2]
μείζους ἐν τοῖσι κοίλοισιν αὐτοῦ καὶ ἐν τοῖσιν
ἄρθροισι, καὶ ὅκου[3] ἐν τοῖσιν ἄλλοισιν | ὑγρηδών,
καὶ κατὰ τὰ αἱματώδεα χωρία· αἱ μὲν ὡς τὸ ἐπιρρέον
ἄνωθεν ἐς τὰ κοῖλα ἐπιδεχόμεναι ἐπὶ σφέας ἕλκωσιν, αἱ δὲ ὥστε τὴν αὖθις γινομένην ὑπὸ τῶν πόνων
ὑγρασίην ἐπιδεχόμεναι, ἐξαρύωσι[4] τὴν πληθύν,
ἥντινα μεθίησι τὰ ἄρθρα. οὕτω πλάδος οὐκ ἔνι ἐν
τῷ σώματι· εἰ γάρ τι καὶ γίνοιτο παραυτίκα, οὐκ ἂν
ἐπιγίνοιτο πλάδος ὀπίσω· καταναισιμοῦται γὰρ καὶ
τὸ πολὺ καὶ τὸ ὀλίγον ἐς τὰς ἀδένας.

4. Καὶ οὕτω τὴν πλεονεξίην τοῦ ἄλλου σώματος
αἱ ἀδένες κέρδος ποιεύμεναι, τροφὴ ξύντροφος αὐτῇ-
σίν ἐστιν· ὥστε ὅκου τελματώδεα, ἐκεῖ καὶ ἀδένες·
σημεῖον, ὅκου ἀδήν, ἐκεῖ καὶ τρίχες· ἡ γὰρ φύσις
ποιέει ἀδένας καὶ τρίχας <καὶ>[5] ἄμφω χρέος τωὐτὸ
λαμβάνουσιν, αἱ μὲν ἐς[6] τὸ ἐπιρρέον, ὡς καὶ ἔμπροσθεν εἴρηται· αἱ δὲ τρίχες τὴν ἀπὸ τῶν ἀδένων
ἐπικαιρίην ἔχουσαι φύονταί τε καὶ αὔξονται, ἀνα-
λεγόμεναι τό τε περισσὸν καὶ ἐκβρασσόμενον ἐπὶ
τὰς ἐσχατιάς. ὅκου δὲ αὖον ἐὸν τὸ σώμα, οὔτε ἀδὴν
οὔτε θρίξ· τὰ δὲ ἁπαλὰ καὶ πονεύμενα καὶ κάθυγρα,
ἀδενώδεα καὶ τρίχας <ἔχει>·[7] ἀδένες δὲ καὶ κατὰ
τοῖν οὐάτοιν ἔνθα καὶ ἔνθα ἑκατέρωθεν κατὰ τὰς
σφαγὰς τοῦ τραχήλου, τρίχες τε ἐνταῦθα ἑκατέρω-

---

[1] Ermerins: -γῶσιν V.   [2] Ermerins: ἢ V.
[3] Ermerins: ὁκόσα V.

fever is kindled, and the glands swell up and become inflamed.

3. Glands are present in greater number and size in the hollow parts of the body and the joints, and where moisture is present in other parts, and in regions rich in blood. Some glands are placed so as to attract and receive the moisture that flows down from above into the hollow parts, others again so as to take over the moisture that arises in exertions, and to drain off most of what the joints secrete. In this way, no collection of fluid is permitted to form in the body; for even if some fluid should suddenly arise, afterwards no collection can follow, since quantities both large and small are consumed into the glands.

4. In this way, too, the glands profiting from the surplus present in the rest of the body, the nourishment of the body is also their nourishment. Thus, in any parts where there is fluid, there are also glands. Here is proof: where there is a gland, there too are hairs, for nature makes glands and hairs, and they both fulfil the same office, the one (i.e. glands) with regard to what flows to them, as was described above, while the hairs, taking advantage of what is provided from the glands, grow and increase by collecting the excess that is cast out to the surface. Where the body is dry, there is neither gland nor hair, whereas the parts that are tender, active in exertions and thoroughly moist are glandular and possess hairs. There are glands scattered about the ears on both sides near the jugular vessels of the neck, and there are hairs

---

⁴ ἐξαρύωσι Littré: ἐν τοῖσιν ἄρθροισιν V.
⁵ Ermerins.　　⁶ Froben: ὡς V.
⁷ Ermerins: ἀδὴν ὧδε καὶ τρίχες V.

111

θεν· ἐπὶ ταῖς μασχάλησιν ἀδένες καὶ τρίχες· βου-
βῶνες καὶ ἐπίσιον ἰκέλως μασχάλησιν [ἀδὴν καὶ
τρίχες].[1] ταῦτα μὲν κοῖλα τῶν ἐν τῷ σώματι καὶ ῥηΐ-
δια ἐς περιουσίην | ὑγροῦ· καὶ γὰρ πονέει ταῦτα καὶ
κινέεται μάλιστα τῶν ἐν τῷ σώματι.

5. Τὰ δ' ἄλλα ὁκόσα ἀδένας ἔχει μοῦνον, οἷον
ἔντερα, ἔχει γὰρ καὶ ταῦτα ἀδένας ἐς τὸ ἐπίπλοον
μείζονας, τρίχας οὐκ ἔχει. καὶ γὰρ ἐν τοῖσι τέλμασι
τῆς γῆς καὶ καθύγροισιν οὐ φύεται τὸ σπέρμα οὔτ'
ἐθέλει ἀναβαίνειν τῆς γῆς ἄνω, ἀλλ' ἀποσήπεται
καὶ ἀποπνίγεται τῇ πλεονεξίῃ· βιῆται γὰρ τὸ
σπέρμα. βιῆται δὲ καὶ ἐν τοῖσιν ἐντέροισιν ἡ πλη-
θὺς καὶ τὸ ὑγρὸν πολύ, καὶ οὐκ ἂν φύσαι τρίχας. αἱ
δὲ ἀδένες μείζονες ἤ κου ἄλλοθι τοῦ σώματος· καὶ
νέμονται αἱ ἀδένες ἐν τοῖσιν ἐντέροισιν ἐκπιεζόμεναι
τὸν πλάδον· τὰ δὲ ἔντερα ἐκ τῶν τευχέων ἐς τὰ ἐπί-
πλοα ἐκδέχεται καὶ καθήσι τὴν ὑγρασίην· τὸ δὲ
ἐπίπλοον διαδιδοῖ τῇσιν ἀδέσιν.

6. Ἔχουσι καὶ οἱ νεφροὶ δὲ ἀδένας· καὶ γὰρ
οὗτοι κορίσκονται πολλῆς ὑγρασίης· μείζους δὲ αἱ
ἀδένες ταύτῃ ἢ αἱ ἄλλαι ἀδένες ἐοῦσαι· οὐ γὰρ
ἐμπίνεται τοῖσι νεφροῖσι τὸ ὑγρὸν τὸ ἐπιρρέον,
ἀλλὰ διαρρέει ἐπὶ κύστιν κάτω, ὥστε ὅ τι ἂν ἀπο-
κερδάνωσιν ἀπὸ τῶν ὀχετῶν, τοῦτο ἕλκουσι πρὸς
σφέας.

7. Καὶ ἄλλαι δέ εἰσιν ἐν τῷ σώματι ἀδένες σμι-
κραὶ [καὶ][2] πάνυ, ἀλλ' οὐ βούλομαι ἀποπλανᾶν τὸν
λόγον· ἐς γὰρ τὰς ἐπικαίρους ἡ γραφή. νῦν δὲ ἀνα-
βήσομαι τῷ λόγῳ, καὶ ἐρέω περὶ ἀδένων οὐλομελίης

there on both sides; in the axillae there are glands and hairs; groins and pubis similarly to the axillae. These are hollow parts of the body and easy for excessive moisture to occupy, especially since they labour and move more than any other parts.

5. Other parts of the body which have only glands, like the intestines (these have quite large glands situated in the omentum), lack hairs. After all, in the earth's marshes and wet places a seed fails to germinate and is unable to come up above the soil, being decomposed and suffocated by the excessive moisture, since the moisture overpowers the seed. So too do the abundance and great moisture in the intestines have an overpowering effect, and fail to bring forth hairs. The glands there are larger than in any other part of the body; they gain their nourishment in the intestines by squeezing out moisture for themselves: the intestines receive moisture from the vessels and send it to the omentum, the omentum in turn passes it on to the glands.

6. The kidneys, too, have glands, since they also are saturated with much moisture. The glands there are larger than the other glands, for the moisture that flows in is not soaked up by the kidneys, but flows through them down to the bladder, so that whatever the glands acquire from the pipes they draw to themselves.

7. In the body there are also other very small glands, but I do not want to wander off from my subject; this treatise, after all, pertains to the essentials. So now I shall go on in my argument and discuss the general nature of the

---

[1] Del. Potter.　　[2] Del. Ermerins.

τραχήλου· τράχηλος τὰ μέρεα αὐτοῦ ἑκάτερα ἔνθα
καὶ ἔνθα ἀδένας ἔχει, καὶ παρίσθμια καλέονται αἱ
ἀδένες αὗται· χρείη τοιήδε· ἡ κεφαλὴ ὑπέρκειται
ἄνω κοίλη ἐοῦσα καὶ περιφερὴς καὶ <πλήρης>[1] τῆς
562   περὶ | αὐτὴν ἀπὸ τοῦ ἄλλου σώματος ὑγρασίης· καὶ
ἅμα ἀναπέμπει τὸ σῶμα ἀτμοὺς ἐς τὴν κεφαλὴν
παντοίους ἄνω, οὓς αὖθις ἡ κεφαλὴ ὀπίσω ἀφίησιν·
οὐ γὰρ δύναται ἐμμένειν τὸ ἐπιρρέον οὐκ ἔχον ἔνθα
ἕδρην, ἢν μὴ τὴν κεφαλὴν πονέῃ, τότε οὐκ ἀνίησιν,
ἀλλ᾽ αὐτοῦ κρατέει· ἐπὴν δὲ ἀνῇ τὴν ἕλξιν, ἐς τὰς
ἀδένας ἡ ῥοὴ γίνεται, καὶ οὐδὲν λυπέει τὸ ῥεῦμα,
ἔστ᾽ ἂν ὀλίγον τε ᾖ καὶ ξύμμετρον καὶ ἐγκρατέες
ἔωσιν αἱ ἀδένες· ἢν γὰρ πολὺ ἐπιρρυῇ δριμύ, ἢν μὲν
ᾖ δριμὺ καὶ κολλῶδες, φλεγμαίνει καὶ ἀνοιδίσκεται
καὶ ξυντείνει ὁ τράχηλος, καὶ οὕτω προίει ἐς οὖς·
κἢν μὲν <ἐς>[2] ἑκάτερα τὰ μέρεα, <ἑκάτερον>·[2] ἢν δὲ
ἐς θάτερον, πονέει θάτερον· ἢν δὲ ᾖ φλεγματῶδες
καὶ πολὺ καὶ ἀργὸν ἡ ῥοή, φλεγμαίνει δὲ καὶ ὧδε·
καὶ ἡ φλεγμονή, στάσιμον ἐὸν ὑγρόν, χοιράδες
ἐγγίνονται· αὗται δ᾽ εἰσὶν[3] αἱ νοῦσοι τραχήλου.

8. Μασχάλῃσι δὲ ξυρρέει μὲν καὶ ἐνταῦθα, ἀλλ᾽
ὅταν πλῆθος ᾖ δριμέος ἰχῶρος,[4] καὶ ὧδε γίνονται
φύματα. κατὰ ταὐτὰ καὶ ἐν τοῖσι βουβῶσιν ἕλκει
τὴν ἀπὸ τῶν ὑπερκειμένων ὑγρασίην ἡ ἀδήν· ἄλλως
<τ᾽>[5] εἰ πλῆθος λάβοι, βουβωνοῦνται καὶ διαπυΐσκε-
ται καὶ φλεγμαίνει ἰκέλως μασχάλῃσί τε καὶ τρα-
χήλῳ· τὰ δ᾽ αὐτὰ οἱ δοκέει παρέχειν ἀγαθὰ καὶ

----

[1] Zwinger in margin.   [2] Aldina.

glands of the throat. The throat has glands here and there on both sides, and these glands are called the paristhmia. This is their function: the head is situated in the superior position, hollow, round, and filled with moisture it has from the rest of the body; at the same time, the body sends up vapours of every sort to the head, which the head in turn sends back; for this influx, having nowhere to settle, cannot remain in the head without making the head ill: in that case the head does not send it away, but it gains the upper hand there. If, however, the head relaxes its attraction, a flux to the glands occurs; this fluid causes no damage, as long as it is small in amount and proportional, and the glands can control it. But if there is a copious sharp afflux, which is sharp and viscous, the throat fills with phlegm, swells up and is stretched tight, and as a consequence there is a discharge to the ear; if this happens on both sides, each ear is involved, if on one side, then one ear suffers. If, on the other hand, the flux is like phlegm, copious and inert, swelling occurs in this case, too, but the swelling, since the moisture tends towards stasis, involves the formation of a scrofula. These are the diseases of the throat.

8. To the axillae, too, there are affluxes; when there is a great amount of sharp serum, tubercles form this way too. In the same way, the gland in the groins draws moisture from the regions above; especially if it takes in a great amount, the groins swell up, suppurate and fill with phlegm, just as happens in the axillae and the throat. In the groins, too, glands appear to produce the same good

---

3 δ' εἰσὶν Potter: χρειήοις V: del. Ermerins.
4 Potter: δριμεῖς ἰχῶρες V.   5 Joly.

κακά. καὶ ταῦτα μὲν ἀμφὶ τῶνδε.

9. Τὰ δὲ ἔντερα ἔχει κόρον πολὺν ἀπό τε σιτίων

564 καὶ ποτῶν· | ἔχει δὲ καὶ τὴν ὑπὸ τοῦ δέρματος ὑγρα-
σίην· αὕτη πᾶσα ἀπαναισιμοῦται ἰκέλη τοῖς πρόσ-
θεν· νούσους δὲ οὐ ποιεῖ τὰ πολλά, ὅκως περ καὶ ἐν
τοῖσιν ἄρθροισι γίνεται· συχναὶ γάρ τοι ἀδένες καὶ
ἀναπετέες, καὶ οὐ κοῖλαι, καὶ οὐ πολὺ πλῆθος ἐπαυ-
ρισκόμεναι ἡ ἑτέρη τῆς ἑτέρης, ἐπεὶ μᾶλλον πλεονε-
κτεῖν ἐθέλουσα οὐδεμίη τότε πλέον[1] ἔχειν δύναται,
ἀλλ᾽ ὀλίγον ἑκάστη[2] ξυρρέον ἐς τὸ ἄρθρον ἐς πολλὰ
διαιρεόμενον· ἰσότης ἐστὶν αὐτῇσιν.

10. Ἡ κεφαλὴ καὶ αὐτὴ τὰς ἀδένας ἔχει, τὸν
ἐγκέφαλον ἴκελον ἀδένι· ἐγκέφαλος γὰρ καὶ λευκὸς
καὶ ψαφαρός, ὅκως περ καὶ ἀδένες, καὶ ταὐτὰ ἀγαθὰ
τῇσιν ἀδέσι ποιεῖ τὴν κεφαλήν· ἐνεοῦσαν[3] <γὰρ>[4]
διὰ τὰ εἰρημένα μοι, τιμωρέων ὁ ἐγκέφαλος ἀποστε-
ρέει τὴν ὑγρασίην, καὶ ἐπὶ τὰς ἐσχατιὰς ἔξω ἀπο-
στέλλει τὸ πλέον ἀπὸ τῶν ῥόων. μείζων δ᾽ ὁ ἐγκέφαλος τῶν ἄλλων ἀδένων, καὶ αἱ τρίχες μείζους ἢ αἱ
ἄλλαι τρίχες· μείζων τε γὰρ ὁ ἐγκέφαλος καὶ ἐν
εὐρυχωρίῃ κέεται τῇ κεφαλῇ.

11. Ποιέει δὲ νούσους καὶ ἥσσονας καὶ μείζονας
ἢ αἱ ἄλλαι ἀδένες· ποιέει δέ, ὁκόταν ἐς τὰ κάτω τοῦ
σώματος τὴν σφετέρην πλεονεξίην ἀποστείλῃ. ῥόοι
δὲ ἀπὸ κεφαλῆς ἐνίοτε ἀποκρίνονται,[5] δι᾽ ὤτων κατὰ

---

[1] Joly: πλήην V.
[2] ὀλίγον ἑκάστη Littré: ἐς ὀλίγον ἑκάστης V.

and bad effects. So much on this subject.

9. The intestines have a great nutritional abundance which comes from the foods and drinks; also they have the moisture under the skin. All this moisture is consumed just as in the cases above, but in most instances, contrary to what happens at the joints, it does not provoke diseases. For in the intestines the glands are thickly distributed and wide open; neither do they have cavities, nor does one gland obtain a great quantity at the expense of another, since none is ever able to have an excessive amount—even though it might wish to have more than its share—but each gets a little of what flows to the part and is divided between many: i.e. there is an equal division among them.

10. The head too has glands, viz. the brain which is like a gland; for the brain is white and friable just like glands, and it renders the same benefits in the head as do glands. For whatever moisture occupies the head in the manner I have described, the brain helps by drawing off, and it sends away to the extremities most of what arises from fluxes. The brain is larger than other glands, and the hairs over it are longer than other hairs; for the brain is of greater size, and it lies in an open space in the head.

11. The brain provokes diseases that are both less and more severe than those provoked by other glands; it does this when it sends its own surplus to the lower parts of the body. Fluxes from the head are sometimes secreted

---

[3] Joly: ἐοῦσαν V.

[4] Ermerins.

[5] ἐνίοτε ἀποκρ. Potter: ἕως ἀποκρίσιος V.

φύσιν, δι᾽ ὀφθαλμῶν, διὰ ῥινῶν· τρεῖς οὗτοι· καὶ
ἄλλοι δι᾽ ὑπερώης ἐς φάρυγγα, ἐς στόμαχον· ἄλλοι
διὰ φλεβῶν ἐπὶ νωτιαῖον, ἐς τὰ ἰσχία,[1] οἱ πάντες
ἑπτά.

12. Οὗτοι τοῦ τε ἐγκεφάλου λύματά εἰσιν ἀπιόν-
566 τες· καὶ εἰ μὴ | ἀπίοιεν, νοῦσος αὐτῷ. οὕτω δὲ καὶ
τῷ ἄλλῳ σώματι, ἢν ἐς τὰ ἔνδον ἀπίωσι καὶ μὴ ἔξω,
καὶ αὐτοῖς ὄχλος πολύς, κἄνδοθεν ἑλκοῖ· καὶ δριμὺ
μὲν εἰ πρόοιτο ὁ ἐγκέφαλος ῥεῦμα, τὰς ἐπιρροὰς
ἐσθίει καὶ ἑλκοῖ· καὶ τὸ μὲν ἐπιὸν ἢν ᾖ πλῆθος
κατιὸν ἁλές, οὐκ ἀνέχει ὁ ῥόος, ἔστ᾽ ἂν ἐξαρύηται
τὴν πληθὺν τοῦ κατιόντος· καὶ τὸ μὲν ἐπιρρέον ἀπο-
πέμπων ἔξω, ἕτερον δὲ ἐσδεχόμενος, ἐς τὸ ὅμοιον
αἰεὶ καθιστάμενος· τά τε ὑγρὰ ἑλκοῖ[2] καὶ ποιέει νούσους. ἄμφω δὲ ἐν ἀηθίῃ[3] καταγυιοῖ τὴν φύσιν· καὶ
ἢν πάθῃ, δυ᾽ ἐστὸν κακία· τὰ μὲν γὰρ πάθη τῆς
φύσεως, οἱ προειρημένοι ῥόοι· δυσφορέουσι τὸ πλῆθος, καὶ ὀδάξον καὶ[4] ἄλογον καὶ οὐ ξύνηθες ὄν· ὁ δὲ
ἐγκέφαλος πῆμα ἴσχει καὶ αὐτὸς οὐχ ὑγιαίνων· ἀλλ᾽
εἰ μὲν δάκνοιτο, τάραχον πολὺν ἴσχει, καὶ ὁ νόος
ἀφραίνει, καὶ ὁ ἐγκέφαλος σπᾶται καὶ ἕλκει τὸν
ὅλον ἄνθρωπον, ἐνίοτε[5] δ᾽ οὐ φωνέει καὶ πνίγεται,
ἀποπληξίῃ τῷ πάθει τοὔνομα. ἄλλοτε δὲ δριμὺ μὲν

---

[1] Littré: τὸ αἷμα V.
[2] Potter: ἕλκη V.
[3] Potter: ἀκηδίῃ V.
[4] Joly: ὀδάξον τὸ Ermerins: ὀδάξονται τὸ V.

through the natural passages of the ears, eyes or nose: these are three possibilities; others flow through the palate into the throat or into the oesophagus, others through the vessels into the spinal marrow, or to the hips: seven possibilities in all.[a]

12. These fluxes are purgations of the brain when they pass off, and if they do not pass off, the brain becomes ill. So too in the rest of the body, if they pass into the interior rather than to the exterior, it means great trouble for those parts, with ulcers forming in the interior. If the brain sends forth a fluxion that is sharp, this corrodes and ulcerates the channels by which it arrives. If what arrives is great in quantity and passes down all in a mass, the flux does not cease until it has drained off most of what is passing down. In continually sending one afflux away while at the same time receiving a new one, a constant state becomes established, and the moisture causes ulcerations and provokes diseases. Both these fluxes in their abnormality weaken the constitution, and if a person becomes ill, the evils are two. For first there are the affections of the constitution, the fluxes mentioned above; these patients bear up poorly under the quantity and the fact that it is irritating, inappropriate and unaccustomed. Secondly, the brain is harmed and is not healthy itself; and if it is irritated, it suffers a great disturbance, the mind is deranged, and the brain pulls and convulses the whole person, who sometimes becomes speechless and is suffocated; the name of this disease is apoplexy. Other

[a] Cf. *Places in Man* 10.

---

[5] Littré after Dietz: ἐν ἑωυτῷ V.

568 οὐ ποιέεται τὸ[1] ῥεῦμα, πλῆθος δ' | ὂν τὸ ἐμπεσὸν
πονέει τοῦτο, καὶ ἡ γνώμη ταράττεται, καὶ περίεισιν
ἀλλοῖα φρονῶν καὶ ἀλλοῖα ὁρέων· φέρων τὸ ἦθος
τῆς νούσου σεσηρόσι μειδιήμασι καὶ ἀλλοκότοισι
φαντάσμασιν.

13. Ἄλλος ῥόος ἐπ' ὀφθαλμούς, ὀφθαλμίαι, καὶ
οἰδέουσιν αἱ ὄψεις. εἰ δὲ ἐπὶ ῥῖνας ὁ κατάρρους,
ὀδάξονται μυκτῆρας, καὶ ἄλλο οὐδὲν δεινόν· αἵ τε
γὰρ ὁδοὶ τούτων εὐρέαι καὶ ἱκαναὶ τιμωρέειν σφίσιν·
πρὸς δὲ καὶ ἀσύστροφον τὸ ἀπιὸν αὐτῇσι. τὰ δὲ
οὔατα σκολιὸς μὲν πόρος καὶ στεινός· ὁ δ' ἐγκέφαλος πλησίον αὐτοῦ αὐτοῖσιν ἐμπιέζεται· νοσέων δὲ
τὴν νοῦσον ταύτην, τὰ πολλὰ ἀποκρίνει κατὰ τὸ οὖς
ἀπὸ τῆς πυκινῆς ῥοῆς <καὶ>[2] ἀνὰ χρόνον κεχώρι-
σταί[3] τε καὶ ῥέει δυσῶδες πῦον. οὕτως ἐς τὰ ἔξω
δῆλοι τῷ ὀφθαλμῷ ῥόοι, καὶ οὐ πάμπαν θανατώδεες.

14. Ἢν δὲ ὀπίσω [τὸ ῥεῦμα][4] ἴῃ δι' ὑπερῴης τὸ
ἀφικόμενον φλέγμα ἐς τὴν κοιλίην, ῥέουσι μὲν καὶ
αἱ κοιλίαι τούτων, νοσέουσι δὲ οὔ· ἀναμένοντος
κάτω τοῦ φλέγματος, εἰλεοὶ καὶ[5] πάθη χρόνια·
ἄλλοις δι' ὑπερῴης ἐπὶ φάρυγγα, ἢν πολὺ ῥυῇ καὶ
ἐπὶ πολύ, αἱ νοῦσοι φθινάδες· κορίσκονται γὰρ τοῦ
570 φλέγματος οἱ πνεύμονες, καὶ γίνεται | τὸ πῦον·
τοῦτο διεσθίει τοὺς πνεύμονας· καὶ οἱ νοσέοντες οὐ
ῥᾷον περιγίνονται· καὶ ἡ γνώμη τοῦ ἰητροῦ, καὶ ἢν
ἀγαθὸς καὶ ἢν ἀγχίνοος, τὰ πολλ' ἀξυνετέει τῆς
προφάσιος.

[1] Potter: ποιέει αὐτὸ V.

120

times the fluxion is not made sharp, but the afflux, being great in quantity, causes pain; the reason is disturbed and the victim goes about thinking and seeing alien things; one bears this kind of disease with grinning laughter and grotesque visions.

13. Another flux is to the eyes—ophthalmia—and the organs of sight swell up. If the defluxion is to the nose, the nostrils become itchy, but nothing else untoward occurs, for these passages are wide and sufficient to look after themselves; furthermore what goes out through the nose does not tend to congeal. In the ears, the channels are curved and narrow; besides, the brain, being close, exerts pressure on them. If a person has this disease, generally he secretes through the ear what comes from the frequent flux; with time this is separated off, and ill-smelling pus flows out. In this way fluxes are revealed to the eye as they flow to the exterior, and such are not altogether mortal.

14. If the phlegm that arrives from above goes backward through the palate into the cavity, these persons have a flux of the cavities, but do not become ill; if, however, the phlegm tarries in the lower parts, ileus and chronic sufferings arise. In other persons, a flux through the palate into the throat, if it is copious and lasts a long time, produces consumptive diseases; for the lungs become saturated with the phlegm, and pus is formed, which eats its way through the lungs. These patients survive only with difficulty; the physician's understanding, even if he is good and he is shrewd, in most cases fails to grasp the cause.

---

² Ermerins.   ³ Ermerins: ὤρισταί V.
⁴ Del. Joly.   ⁵ Joly: τὰ V.

121

Ἄλλη νοῦσος· διὰ φλεβῶν ἐπὶ νωτιαῖον ἀπὸ
καταρρόου κεφαλῆς· ἀΐσσει δὲ ἐντεῦθεν ἐπὶ ἱερὸν
ὀστέον, ἄγων τὴν ἐπιρροὴν ὁ νωτιαῖος, καὶ ἐναπέ-
θετο τῆσι κοτύλῃσι τῶν ἰσχίων· [ἰσχία]¹ καὶ ἦν
ποιέῃ φθίσιν, [καὶ]² μαραίνεται ὁ ἄνθρωπος ὧδε καὶ
ὧδε καὶ ζώειν οὐκ ἐθέλει· ταχὺ γὰρ πονέει τὴν
σπάθην, καὶ ἄμφω τὼ πόδε καὶ μηρὼ παρέπονται,
καὶ αἰεὶ τελέως ὄλλυνται χρόνῳ πολλῷ μελεδαινόμε-
νοι, καὶ οὕτως ἀπηύδηκε καὶ θνήσκει. ταῦτά μοι
<περὶ>³ ῥόων ἀπὸ κεφαλῆς εἴρηται.

15. Καὶ πάθεα ἐγκεφάλου καὶ ἄλλαι νοῦσοι,
παραφροσύναι καὶ μανίαι, καὶ πάντα ἐπικίνδυνα,
καὶ πονέει ὁ ἐγκέφαλος καὶ αἱ ἄλλαι ἀδένες· ἔχει
γὰρ καὶ τόνον καὶ ἄλλη ξύνοδος ἐνταῦθα πάλιν τοῦ
σώματος.

16. Ἀλλὰ καὶ ἀδένες ἐν τοῖσι στήθεσι μαζοὶ
καλέονται, καὶ | διαίρονται γάλα ποιέουσιν· τοῖς δὲ
οὐ ποιέουσι γάλα, <οὔ>·⁴ ποιέουσι μὲν αἱ γυναῖκες,
οἱ δὲ ἄνδρες οὐ ποιέουσι. τῆσι μὲν γυναιξὶν ἀραιή
τε ἡ φύσις κάρτα⁵ τῶν ἀδένων, ὥσπερ τὸ ἄλλο
σῶμα, καὶ τὴν τροφήν, ἥντινα ἕλκουσιν ἐπὶ σφᾶς,
ἀλλοιοῦσιν ἐς τὸ γάλα· καὶ ἀπὸ τῆς μήτρης παρα-
γίνεται ἐπὶ τοὺς μαζοὺς ἐς τὴν μετὰ τὸν τόκον τῷ
παιδίῳ τροφήν, ἥν τινα ἀποπιέζει καὶ ὑπερβάλλει
τὸ ἐπίπλοον ἐς τὰ ἄνω, στενοχωρούμενον ὑπὸ τοῦ
ἐμβρύου.

Τοῖσι δὲ ἄρρεσι καὶ ἡ στενοχωρίη καὶ ἡ πυκνό-
της τοῦ σώματος μέγα συμβάλλεται μὴ εἶναι μεγά-

572

Another disease: a flux down from the head through the vessels to the spinal marrow; from there the afflux darts along the spinal marrow to the sacrum, to be deposited in the cups of the hips. If the flux leads to consumption, the person wastes away bit by bit and does not care to live; for soon he is affected in the shoulder blades, his two legs and thighs follow—in the end such patients always perish after a long period of care—and so he fails and dies. This is what I have to say about fluxes from the head.

15. Other diseases too are affections of the brain—derangements and delirious states—and all these things are dangerous. Both the brain and the other glands suffer; for the brain has a band connecting it to the other glands, and also there is another conjunction there coming back from the body.

16. The glands on the chest are called breasts, and these develop in individuals that make milk, not in those that do not make milk: women make milk, but men do not. In women the substance of the glands is very rarefied, just like the rest of their bodies, and the nourishment these glands draw to themselves they alter into milk. This passes from the uterus to the breasts as nourishment for the baby after its birth, being squeezed out by the omentum and cast up to the higher regions of the body when it becomes cramped by the fetus.

In males it is largely the compactness and density of their bodies that contribute to the smallness of these

---

[1] Del. Ermerins.   [2] Del. Linden.
[3] Foes after Cornarius' *de*.
[4] Littré.   [5] Zwinger in margin: καὶ κατὰ V.

λας τὰς ἀδένας· τὸ γὰρ ἄρρεν ναστόν ἐστι καὶ οἷον
εἶμα πυκνὸν καὶ ὀρέοντι καὶ ἐπαφωμένῳ· τὸ δὲ θῆλυ
ἀραιὸν καὶ χαῦνον καὶ οἷον εἴριον ὀρέοντι καὶ ἐπα-
φωμένῳ· ὥστε τὴν ὑγρασίην οὐ μεθίησι τὸ ἀραιὸν
καὶ μαλθακόν· τὸ δὲ ἄρσεν οὐκ ἄν τι προσδέξαιτο,
πυκνόν τε ἐὸν καὶ ἀστεργές, καὶ ὁ πόνος κρατύνει
αὐτοῦ τὸ σῶμα, ὥστε οὐκ ἔχει δι' οὗ λήψεταί τι[1] τῶν
περισσῶν. οὕτως ἀναγκάζει ὅδε ὁ λόγος καὶ στήθεα
καὶ μαζοὺς καὶ τὸ ἄλλο σῶμα τῆσι γυναικὶ χαῦνα
καὶ μαλακὰ εἶναι καὶ διὰ τὴν ἀργίην καὶ διὰ τὰ
προειρημένα· τοῖσι δὲ ἀνδράσι τὰ ἐναντία.

17. Ποιέουσι καὶ μαζοὶ φύματα, φλεγμονάς, τὸ
574 γάλα ἀποσή|ποντες· ἀγαθὰ δὲ ἔχουσι τοῖσιν ἔμπροσ-
θεν ὅμοια· ἀποστερίζουσι τὴν πλεονεξίην τοῦ ἄλλου
σώματος. μαρτύριον τῆσι γυναιξὶν ᾗσιν ἀφαιρεῖται
νούσῳ ἢ ἄλλῃ τινὶ ξυμφορᾷ[2] μαζός· καὶ ἡ φωνὴ
θρασεῖα, καὶ ὑγρὰ εἰς στόμαχον, καὶ πτυελίζουσι,
καὶ τὴν κεφαλὴν ἀλγέουσι· καὶ ἀπὸ τῶνδε νοσέουσιν·
ἰὸν γὰρ ἀπὸ τῆς μήτρης καὶ ἐπιρρέον τὸ γάλα ὥσπερ μεθίετο[3] καὶ ἔμπροσθεν ἐς τὰ ἄνω [τεύχεα],[4] τὰ
οἰκεῖα οὐκ ἔχον τεύχεα, συντυγχάνει τοῖσι κυρίοις
τοῦ σώματος, καρδίῃ, πνεύμονι, καὶ ἀποπνίγονται.

---

[1] Ermerins: τὸ V.
[2] Froben: νοῦσος ἢ ἄλλη τις ξυμφορὰ V.
[3] Joly: μεθίη V.
[4] Del. Joly.

glands; for the male is close-pressed like a thick carpet both in appearance and to the touch. The female, on the other hand, is rarefied and porous like a flock of wool in appearance and to the touch: it follows that this rarefied and soft tissue does not reject moisture. The male cannot accept anything, first because of his naturally dense and unyielding substance, and then exercise, too, strengthens his body, so that it has no way of taking up any excess. Thus, this reasoning proves that the chest, breasts and rest of the body in females are porous and soft, both because of the sex's inactivity and because of what has been said above, whereas in males the opposite is true.

17. The breasts develop tubercles and inflammations when their milk turns bad. Their positive actions are the same as noted above: to carry off the rest of the body's surplus. Proof of this is provided by women that lose a breast either in a disease or because of some other accident: their voice becomes bold, fluid enters their oesophagus, they secrete excessive saliva, and they have pains in the head. They suffer these things because of the following: since milk continues to leave the uterus and flow out towards the upper regions of the body as it did before, not having a proper receptacle it lights upon the principal parts of the body such as the heart and lung, and the women suffocate.

FLESHES

INTRODUCTION

The only trace of this work to be found in ancient writings appears in the Hippocratic glossaries of Erotian[1] and Galen, both of which contain the uniquely occurring word ἀνακῶς (*Fleshes* ch. 19). One or more of three other entries in Galen's *Glossary* may also derive from *Fleshes*, although not necessarily so, since they do occur elsewhere in the Corpus: αἰών (*Fleshes* ch. 19 bis); συριγγώδη (chs. 3, 7, 15); τέως (chs. 8, 9, 13).

The theory put forth in the first part of the treatise, and developed with such energy and care, marks its author as one of the most extraordinary scientific minds of the Corpus; the reasoning here is comparable in power and originality to anything we find in *Generation-Nature of the Child*, *Diseases IV*, *Regimen*, or the *Sacred Disease*.

Beginning from a division of the macrocosm into aether = heat, earth (cold and dry), and air (thick and moist), the author traces the effects of heat and cold on the various elements and their combinations, as they give rise to the tissues of the human body. His basic "chemical" principles are:

[1] Cf. Nachmanson p. 387, note 2 and p. 411.

Heat: {
burns the fatty to form bone
bakes the gluey to form membranes
dries the moist to make it disappear
melts the cold to make it flow away.

Cold: congeals the liquid to make it stand still.

Consistent application of these principles results in cogent explanations of the formation of the hollow parts and cords (ch. 3), brain (ch. 4), heart and vessels (chs. 5, 6), lung (ch. 7), liver (ch. 8), spleen and kidney (ch. 9), joints (ch. 10), nails (ch. 11), and hair (ch. 14). Only in the case of the teeth (chs. 12, 13) does the author fall back on the traditional "like to like" principle of other Hippocratic treatises.[2]

The second part of the treatise explores the mechanisms of hearing (ch. 15), smell (ch. 16), sight (ch. 17), and speech (ch. 18), generally achieving a judicious synthesis of common Hippocratic anatomical beliefs and physiological principles (e.g. hearing as echo; seeing as reflection; phonation as resonance within the chest) with ideas otherwise unrecorded in the Corpus (e.g. smell as the perception of the drier by the moister; the cornea as a skin formed on the surface of moisture flowing from the brain when it comes into contact with external cold; the articulation of speech as the act of the tongue touching the palate and teeth).

Chapter 19 takes up a theme remote from those of the first two sections, setting out to illustrate the special significance of the seven-day period in biology. Examples,

[2] Cf. C. W. Müller, *Gleiches zu Gleichem*, Wiesbaden, 1965, pp. 112–50.

many of which are also alluded to in other Hippocratic treatises, include the time taken for the first formation of the fetus, the fact that the seven-month, but not eight-month, fetus is viable, and the special importance of the seven-day period, or fraction of it, in the course of acute diseases.[3]

Several emendations to the text of *Fleshes* are proposed in: W. A. Heidel, "Hippocratea I", *Harvard Studies in Classical Philology*, 25 (1914), 178–187 (= Heidel). An edition, German translation, and commentary of *Fleshes* appeared as: K. Deichgraeber, *Über Entstehung und Aufbau des menschlichen Körpers ... mit einem sprachwissenschaftlichen Beitrag von Ed. Schwyzer*, Leipzig and Berlin, 1935 (= Deichgraeber or Schwyzer). *Fleshes* is also included in Joly's Budé volume XIII (1978).

---

[3] Cf. Wilhelm Roscher, *Die Hebdomadenlehren der griechischen Philosophen und Ärzte*, Leipzig, 1906, pp. 44–86.

ΠΕΡΙ ΣΑΡΚΩΝ

VIII 584
Littré

1. Ἐγὼ τὰ μέχρι τοῦ λόγου τούτου κοινῇσι γνώμῃσι χρέομαι ἑτέρων τε τῶν ἔμπροσθεν, ἀτὰρ καὶ
ἐμεωυτοῦ· ἀναγκαίως γὰρ ἔχει κοινὴν ἀρχὴν ὑποθέ-
σθαι τῇσι γνώμῃσι βουλόμενον ξυνθεῖναι τὸν λόγον
τόνδε περὶ τῆς τέχνης τῆς ἰητρικῆς. περὶ δὲ τῶν
μετεώρων οὐδὲν[1] δέομαι λέγειν, ἢν μὴ τοσοῦτον ἐς
ἄνθρωπον ἀποδείξω καὶ τὰ ἄλλα ζῷα, ὅκως[2] ἔφυ καὶ
ἐγένετο, καὶ ὅ τι ψυχή ἐστιν, καὶ ὅ τι τὸ ὑγιαίνειν,
καὶ ὅ τι τὸ κάμνειν, καὶ ὅ τι τὸ ἐν ἀνθρώπῳ κακὸν
καὶ ἀγαθόν, καὶ ὅθεν ἀποθνήσκει. νῦν δὲ ἀποφαίνο-
μαι αὐτὸς <τὰς>[3] ἐμεωυτοῦ γνώμας.

2. Δοκέει δέ μοι ὃ καλέομεν θερμόν, ἀθάνατόν τε
εἶναι καὶ νοέειν πάντα καὶ ὁρῆν καὶ ἀκούειν καὶ εἰδέναι πάντα καὶ τὰ ἐόντα καὶ τὰ μέλλοντα ἔσεσθαι·
τούτου[4] οὖν τὸ πλεῖστον, ὅτε ἐταράχθη πάντα, ἐξε-
χώρησεν εἰς τὴν ἀνωτάτω περιφορήν· καὶ ὀνομῆναί
μοι αὐτὸ δοκέουσιν οἱ παλαιοὶ αἰθέρα. ἡ δευτέρα
μοῖρα κάτωθεν αὐτῆς[5] καλέεται μὲν γῆ, ψυχρὸν καὶ
ξηρὸν καὶ πολὺ κινοῦν· καὶ ἐν τούτῳ ἔνι δὴ πολὺ τοῦ

[1] Joly after Heidel: οὐδὲ V.

FLESHES

1. In this treatise I shall employ assumptions that are generally held—both those of my predecessors and my own—since it is necessary to establish a common starting point, if I wish to compose this treatise about medicine. About what is in the heavens I have no need to speak, except insofar as is necessary in order to explain how man and the other animals are formed and come into being, what the soul is, what health and sickness are, what in man is evil and what good, and where his death comes from. From here on, then, I present opinions that are my own.

2. I believe that what we call heat is in fact immortal, that it perceives all things, and sees, hears and knows all that is and all that will be. Now at the time that the universe was in a state of turbulence, the greatest part of this heat separated off into the uppermost revolving vault of heaven. This the ancients, I believe, called the "aether". The second portion of material below this is called earth; it is cold, dry and in great motion, although it too contains

[2] Ermerins: ὁκόσα V.   [3] Ermerins.

[4] Potter: τοῦτο V.

[5] Deichgraeber: αὐτῇ V.

θερμοῦ. ἡ δὲ τρίτη μοῖρα ἡ τοῦ ἠέρος τοῦ ἐγγυτάτω
πρὸς τῇ γῇ, ὑγρότατόν τε καὶ παχύτατον.

3. Κυκλεομένων δὲ τούτων, ὅτε συνεταράχθη,
586 ἀπελείφθη τοῦ | θερμοῦ πολὺ ἐν τῇ γῇ ἄλλοθι <καὶ
ἄλλοθι>,[1] τὰ μὲν μεγάλα, τὰ δὲ ἐλάσσω, τὰ δὲ καὶ
πάνυ σμικρά, πλῆθος πολλά. καὶ τῷ χρόνῳ ὑπὸ τοῦ
θερμοῦ ξηραινομένης τῆς γῆς, ταῦτα <τὰ>[2] κατα-
λειφθέντα περὶ αὐτὰ σηπεδόνας ποιέει οἷόν περ[3]
χιτῶνας. καὶ πολλῷ χρόνῳ θερμαινόμενον, ὅσον
μὲν ἐτύγχανεν ἐκ τῆς γῆς σηπεδόνος λιπαρόν τε ἐὸν
καὶ ὀλίγιστον τοῦ ὑγροῦ ἔχον, τάχιστα ἐξεκαύθη
καὶ ἐγένετο ὀστέα. ὁκόσα δὲ ἐτύγχανε κολλωδέ-
στερα ἐόντα καὶ τοῦ ψυχροῦ μετέχοντα, ταῦτα δὲ
θερμαινόμενα οὐκ ἠδύνατο ἐκκαυθῆναι, οὐδὲ ξηρὰ
γενέσθαι· οὐ γὰρ ἦν τοῦ λιπαροῦ ὡς ἐκκαυθῆναι,
οὐδὲ μὴν τοῦ ὑγροῦ ὡς ἐκκαυθὲν ξηρὸν γενέσθαι·
διὰ τοῦτο εἰδέην ἀλλοιοτέρην ἔλαβε τῶν ἄλλων καὶ
ἐγένετο νεῦρα καὶ φλέβες. αἱ μὲν φλέβες κοῖλαι, τὰ
δὲ νεῦρα στερεά· οὐδὲ γὰρ ἐνῆν πολὺ τοῦ ψυχροῦ
αὐτοῖσιν.[4] αἱ δὲ φλέβες τοῦ ψυχροῦ εἶχον· καὶ τούτου τοῦ ψυχροῦ τὸ μὲν πέριξ ὅσον κολλωδέστατον
ἦν, ὑπὸ τοῦ θερμοῦ ἐξοπτηθὲν μῆνιγξ ἐγένετο, τὸ δὲ
ψυχρὸν ἐνεὸν[5] κρατηθὲν ὑπὸ τοῦ θερμοῦ διελύθη καὶ
ἐγένετο ὑγρὸν διὰ τοῦτο.

Κατὰ δὲ τὸν αὐτὸν λόγον καὶ ὁ φάρυγξ καὶ ὁ
στόμαχος καὶ ἡ γαστὴρ καὶ τὰ ἔντερα ἐς τὸν ἀρχὸν

---

[1] Littré.   [2] Ermerins.

much heat. The third portion is the air closest to the earth; it is moistest and thickest.

3. Now while these things were mingled with one another in a state of turbulence as they rotated, much heat was left behind at various places in the earth, in some places great amounts, in others lesser amounts and in still others very small amounts, but these many in number. As with time the earth was dried out by this heat, the materials left behind engendered putrefactions about themselves, which had the form of tunics. Now what was heated for a great time and happened to arise from the putrefaction of the earth as fat, and containing the least moisture, quickly burnt up and became bones. That, on the other hand, which happened to be more gluey and to contain cold could not be burnt up on being heated or become dry, since it contained neither any fat that could be burnt up nor any moisture that, on being burnt up, could become dry; for this reason it took a form rather different from the other things, and became cords and vessels. The vessels were hollow and the cords solid owing to the fact that in the cords there was not much cold; in the vessels, on the other hand, cold was present, so that in the case of the vessels the part of this cold around the outside that was most gluey was baked through by the heat and became membraneous, while the cold in the inside, on being mastered by the heat, dissolved and became liquid.

In this same way the pharynx, oesophagus, stomach and intestines as far down as the rectum too became

---

[3] Froben: περὶ V.   [4] Ermerins: αὐτὸ V.

[5] Zwinger: ἐὸν V[2].

κοῖλα ἐγένοντο· τοῦ γὰρ ψυχροῦ αἰεὶ θερμαινομένου
τὸ μὲν πέριξ ἐξωπτήθη ὅσον αὐτὸ κολλῶδες ἦν, καὶ
ἐγένετο χιτὼν ὁ περὶ αὐτὸν μῆνιγξ, τὸ δὲ ἐντὸς τοῦ
ψυχροῦ,[1] οὐ γὰρ ἐνῆν ἐν αὐτῷ οὔτε λιπαρὸν οὔτε
κολλῶδες πολύ, διετάκη καὶ ἐγένετο ὑγρόν. κατὰ δὲ
τὸν αὐτὸν λόγον καὶ ἡ κύστις, πολὺ ψυχρὸν ἀπολει-
φθέν, τὸ πέριξ αὐτοῦ ὑπὸ τοῦ θερμοῦ θερμαινόμενον
χιτὼν ἐγένετο μῆνιγξ, τὸ δὲ μέσον πολὺν χρόνον
θερμαινόμενον διελύθη καὶ ἐγένετο ὑγρόν· οὐ γὰρ |
588   ἐνῆν ἐν αὐτῷ οὔτε τοῦ λιπαροῦ οὔτε τοῦ κολλώδεος·
ὅσον δὲ περιῆν χιτὼν ἐγένετο.

Ἀτὰρ καὶ περὶ τῶν ἄλλων, ὅσα κοῖλα, τὸν αὐτὸν
ἔχει τρόπον· ὅκου μὲν ἦν τοῦ κολλώδεος πλέον ἢ τοῦ
λιπαροῦ, χιτὼν μῆνιγξ ἐγένετο· ὅκου δὲ τοῦ λιπαροῦ
πλέον ἢ τοῦ κολλώδεος, ὀστέα ἐγένετο. ὡϋτὸς δὲ
λόγος καὶ τῶν ὀστέων· ὅκου μὲν μὴ ἐνῆν τοῦ κολλώδεος, τοῦ δὲ λιπαροῦ καὶ τοῦ ψυχροῦ, ἐξεκαίετο θᾶσσον διὰ τὸ λιπαρόν, καὶ ταῦτα τῶν ὀστέων καὶ
σκληρότατα καὶ στριφνότατα· ὅκου δὲ λιπαρὸν καὶ
κολλῶδες παραπλήσια, ταῦτα δὲ τῶν ὀστέων
σηραγγώδεα. περὶ μὲν τούτων οὕτως· τὸ μὲν
ψυχρὸν πήγνυσιν· τὸ δὲ θερμὸν διαχέει, ἐν δὲ τῷ
πολλῷ καὶ ξηραίνει χρόνῳ· ὅκου δὲ τοῦ λιπαροῦ
ξυνῇ τι τούτοισι, θᾶσσον ἐκκαίει καὶ ξηραίνει· ὅκου
δὲ ἂν τὸ κολλῶδες ξυνῇ τῷ ψυχρῷ ἄνευ τοῦ λιπαροῦ, οὐκ ἐθέλει ἐκκαίεσθαι, ἀλλὰ τῷ χρόνῳ θερμαινόμενον πήγνυται.

hollow: for as the cold in each case was heated, the part around the outside, being gluey, was baked through and became a tunic, the membrane around it, while the cold in the inside, not containing much that was fatty or gluey, melted completely and became liquid. In the same way the bladder too was formed: for when much cold was left behind, the part around the outside was warmed by the heat and became a tunic, a membrane, while the part in the inside, on being heated for a long time, dissolved and became liquid, since it contained nothing that was fatty or gluey, and what was around the outside became a tunic.

In the case of the other hollow parts it happened in the same way: where there was more gluey material than fatty, a tunic, membrane, was formed, where there was more fatty than gluey, bones were formed. Concerning the bones the explanation is the same. Where fatty and cold material was present, but no gluey, it was burnt more quickly because it was fatty, and these were the hardest and solidest of the bones. But where fatty and gluey components were in equal proportions, the bones formed were porous.[a] These processes occur according to the following principles. Cold condenses, heat melts and, over a long period of time, also dries. Where any fat is present with other materials, heat burns it up and dries it very quickly; where, on the other hand, gluey material is present with cold but without fat, what is gluey refuses to be burnt up, but, on being heated for a time, condenses (i.e. into a membrane).

[a] I.e. a mixture of bone and membrane.

[1] Froben: ὑγροῦ V.

4. Ὁ δὲ ἐγκέφαλός ἐστι μητρόπολις τοῦ ψυχροῦ
καὶ τοῦ κολλώδεος, τὸ δὲ θερμὸν τοῦ λιπαροῦ
μητρόπολις· θερμαινόμενον γάρ, τὸ πρῶτον πάντων
διαχεόμενον λιπαρὸν γίνεται· καὶ διὰ τοῦτο ὁ ἐγκέφαλος ὅτι ὀλίγιστον ἔχει τοῦ λιπαροῦ, τοῦ δὲ
κολλώδεος πλεῖστον, οὐ δύναται ἐκκαυθῆναι ὑπὸ τοῦ
θερμοῦ, ἀλλ᾽ ἐν τῷ χρόνῳ χιτῶνα μήνιγγα παχείην
ἔλαβε· περὶ δὲ τὴν μήνιγγα ὀστέα ὁκόσον τὸ θερμὸν
ἐκράτησε, καὶ ἐν ὅσοισι τοῦ λιπαροῦ ἐνῆν. καὶ ὁ
μυελὸς ὁ καλεόμενος νωτιαῖος καθήκει ἀπὸ τοῦ ἐγκε-
φάλου· καὶ οὐκ ἔστιν ἐν αὐτῷ τοῦ λιπαροῦ [ἢ τοῦ
κολλώδεος]¹ πολύ, ὥσπερ καὶ τῷ ἐγκεφάλῳ· διὰ
τοῦτο οὐκ ἂν δικαίως καὶ αὐτῷ εἴη μυελὸς οὔνομα·
οὐ γὰρ ὅμοιος τῷ ἄλλῳ μυελῷ, ὡς ἐν τοῖσιν ἄλλοισιν ὀστοῖσιν ἔνι· μοῦνος γὰρ μήνιγγας ἔχει, ὁ δὲ
590 ἄλλος οὐκ ἔχει. τεκμήρια δὲ τούτων | σαφέα, εἴ τις
ἐθέλοι ὀπτᾶν νευρώδεά τε καὶ κολλώδεα, καὶ τὰ
ἄλλα δέ· τὰ μὲν ἄλλα ταχὺ ὀπτᾶται, τὰ δὲ νευρώδεά
τε καὶ κολλώδεα οὐκ ἐθέλει ὀπτᾶσθαι· ἐλάχιστον
γὰρ ἔχει τοῦ λιπαροῦ· τὸ δὲ πιότατον καὶ λιπαρὸν
τάχιστα ὀπτᾶται.

5. Τὰ δὲ σπλάγχνα ὧδέ μοι δοκέει ξυστῆναι·
περὶ μὲν οὖν τῶν φλεβῶν εἴρηταί μοι πρότερον· ἡ δὲ
καρδίη πολὺ τοῦ κολλώδεος καὶ τοῦ ψυχροῦ ἔχει·
καὶ ὑπὸ τοῦ θερμοῦ θερμαινόμενον, κρέας ἐγένετο
σκληρὸν καὶ γλίσχρον, καὶ μῆνιγξ περὶ αὐτήν, καὶ
ἐκοιλώθη [οὐχ]² ὥσπερ φλέβες· καί ἐστιν ἐπὶ τῆς

4. The brain is the metropolis[a] of the cold and gluey, and heat the metropolis of the fatty, since the first thing of all to melt, on being heated, becomes fat. Thus the brain, having the least fat but the most gluey material, cannot be burnt up by heat, but after a time engenders a tunic, the thick membrane (*dura mater*) and, around the membrane, bones, to the degree that heat gains mastery and any fat is present. The marrow called spinal extends down from the brain, and it, just like the brain, has little fat [or gluey material]. Thus, it does not rightly bear the name "marrow", since it is not like the other marrow in the bones, it alone having a membrane, and they none. Proof of this difference would be obvious if you were to burn such cords, then something gluey, and then anything else. Whereas anything else is rapidly burnt through, the cords and whatever is gluey refuse to be burnt through because they have so little fat. And fat and what is very rich are quickest to burn.

5. The internal parts arise, in my opinion, as follows— about the vessels I have spoken above.[b] The heart has much that is cold and gluey; as it was warmed by the heat, it became hard, tough flesh, and a membrane developed around it, and it was hollowed out [but not] like the

[a] I.e. source, as the "mother-city" was the source of settlers in a Greek colony.

[b] See chapter 3 above.

---

[1] Del. Deichgraeber.

[2] Del. Ermerins.

κεφαλῆς τῆς φλεβὸς τῆς κοιλοτάτης. δύο γάρ εἰσι
κοῖλαι φλέβες ἀπὸ τῆς καρδίης· τῇ μὲν οὔνομα
ἀρτηρίη· τῇ δὲ κοίλη φλέψ, πρὸς ᾗ ἡ καρδίη ἐστίν·
καὶ πλεῖστον ἔχει τοῦ θερμοῦ ἡ καρδίη καὶ[1] ἡ κοίλη
φλέψ, καὶ ταμιεύει τὸ πνεῦμα. πρὸς δὲ ταύταιν ταῖν
φλεβοῖν ἄλλαι κατὰ τὸ σῶμα· ἡ δὲ κοιλοτάτη φλέψ,
πρὸς ᾗ ἡ καρδίη, διὰ τῆς κοιλίης ἁπάσης διήκει
καὶ διὰ τῶν φρενῶν, καὶ σχίζεται ἐς ἑκάτερον τῶν
νεφρῶν· καὶ ἐπὶ τῇ ὀσφύϊ σχίζεται, καὶ ἀΐσσει ἐπί
τε τὰ ἄλλα καὶ ἐς ἑκάτερον σκέλος, ἀτὰρ καὶ ἄνωθεν
τῆς καρδίης πρὸς τῷ αὐχένι, τὰ μὲν ἐπὶ δεξιά, τὰ δ'
ἐπ' ἀριστερά· καὶ τότε ἐπὶ τὴν κεφαλὴν ἄγει καὶ
ἐν τοῖς κροτάφοισι σχίζεται ἑκατέρη· ἔστι δὲ καὶ
ἀριθμῷ εἰπεῖν τὰς φλέβας τὰς μεγίστας· ἑνὶ δὲ
λόγῳ ἀπὸ τῆς κοίλης φλεβὸς καὶ ἀπὸ τῆς ἀρτηρίης
ἄλλαι φλέβες ἐσχισμέναι εἰσὶ κατὰ πᾶν τὸ σῶμα·
κοιλόταται δὲ αἱ πρὸς τῇ καρδίῃ καὶ τῷ αὐχένι καὶ
ἐν τῇ κεφαλῇ καὶ κάτωθεν τῆς καρδίης μέχρι τῶν
ἰσχίων.

592   6. Καὶ τὸ θερμὸν πλεῖστον ἔνι ἐν τῇσι φλεψὶ καὶ
τῇ καρδίῃ, καὶ διὰ τοῦτο πνεῦμα ἡ καρδίη ἔχει
θερμὴ ἐοῦσα μάλιστα τῶν ἐν τῷ ἀνθρώπῳ. ῥηΐδιον
δὲ τοῦτο καταμαθεῖν, ὅτι θερμόν ἐστι τὸ πνεῦμα· ἡ
καρδίη καὶ αἱ κοῖλαι φλέβες κινέονται αἰεὶ [καὶ τὸ
θερμότατον πλεῖστον ἔνι ἐν τῇσι φλεψίν· καὶ διὰ
τοῦτο πνεῦμα ἡ καρδίη ἔχει θερμὴ ἐοῦσα μάλιστα

---

[1] Deichgraeber: ἢ V.

140

vessels. The heart is at the head of the hollowest vessel (*vena cavissima*); for two hollow vessels come out of it: one has the name artery, the other hollow vessel (*vena cava*), and it is against this that the heart sits. The heart and the hollow vessel have the most heat, and control the breath. Attached to these two vessels are others to the body. The hollowest vessel, against which the heart sits, traverses the whole cavity and the diaphragm, and branches to each of the kidneys; it also branches at the loins, and proceeds to the other parts, and to each leg. Furthermore, above the heart in the region of the throat there are branches to the right and to the left. Then one of these on each side goes to the head, and branches at the temples. It is possible to enumerate the largest vessels: in a word, from the hollow vessel and the artery the other vessels branch off through the whole body; the most capacious ones are those at the heart and throat and in the head, and below the heart as far down as the hips.

6. The greatest amount of heat is in the vessels and heart, and for this reason the heart, being the hottest part in a person, holds the breath, which is very easy to understand, given that the breath is hot.[a] The heart and the hollow vessels are in constant movement [and the greatest amount of heat is mostly in the vessels, and for this reason the heart, being the hottest part in a person, holds the

[a] Cf. *Nature of the Child* 12.

τῶν ἐν τῷ ἀνθρώπῳ].[1] ἔστι δὲ καὶ ἄλλως γνῶναι·
πῦρ εἴ τις θέλει καίειν ἐν οἰκήματι ὁκόταν ἄνεμος μὴ
εἰσπνέῃ, φλὸξ κινέεται τοτὲ μὲν μᾶλλον, τοτὲ δὲ
ἧσσον· καὶ λυχνὸς[2] καιόμενος τὸν αὐτὸν τρόπον
κινέεται, τοτὲ μὲν μᾶλλον, τοτὲ δὲ ἧσσον, ἀνέμου
οὐδενὸς κινέοντος, ὅντινα καὶ ἡμεῖς οἷοί τέ ἐσμεν
γινώσκειν πνέοντα· καὶ τροφή ἐστι τῷ θερμῷ τὸ
ψυχρόν. τὸ δὲ παιδίον ἐν τῇ γαστρὶ συνέχον τὰ χεί-
λεα μύζει ἐκ τῶν μητρέων τῆς μητρὸς καὶ ἕλκει τήν
τε τροφὴν καὶ τὸ πνεῦμα τῇ καρδίῃ εἴσω· τοῦτο γὰρ
θερμότατόν ἐστιν ἐν τῷ παιδίῳ, ὅταν περ ἡ μήτηρ
ἀναπνέῃ· τούτῳ δὲ καὶ τῷ ἄλλῳ σώματι τὴν κίνησιν
παρέχει τὸ θερμὸν καὶ τοῖς ἄλλοις πᾶσιν. εἰ δέ τις
ἐρωτοίη πῶς τοῦτο οἶδέ τις, ὅτι ἐν τῇ μήτρῃ τὸ παιδίον ἕλκει καὶ μύζει, τάδε αὐτῷ ἔστιν ἀποκρίνασθαι·
594 κόπρον ἔχον | ἐν τοῖσιν ἐντέροισιν γίνεται, καὶ ἀπο-
πατέει ἐπειδὰν γένηται τάχιστα, καὶ οἱ ἄνθρωποι
καὶ τὰ πρόβατα· καίτοι οὐκ ἂν εἶχε κόπρον, εἰ μὴ ἐν
τῇσι μήτρῃσιν ἔμυζεν, οὐδ' ἂν θηλάζειν τὸν μασθὸν
ἠπίστατο γεννώμενον αὐτίκα, εἰ μὴ καὶ ἐν τῇ μήτρῃ
ἔμυζε. καὶ περὶ μὲν τῆς κινήσιος τῆς καρδίης καὶ
τῶν φλεβῶν οὕτως ἔχει.

7. Ὁ δὲ πνεύμων πρὸς τῇ καρδίῃ ἐγένετο ὧδε·
τοῦ ὑγροῦ ὁκόσον ἦν κολλωδέστατον, ἡ καρδίη

---

[1] Del. Potter.
[2] Froben: αὐχμὸς V.

breath].[a] The origin of this movement can also be understood from another example: if you light a fire inside a
room when no wind is blowing in, the flame moves, sometimes more, sometimes less. Also a lighted lamp moves in
the same way, sometimes more, sometimes less, although
no wind is stirring that we are able to perceive. And we
know that the nourishment of hot is cold.[b] For example
the fetus in the belly continually sucks with its lips from
the uterus of the mother and draws nourishment and
breath to its heart inside, for this breath is hottest in the
fetus just at the time that the mother is inspiring.[c] To this
(the fetus's heart) and to the rest of the body, too, heat
gives movement, and to all other things. If anyone asks
you how you know that the fetus draws and sucks in the
uterus, you may reply as follows. Both humans and animals have faeces in their intestines at the time of birth,
and immediately at birth pass stools—but there would be
no faeces, if the fetus did not suck in the uterus. Nor
would a baby know how to suck from the breast immediately at birth, if it did not also suck in the uterus. So it is
with the movement of the heart and vessels.

7. The lung arose next to the heart as follows. The
heart, warming that part of the moisture that was most

[a] I do not find the repetitiousness of this passage merely "très
typique de la manière archaique" (Joly ad loc.), but suspect some
sort of textual disturbance.

[b] The connection of thought between this and the preceding
sentence is that the cold, which is "the nourishment of the hot",
is drawn in from the outside by the hot and thus gives rise to
movement.

[c] Presumably the model here is that of a fireplace with the
damper open.

θερμαίνουσα ταχὺ ἐξήρανεν ὅκωσπερ ἀφρὸν καὶ
ἐποίησε σηραγγῶδες, καὶ φλέβια πολλὰ ἐν αὐτῷ.
διὰ δὲ τοῦτο ἐποίησε τὰ φλέβια· ὁκόσον ἐν τῷ κολλώδει ἐνῆν ψυχρόν, τοῦτο μὲν ὑπὸ τοῦ θερμοῦ διετάκη καὶ ἐγένετο ὑγρόν· τὸ δὲ ἀπὸ τοῦ κολλώδεος
αὐτοῦ[1] ὁ χιτών.

8. Τὸ δὲ ἧπαρ ὧδε ξυνέστη· ξὺν τῷ θερμῷ πολὺ
τοῦ ὑγροῦ ἀπολειφθὲν ἄνευ τοῦ κολλώδεος καὶ τοῦ
λιπαροῦ, ἐκράτησε τὸ ψυχρὸν τοῦ θερμοῦ, καὶ
ἐπάγη. τεκμήριον δέ μοι τόδε· ὁκόταν σφάξῃ τις
ἱερεῖον, τέως μὲν ἂν θερμὸν ᾖ, ὑγρόν ἐστι τὸ αἷμα·
ἐπειδὰν δὲ ψυχθῇ, ἐπάγη· ἢν δέ τις αὐτὸ τινάσσῃ,
οὐ πήγνυται· αἱ γὰρ ἶνες εἰσι ψυχραὶ καὶ κολλώδεες.

9. Ὁ δὲ σπλὴν συνέστη ὧδε· σὺν τῷ ὑγρῷ[2] καὶ
κολλώδει ἐκάετο[3] θερμοῦ πλεῖστον, τοῦ δὲ ψυχροῦ
ἐλάχιστον, τοσοῦτον μόνον ὁκόσον πῆξαι τὸ κολλῶδες αὐτό, ὅ εἰσιν αἱ ἶνες αἱ ἐνοῦσαι ἐν τῷ σπληνί·
καὶ διὰ τὰς ἶνας ταύτας μαλακός ἐστιν ὁ σπλὴν καὶ
ἰνώδες.

Οἱ δὲ νεφροὶ ξυνέστησαν ὧδε· ὀλίγον τοῦ κολλώ-
δεος, τοῦ ὑγροῦ[4] πλεῖστον | τοῦ ψυχροῦ πλεῖστον,
καὶ ἐπάγη ὑπὸ τούτου, καὶ ἐγένετο σκληρότατον τὸ
σπλάγχνον καὶ ἥκιστα ἐρυθρόν, ὅτι οὐ πολὺ τοῦ
θερμοῦ ξυνέστη.

Ὁ δὲ αὐτὸς λόγος καὶ περὶ τῶν σαρκῶν· τὸ μὲν

---

[1] Ermerins: αὐτὸς V.
[2] Ermerins: θερμῷ V.

gluey, quickly dried it out into a kind of foam, making it porous and filling it with many small vessels. This is how the heart formed the small vessels: any cold that was in the gluey material was melted by its heat and became liquid, while from the gluey part itself the membrane of the vessel arose.[a]

8. The liver was formed as follows. Where much moisture was left behind together with heat and in the absence of anything gluey or fatty, cold gained mastery over the heat, and the moisture congealed. This is my proof. When you slaughter a sacrificial animal, as long as its blood remains warm it is liquid, but when it becomes cold, it congeals. If, however, you shake it, the blood does not solidify, for its fibres are cold and gluey.

9. The spleen was formed as follows. With what was moist and gluey was kindled very much heat, and very little cold, just the right amount to condense the gluey material itself,[b] which constitutes the fibers that are in the spleen; it is because of these fibres that the spleen is soft and stringy.

The kidneys were formed as follows. Little gluey material, a great amount of moist material, a great amount of cold: the viscus congealed on account of this, and became the toughest and the least red, since not much heat was present.

About muscles the explanation is the same: cold

[a] Cf. chapter 3 above.

[b] For the condensing effect of heat on gluey material see chapter 3 *fin.* above.

---

[3] Potter: καὶ τοῦ V.   [4] Potter: θερμοῦ V.

ψυχρὸν ἔστησε καὶ ξυνέπηξε καὶ ἐποίησε σάρκα, τὸ
δὲ κολλῶδες τρώγλαι ἐγένοντο· ἐν δὲ τῇσι τρώγλησι
ταύτῃσι τὸ ὑγρὸν ὥσπερ καὶ ἐν τῇσι φλεψὶ τῇσι
μεγάλῃσιν.

Τὸ δὲ θερμὸν ἐν παντὶ τῷ σώματι, πλεῖστον δὲ τῷ
σώματι· καὶ τοῦ ψυχροῦ πολὺ ἐν τῷ ὑγρῷ· τοσοῦτο
δέ ἐστι τοῦ ψυχροῦ ὁκόσον δύναται πῆξαι τὸ ὑγρόν·
ἀλλὰ νενίκηται, ὥστε διακέχυται ὑπὸ τοῦ θερμοῦ.

Ἡ δὲ ἀπόδειξις τοῦ ὑγροῦ ὅτι θερμόν ἐστιν, εἴ τις
ἐθέλοι τάμνειν τοῦ ἀνθρώπου τοῦ σώματος, ὅκου
ἐθέλοι, ῥεύσεται[1] αἷμα θερμόν, καὶ τέως μὲν ἂν θερμὸν ᾖ, ὑγρὸν ἔσται· ἐπειδὰν δὲ ψυχθῇ ὑπό τε τοῦ
ἐνεόντος ψυχροῦ καὶ τοῦ ἐκτός, ἐγένετο δέρμα καὶ
ὑμήν, καὶ εἴ τις ἀφελὼν τοῦτο τὸ δέρμα ἐάσειεν ὀλίγου χρόνου, ὄψεται ἄλλο δέρμα γινόμενον· εἰ δέ τις
τοῦτο αἰεὶ ἀφαιρεῖ, ἄλλο δέρμα γίνοιτ' ἂν πρὸς τοῦ
ψυχροῦ. τούτου δὲ εἵνεκα πλείω ἔλεξα, ὅκως[2] ἀποδείξω ὅτι τὸ ἔσχατον τοῦ σώματος πρὸς τοῦ ἠέρος
ἀναγκαίως ἔχει δέρμα γενέσθαι ὑπὸ τοῦ ψυχροῦ καὶ
τῶν πνευμάτων προσβαλλόντων.

10. Τὰ δ' ἄρθρα ὧδε ἐγένετο· ὅτε τὰ ὀστέα ξυνί-
στατο, ὁκόσα μὲν αὐτῶν λιπαρὰ ἦν, τάχιστα ἐξεκαύθη, ὥσπερ πρόσθεν λέλεκται ἐν τῷ προτέρῳ
λόγῳ· ὁκόσον δ' αὐτῶν κολλῶδες ἦν, τοῦτο δὲ οὐκ
ἠδύνατο ἐκκαυθῆναι, ἀλλὰ μεταξὺ ἀπολειφθὲν τοῦ
καιομένου <καὶ>[3] ξηραινομένου ὑπὸ τοῦ θερμοῦ
598 νεῦρα καὶ σίαλον ἐγένετο· τὸ δὲ σίαλον, ὁκόσον τοῦ
κολλώδεος ὑγρότατον ἦν, τόδε θερμαινόμενον παχύ-

brought what was moist to a stand-still, congealed it, and turned it into muscle, while what was gluey formed canals; in these canals there is moisture just as in the large vessels.

There is heat in the whole body, in fact it constitutes the greatest part in the body; in the moisture of the body, again, there is much cold; in fact so much cold that it could congeal this moisture, if it were not overwhelmed and so liquified by the whole body's heat.

Demonstration that the moist part in the body is kept hot. If someone should make a cut in a person's body anywhere he wishes, hot blood will flow out and, as long as it is hot, it will be liquid. But when this blood is cooled by the internal cold and the external cold, a skin or membrane forms; if you remove this and let a little time pass, another skin will be seen to form; and if you keep on removing this, still another skin will form, because of the effect of the cold. I have dilated on this point in order to show that at the outer confines of the body next the air a skin must of necessity arise on account of the cold and of the winds that strike it.

10. The joints came into being as follows. When bones arose what was fat in them was very rapidly burnt up, as has been said above (ch. 3); what was gluey could not be burnt up, but, being left behind between what was burnt up and dried by the heat, became cords and fluid. As to the fluid, the most liquid part of what was gluey, this

---

[1] Schwyzer: ῥεύσει τε V.

[2] Ermerins: τέως V.

[3] Zwinger in margin after Calvus' *perustum desiccatumque*.

τερον ἐγένετο ὑγρὸν ἐόν· καὶ ἀπὸ τούτου σίαλον ἐγένετο.

11. Οἱ δὲ ὄνυχες ἀπὸ τούτου ἐγένοντο τοῦ κολλώ-
δεος· ἀπὸ γὰρ τῶν ὀστέων καὶ τῶν ἄρθρων αἰεὶ τὸ
ὑγρότατον αὐτοῦ ἀπιὸν κολλῶδες[1] γίνεται ἀπὸ τοῦ
θερμοῦ ξηραινόμενον καὶ ἐξαυαινόμενον θύραζε
ὄνυχες.

12. Οἱ δὲ ὀδόντες [ὕστερον][2] γίνονται διὰ τόδε·
ἀπὸ τῶν ὀστέων τῶν ἐν τῇ κεφαλῇ καὶ ταῖν γνάθοιν
ἡ αὔξησις γίνεται· τοῦ κολλώδεος <καὶ>[3] τοῦ λιπαροῦ τὸ ἐνεὸν ὑπὸ τοῦ θερμοῦ ξηραινόμενον ἐκκαίε-
ται, καὶ γίνονται ὀδόντες σκληρότεροι τῶν ἄλλων
ὀστέων, ὅτι οὐκ ἔνεστι τοῦ ψυχροῦ. καὶ οἱ μὲν
πρῶτοι ὀδόντες φύονται ἀπὸ τῆς διαίτης τῆς ἐν τῇ
μήτρῃ, καὶ, ἐπὴν γένηται, ἀπὸ τοῦ γάλακτος θηλά-
ζοντι τῷ παιδίῳ· ἐπειδὰν δὲ οὗτοι ἐκπέσωσιν ἀπὸ
τῶν σιτίων καὶ τῶν ποτῶν· ἐκπίπτουσι δὲ ἐπειδὰν
ἑπτὰ[4] ἔτεα ἔῃ τῆς πρώτης τροφῆς· ἔστι δὲ καὶ οἷς
πρότερον, ἢν ἀπὸ νοσερῆς τροφῆς φυέωσι· τοῖσι δὲ
πλείστοισιν, ἐπειδὰν ἑπτὰ ἔτεα γένηται· οἱ δὲ μετα-
φυέοντες συγκαταγηράσκουσιν, ἢν μὴ ὑπὸ νόσου
διαφθαρῶσι.

13. Διὰ δὲ τοῦτο φύονται οἱ ὀδόντες ὕστερον τῶν
ἄλλων· ἐν τῇ | γνάθῳ φλέβες εἰσὶ καὶ ἐκ τῆς κάτω
κοιλίης[5] μούνοισι τῶν ὀστέων αὗται τὴν τροφὴν
παρέχουσι [τῷ ὀστέῳ]·[6] τὰ δὲ ὀστέα τοιαύτην αὔξη-

600

---

[1] Potter: κολλῶδες V: del. Deichgraeber, Joly.

thickened on being heated, but still remained liquid; this
is the origin of the joint fluid.

11. The nails too arose from this gluey material; for
the most liquid part of the gluey material which is continually being discharged from the bones and joints is dried
up and desiccated by heat to become nails at the exterior.

12. The teeth came into being for the following reason. The teeth grow out of the bones of the skull and
jaws, when what is gluey and fatty in them is dried and
burnt up by the heat. Teeth are harder than other bones
because they contain no cold. The primary teeth take
their growth from the regimen in the uterus,[a] and after
birth from milk, in the child at the breast; when these
have fallen out, the permanent teeth take their growth
from foods and drinks. The primary teeth fall out when
there have been seven years of the first regimen; in some
cases this occurs earlier, if the children are being fed a
regimen that is sickly—but in most cases when they are
seven years old. The teeth that grow in afterwards grow
old with the person, unless they are lost as the result of
disease.

13. Teeth are formed later than the other bones for
the following reason. In the jaws there are vessels which
provide only them among bones with nourishment from
the lower cavity. Now bones increase, as do all other

[a] Cf. chapter 6 above.

---

[2] Del. Deichgraeber.   [3] Zwinger in margin.

[4] Littré after Coray: ἐπὶ τὰ V.

[5] Foes in note 33 after Cornarius' *ex inferno ventre*: ἐν τῇ —
κοιλίῃ V.   [6] Del. Ermerins.

σιν ὁποῖά πέρ ἐστι, καὶ τὰ ἄλλα δὲ πάντα τοιαύτην
ἀποδίδωσιν αὔξησιν, ὁποῖα αὐτά ἐστιν· καὶ γὰρ αἱ
φλέβες αἱ ἐκ τῆς νηδύος <καὶ>[1] τῶν ἐντέρων, εἰς ἃ
συλλέγεται τὰ σιτία καὶ τὰ ποτά, ἐπειδὰν θερμανθῇ
ταῦτα, ἕλκουσι τὸ λεπτότατον καὶ τὸ ὑγρότατον·
τὸ δὲ παχύτατον αὐτοῦ καταλείπεται, καὶ γίνεται
κόπρος ἐν τοῖσιν ἐντέροισι τοῖσι κάτω· τὸ δὲ λεπτότατον αἱ φλέβες ἕλκουσιν ἐκ τῆς νηδύος καὶ τῶν
ἐντέρων τῶν ἄνωθεν τῆς νήστιος, θερμαινομένων
τῶν σιτίων· ἐπὴν δὲ περήσῃ τὴν νῆστιν ἐς τὰ κάτω
ἔντερα, ξυνεστράφη καὶ κόπρος ἐγένετο· ἡ δὲ τροφὴ
ἐπειδὰν ἀφίκηται ἐς ἕκαστον, τοιαύτην ἀπέδωκε τὴν
ἰδέην ἑκάστου ὁκοία περ ἦν· ἀρδόμενα γὰρ ὑπὸ τῆς
τροφῆς αὔξεται ἕκαστα, τὸ θερμὸν καὶ τὸ ψυχρὸν
καὶ τὸ κολλῶδες καὶ τὸ λιπαρὸν καὶ τὸ γλυκὺ καὶ τὸ
πικρὸν καὶ τὰ ὀστέα καὶ τὰ ἄλλα ξύμπαντα ὁκόσα
ἐν τῷ ἀνθρώπῳ ἔνι.

Διὰ τοῦτο ὕστερον οἱ ὀδόντες φύονται· εἴρηται δέ
μοι καὶ πρότερον ὅτι μοῦναι τῶν ὀστέων αἱ γνάθοι
602 φλέ|βας ἔχουσιν αὐταὶ ἐν ἑωυταῖσι· καὶ διὰ τοῦτο
τροφὴ ἕλκεται πλέον ἢ ἐς τὰ ἄλλα ὀστέα. καὶ
πλέονα τὴν τροφὴν ἔχοντα καὶ ἀθροωτέρην τὴν
ἐπιρροήν, ἀποτίκτει αὔξησιν αὐτὰ ἀφ' ἑαυτῶν τοιαύτην οἷά πέρ ἐστιν αὐτά, τέως ἄν περ καὶ ἄνθρωπος
ὅλος[2] αὐξάνηται. αὐξάνεται δὲ ἐπὴν γένηται ἐπίδη-
λος· ἐπίδηλος δὲ μάλιστα γίνεται ἀπὸ ἑπταετέος
μέχρι τεσσαρεσκαιδεκαετέος, καὶ ἐν τούτῳ τῷ χρόνῳ
οἵ τε μέγιστοι τῶν ὀδόντων φύονται καὶ ἄλλοι

parts, by adding what is similar to their own particular quality. For when food and drink collect in the stomach and intestines, and are heated, the vessels arising there draw off the finest and moistest part, leaving the thickest part behind, which turns to faeces in the lower intestines. As the foods are heated, then, these vessels draw off the finest part from the stomach and the upper intestines— the part above the jejunum—and as the foods pass through the jejunum into the lower intestines, they solidify and become faeces. When the nourishment arrives, it gives up the particular quality corresponding to each part, for it is by being watered by this nourishment that every part increases, the hot, the cold, the gluey, the fat, the sweet, the bitter, the bones, and all the other parts that are in a person.

Here is why teeth are formed at a later time. I mentioned above that the jaws alone of the bones have vessels inside themselves, and this is why more nourishment is drawn to them than to other bones. Since they have more nourishment coming in a more massive afflux, they continue to increase by adding what is similar to their own particular quality, until a person has reached adulthood. A person reaches adulthood when he has acquired his definitive form, and this generally occurs between seven and fourteen years of age. In that time all the teeth,

---

[1] Zwinger in margin after Carnarius' *ac.*

[2] Joly: οὖλος V.

πάντες, ἐπὴν ἐκπέσωσιν οἳ ἐγένοντο ἀπὸ τροφῆς τῆς
ἐν τῇ μήτρῃ. αὐξάνεται δὲ καὶ ἐς τὴν τρίτην ἑβδο-
μάδα, ἐν ᾗ νεηνίσκος γίνεται, <καὶ>[1] μέχρι τεσσά-
ρων καὶ πέντε ἑβδομάδων· καὶ ἐν τῇ τετάρτῃ δὲ
ἑβδομάδι ὀδόντες φύονται δύο τοῖσι πολλοῖσι τῶν
ἀνθρώπων, οὗτοι καλέονται σωφρονιστῆρες.

14. Αἱ δὲ τρίχες φύονται ὧδε· ὀστέα εἰσὶ καὶ
ἐγκέφαλος, ἀφ᾽ ὧν ἡ τοιαύτη αὔξησις, ὅτι τὸ πέριξ
κολλῶδες,[2] ὅπωσπερ τοῖσι νεύροισι, καὶ τοῦ λιπαροῦ
οὐκ ἔνεστιν· εἰ γὰρ ἐνῆν τοῦ λιπαροῦ, ἐξεκαίετο ἂν
ἐκ τοῦ θερμοῦ. τάχα δὲ θαυμάσειεν ἄν τις ὅτι καὶ ἐν
τῇσι μασχάλῃσι καὶ ἐν τῷ ἐπισίῳ τρίχες πολλαὶ καὶ
ἐν τῷ σώματι παντί εἰσιν· ωὑτὸς λόγος περὶ τούτου·
ὅπου τυγχάνει τοῦ σώματος τὸ κολλῶδες ὄν,
ἐνταῦθα αἱ τρίχες γίνονται ὑπὸ τοῦ θερμοῦ.

15. Ἀκούει δὲ διὰ τόδε· τὰ τρήματα τῶν οὐάτων
προσήκει πρὸς ὀστέον σκληρόν τε καὶ ξηρὸν ὅμοιον
λίθῳ· τοῦτο δὲ πέφυκε πρὸς ὀστέον κοίλωσις
σηραγγώδης· οἱ δὲ ψόφοι ἀπερείδονται πρὸς τὸ
σκληρόν· τὸ δὲ ὀστέον τὸ κοῖλον ἐπηχεῖ διὰ τοῦ
σκληροῦ· τὸ δὲ δέρμα τὸ πρὸς τῇ ἀκοῇ πρὸς τῷ
ὀστέῳ τῷ σκληρῷ λεπτόν ἐστιν ὥσπερ ἀράχνιον,
ξηρότατον τοῦ ἄλλου δέρματος. τεκμήρια δὲ πολλὰ
604 | ὅτι τὸ ξηρότατον ἠχεῖ μάλιστα· ὅταν δὲ μέγιστον
ἠχήσῃ, τότε μάλιστα ἀκούομεν. καὶ εἰσί τινες οἳ
ἔλεξαν φύσιν ξυγγράφοντες ὅτι ὁ ἐγκέφαλός ἐστιν ὁ

---

[1] Potter.   [2] Ermerins: τοῦ — κολλώδεος V.

including the largest, are formed, once those that came into being from the nourishment in the uterus have fallen out. He continues to grow into the third seven-year period, in which he becomes a young man, and even until the fourth and fifth seven-year periods. In the fourth seven-year period two more teeth are formed in many persons, and these are called the wisdom teeth.

14. Hair grows as follows.[a] It is the bones and the brain from which this growth occurs, since their exterior is gluey—just as it is in cords—and lacks fat,[b] for if any fat were present, it would be burnt out by the heat. You may perhaps wonder that there is much hair in the axillae and on the pubis, and in fact over the whole body. The same explanation holds there: in any part of the body where anything gluey happens to be present, there hair occurs because of the effect of heat.

15. Hearing occurs for the following reason. The openings of the ears lead to a bone that is hard and dry like a stone, and besides there is a cavernous hollow next the bone. Sounds are directed towards this hardness, and through its hardness the hollow bone resounds. The skin in the ear next the hard bone is thin like a spider's web, and drier than other skin. There are many proofs that what is driest echoes best; and when a thing echoes best, we hear best. There are some writers on nature who have

[a] Cf. *Glands* chapters 4 and 10.
[b] Cf. chapter 4 above.

ἠχέων· τοῦτο δὲ οὐκ ἂν γένοιτο·[1] αὐτός τε γὰρ ὁ
ἐγκέφαλος ὑγρός ἐστι, καὶ μῆνιγξ περὶ αὐτόν ἐστιν
ὑγρὴ καὶ παχείη, καὶ περὶ τὴν μήνιγγα ὀστέα· οὐδὲν
οὖν τῶν ὑγρῶν ἠχεῖ, ἀλλὰ ξηρά· τὰ δὲ ἠχέοντα
ἀκοὴν ποιεῖ.

16. Ὀσφραίνεται δ' ὁ ἐγκέφαλος ὑγρὸς ἐὼν τῶν
ξηρῶν αὐτός, ἕλκων τὴν ὀδμὴν ξὺν τῷ ἠέρι διὰ τῶν
βρογχίων ξηρῶν ἐόντων· προήκει γὰρ ὁ ἐγκέφαλος
τῆς ῥινὸς ἐς τὰ κοῖλα· καὶ ταύτῃ αὐτῷ οὐκ ἔστι
ἐπίπροσθεν <οὔτε κρέας οὔτε>[2] ὀστέον οὐδέν, ἀλλὰ
χονδρίον μαλακὸν ὅκως περ σπόγγος [οὔτε κρέας
οὔτε ὀστέον].[2] καὶ ὅταν μὲν ξηρὰ ᾖ τὰ κοῖλα τῆς
ῥινός, ὀσμᾶσθαι τῶν ξηροτέρων αὐτὸς ἑωυτοῦ ἀκρι-
βέστερός ἐστιν· ὕδατος γὰρ οὐκ ὀδμᾶται· ὑγρότερον
γάρ ἐστι τοῦ ἐγκεφάλου, ἐὰν μὴ σαπῇ· σηπόμενον
γὰρ τὸ ὕδωρ παχύτερον γίνεται καὶ τὰ ἄλλα πάντα·
ὁκόταν δὲ ὑγρανθέωσιν αἱ ῥῖνες, οὐ δύναται ὀσφραί-
νεσθαι· οὐ γὰρ τὸν ἠέρα ἕλκει αὐτὸς πρὸς ἑωυτόν.
ταύτῃ δὲ καὶ ὅταν ἀποτήκῃ ὁ ἐγκέφαλος πλεῖστον
αὐτὸς ἀφ' ἑωυτοῦ ἐς τὴν ὑπερῴην καὶ τὴν φάρυγγα
καὶ τὸν πνεύμονα καὶ ἐς τὴν ἄλλην κοιλίην, γινώσκουσιν οἱ ἄνθρωποι καί φασι καταρρέειν ἐκ τῆς
κεφαλῆς· καταρρέει δὲ καὶ ἐς τὸ ἄλλο σῶμα· καὶ
ἐστι τροφὴ[3] τῷ θερμῷ.

17. Ὁρῇ δὲ διὰ τοῦτο· ἀπὸ τοῦ ἐγκεφάλου τῆς
μήνιγγος φλὲψ καθήκει ἐς τὸν ὀφθαλμὸν διὰ τοῦ
ὀστέου ἑκάτερον· διὰ ταύταιν ταῖν φλεβοῖν ἀπὸ τοῦ
ἐγκεφάλου διηθέεται τὸ λεπτότατον τοῦ κολλωδεσ-

said that it is the brain which echoes, but this cannot be so; for the brain itself is moist, and the membrane around it is moist and thick, and around the membrane there are bones: now nothing moist echoes, but rather what is dry, and it is what echoes that gives rise to hearing.

16. The brain, being itself moist, perceives the smell of dry things, by drawing the odour along with air through the bronchial tubes which are dry; for the brain extends towards the hollows of the nose, and at that point there is no flesh or bone in the way, but soft cartilage like a sponge. When the hollows of the nose are dry, the perception of the smell of drier things is more acute. The brain does not perceive the smell of water, since water is moister than the brain, unless the water is putrefying—on putrefying, of course, water becomes thicker, as does everything else. When the nostrils become moist, the brain loses its sense of smell, since it cannot draw any air to itself. Also, when the brain melts off copious material from itself along that same path to the palate, throat, lung and the rest of the cavity, people recognize this, and say that there is a catarrh from the head; there is a catarrh to the rest of the body, too, and this provides nourishment for heat.

17. Seeing occurs because of the following. From the membrane of the brain a vessel descends into the eye through the bone on each side. Through these two vessels the finest of the gluey part filters from the brain, and

---

[1] Deichgraeber: -ηται V.
[2] Deichgraeber.   [3] Heidel: τροπὴ V.

τάτου· καὶ διὰ τοῦτο αὐτὸ περὶ ἑωυτὸ δέρμα ποιέει
606 τοιοῦτον | οἷόν περ [περὶ]¹ αὐτό ἐστι, τὸ διαφανὲς
τοῦ ὀφθαλμοῦ τὸ πρὸς τοῦ ἠέρος, πρὸς ὃ προσβάλ-
λει τὰ πνεύματα, κατὰ τὸν αὐτὸν λόγον ὥσπερ περὶ
τοῦ ἄλλου δέρματος ἔλεξα. πολλὰ δὲ ταῦτ' ἐστὶ τὰ
δέρματα πρὸ τοῦ ὀρέοντος διαφανέα, ὁκοῖόν περ
αὐτό ἐστιν· τούτῳ γὰρ τῷ διαφανεῖ ἀνταυγέει τὸ
φῶς καὶ τὰ λαμπρὰ πάντα· τούτῳ οὖν ὁρῇ τῷ ἀνταυ-
γέοντι· ὅ τι δὲ μὴ λαμπρόν ἐστι μηδὲ ἀνταυγεῖ,
τούτῳ οὐχ ὁρῇ· τὸ δὲ ἄλλο τὸ περὶ τοὺς ὀφθαλμοὺς
λευκὸν κρέας ἐστίν. ἡ δὲ κόρη καλεομένη τοῦ
ὀφθαλμοῦ μέλαν φαίνεται διὰ τοῦτο ὅτι ἐν βάθει
ἐστὶ καὶ χιτῶνες περὶ αὐτό εἰσι μέλανες· χιτῶνα δὲ
καλέομεν τὸ ἐνεὸν ὥσπερ δέρμα· ἔστι δὲ οὐ μέλαν
ὄψει, ἀλλὰ λευκὸν διαφανές. τὸ δὲ ὑγρὸν κολλῶδες·
πολλάκις γὰρ ὀπώπαμεν ἐπὶ συρραγέντος ὀφθαλμοῦ
ἐξιὸν ὑγρὸν κολλῶδες· κἢν μὲν ᾖ ἔτι θερμόν, ὑγρόν
ἐστιν· ἐπειδὰν δὲ ψυχθῇ, ἐγένετο ξηρὸν ὥσπερ
λιβανωτὸς διαφανής, καὶ τῶν ἀνθρώπων καὶ τῶν
θηρίων ὅμοιόν ἐστι. τὸν δὲ ὀφθαλμὸν ἀνιᾷ πᾶν ὅ τι
ἂν ἐμπέσῃ, καὶ τὰ πνεύματα προσβάλλοντα, καὶ τὰ
ἄλλα ὅσα λαμπρότερα ἢ κατ' αὐτόν ἐστι, καὶ εἴ τις
ἐγχρίσει, διὰ τόδε, ὅτι ὁμόχροια <ἔνυγρός>² ἐστιν,
ὥσπερ τὸ στόμα καὶ ἡ γλῶσσα καὶ τὰ ἄλλα χείλεα³
ἔνυγρά ἐστιν.

18. Διαλέγεται δὲ διὰ τὸ πνεῦμα ἕλκων εἴσω πᾶν
608 τὸ σῶμα, | τὸ πλεῖστον δὲ ἐς τὰ κοῖλα αὐτὸς ἑωυτῷ·
αὐτὸ δὲ θύραζε ὠθεόμενον διὰ τὸ κενὸν ψόφον

because this happens it forms a skin about itself which is like it itself, the transparent part of the eye next the air which the winds strike, this occurring in the same manner as I have described above for the rest of the skin.[a] In fact, these transparent membranes situated in front of the part of the eye that sees are many, and like it itself, for it is in this transparency that light and all bright things reflect; through the agency of this reflection, then, the person sees. And whatever is not bright and does not reflect, the person does not see with. The rest of what is around the eyes is a white tissue. What is called the pupil of the eye appears black because it is deep within, and because the tunics around it are black—we call "tunic" the part inside like a skin; this skin is not black to look at, but rather a transparent white. The liquid in the eye is gluey, for often we have seen, when an eye is ruptured, a gluey fluid coming out; if this material is still warm, it is fluid, but when it has cooled, it is dry like transparent frankincense. It is the same in both humans and animals. The eye is distressed by anything that falls upon it, both winds that strike it and things that are too bright for it; also, if you anoint it; this is because its surface is moist, like the mouth, the tongue and the rest of the lips.

18. Speech takes place through air, by a person drawing it inside his whole body, but mostly into his hollow parts; as this air is forced out through the empty space, it

[a] See chapter 9 *fin.* above.

---

[1] Del. Zwinger.   [2] Linden.
[3] Joly: ἡ ἄλλη κοιλίη V.

ποιέει· ἡ κεφαλὴ γὰρ ἐπηχεῖ. ἡ δὲ γλῶσσα ἀρθροῖ
προσβάλλουσα· ἐν τῷ φάρυγγι ἀποφράσσουσα καὶ
προσβάλλουσα πρὸς τὴν ὑπερῴην καὶ πρὸς τοὺς
ὀδόντας ποιέει σαφηνίζειν· ἢν δὲ μὴ ἡ γλῶσσῃ
ἀρθροῖ προσβάλλουσα ἑκάστοτε, οὐκ ἂν σαφέως
διαλέγοιτο, ἀλλ' ἢ ἕκαστα φύσει τὰ μονόφωνα.
τεκμήριον δέ ἐστι τούτῳ, οἱ κωφοὶ οἱ ἐκ γενεῆς οὐκ
ἐπίστανται διαλέγεσθαι, ἀλλὰ τὰ μονόφωνα μοῦνον
φωνέουσιν. οὐδ' εἴ τις τὸ πνεῦμα ἐκπνεύσας πειρῷτο
διαλέγεσθαι· δῆλον δὲ τόδε· οἱ ἄνθρωποι ὁκόταν
βούλωνται μέγα φωνῆσαι, ἕλκοντες τὸ πνεῦμα τὸ
ἔξω ὠθέουσι θύραζε καὶ φθέγγονται μέγα ἕως ἂν[1]
ἀντέχῃ τὸ πνεῦμα, ἔπειτα δὲ καταμαραίνεται τὸ
φθέγμα· καὶ οἱ κιθαρῳδοί, ὁκόταν δέῃ αὐτοῖς
μακροφωνέειν, ἐπ' ἄκρον ἑλκύσαντες τὸ πνεῦμα ἔσω
πολὺ ἐκτείνουσι τὴν ἐκφορὰν καὶ φωνοῦσι καὶ φθέγγονται μέγα ἕως ἂν[1] ἀντέχωσι τῷ πνεύματι, ἐπὴν δὲ
τὸ πνεῦμα ἐπιλίπῃ, καταπαύονται· τούτοισι δῆλον
ὅτι τὸ πνεῦμά ἐστι τὸ φθεγγόμενον. εἶδον δὲ ἤδη οἳ
σφάξαντες ἑωυτοὺς ἀπέταμον τὸν φάρυγγα παντά-
πασιν· οὗτοι ζῶσι μέν, φθέγγονται δὲ οὐδέν, εἰ μή
τις συλλάβῃ τὸν φάρυγγα· οὕτω δὲ φθέγγονται·
δῆλον δὲ καὶ τούτῳ, ὅτι τὸ πνεῦμα οὐ δύναται, διατε-
τμημένου τοῦ φάρυγγος, ἕλκειν ἔσω ἐς τὰ κοῖλα,
ἀλλὰ κατὰ τὸ διατετμημένον ἐκπνέει. οὕτως ἔχει
περὶ φωνῆς ἴσως καὶ διαλέξιος.

19. Ὁ δὲ αἰὼν ἐστι τοῦ ἀνθρώπου ἑπταήμερος.
610   πρῶτον μὲν | ἐπὴν ἐς τὰς ὑστέρας ἔλθῃ ὁ γόνος, ἐν

produces sound through the resonance of the head. The tongue articulates by touching: as the tongue encloses the air in the throat and touches the palate and the teeth, it gives the sound clarity. If the tongue does not articulate by touching each time, the person does not speak clearly, but utters, as they all are by nature, mere sounds.[a] A proof of this is that persons deaf from birth do not understand how to speak, but utter only mere sounds. Nor is it possible to speak, if you try to after breathing out. This is clear: when people wish to speak loudly, they draw in breath which they then force out and shout loudly as long as their breath lasts; then their voice dies away. Those who sing to the accompaniment of the lyre, when they must sing an extended passage, first forcefully draw in their breath as much as possible, and then greatly extend their delivery, and recite and sing loudly as long as they still have breath; when their breath is gone, they stop. From these examples, it will be clear that it is the breath that speaks. I have seen persons that have cut their own throats sever the throat completely: these may live, but they cannot speak, unless someone closes their throat: then they can speak. From this, too, it is clear that the person cannot draw his breath into the hollow spaces, because his throat has been cut, but instead sends his breath out through the cut. So it is with the voice, and likewise with speech.

19. The period of life of man is seven days. To begin with, when the seed arrives in the uterus, in seven days it

[a] On this difficult passage see Schwyzer's explanation (p. 96), which I follow.

[1] Ermerins: ὡς V.

ἑπτὰ ἡμέρῃσιν ἔχει ὁκόσα περ ἔστιν ἔχειν τοῦ
σώματος· τοῦτο δέ τις ἂν θαυμάσειεν ὅκως ἐγὼ
οἶδα· πολλὰ δὲ εἶδον τρόπῳ τοιῷδε· αἱ ἑταῖραι αἱ
δημόσιαι, αἵτινες αὐτῶν πεπείρηνται πολλάκις, ὁκόταν παρὰ ἄνδρα ἔλθῃ, γινώσκουσιν ὁκόταν λάβω-
σιν ἐν γαστρί, κἄπειτεν διαφθείρουσιν· ἐπειδὰν δὲ
ἤδη διαφθαρῇ, ἐκπίπτει ὥσπερ σάρξ· ταύτην τὴν
σάρκα ἐς ὕδωρ ἐμβαλών, σκεπτόμενος ἐν τῷ ὕδατι,
εὑρήσεις ἔχον πάντα τὰ μέλεα καὶ τῶν ὀφθαλμῶν
τὰς χώρας καὶ τὰ οὔατα καὶ τὰ γυῖα· καὶ τῶν χειρῶν
οἱ δάκτυλοι καὶ τὰ σκέλεα καὶ οἱ πόδες καὶ οἱ δάκτυλοι τῶν ποδῶν, καὶ τὸ αἰδοῖον καὶ τὸ ἄλλο πᾶν
σῶμα δῆλον. εὔδηλον δὲ καὶ ὅταν λάβῃ ἐς γαστέρα
τῇσιν ἐπισταμένῃσιν· αὐτίκα ἔφριξε, καὶ θέρμη καὶ
βρυγμὸς καὶ σπασμὸς ἔχει, καὶ τὸ ἄρθρον καὶ τὸ
σῶμα πᾶν καὶ τὴν ὑστέρην ὄκνος· καὶ ὁκόσαι καθα-
ραί εἰσι καὶ μὴ ὑγραί, τοῦτο πάσχουσιν· ὁκόσαι δὲ
[πάσχουσι][1] παχεῖαι καὶ βλεννώδεις, οὐ γινώσκουσι
τούτων τῶν γυναικῶν πολλαί· ᾗ δέ μοι ἔδειξαν, κατὰ
τοῦτο δὴ καὶ ἐπίσταμαι εἰδέναι.

Δῆλον δὲ καὶ τῷδε, ὅτι ἑπταήμερος ὁ αἰών, εἴ τις
ἐθέλοι ἑπτὰ ἡμέρας φαγέειν ἢ πιέειν μηδέν, οἱ μὲν
πολλοὶ ἀποθνήσκουσιν ἐν αὐτῇσιν· εἰσὶ δέ τινες καὶ
οἳ ὑπερβάλλουσιν, ἀποθνήσκουσι δ' ὅμως· εἰσὶ δέ
τινες οἳ καὶ ἐπείσθησαν ὥστε μὴ ἀποκαρτερῆσαι,
ἀλλὰ φαγέειν τε καὶ πιέειν· ἀλλ' ἡ κοιλίη οὐκέτι
καταδέχεται· ἡ γὰρ νῆστις συνεφύη ἐν ταύτῃσι
τῇσιν ἡμέρῃσιν· ἀλλὰ θνήσκουσι | καὶ οὗτοι. ἔστι

has all the parts the body is to have. You might wonder how I know this: well, I have learned much in the following way. The common prostitutes, who have frequent experience in these matters, after having been with a man know when they have become pregnant, and then destroy the child. When it has been destroyed, it drops out like a piece of flesh. If you put this flesh into water and examine it in the water,[a] you will see that it has all the parts: the orbits of the eyes, the ears and the limbs; the fingers, the legs, the feet, the toes, and the genital parts, and all the rest of the body is distinct. It is also very obvious to women of experience when a woman becomes pregnant: at once she feels a chill; heat, shivering and tension set in, and she feels a sluggishness in her joints, whole body and uterus. Women that are in a clean state and not moist experience these things, but many of those that are stout and phlegmy do not recognize them. It is to the extent that these women have instructed me that I know about these things.

It is also clear from the following that the period of life is seven days: if a person goes seven days without eating or drinking anything, in this period most die; some even survive that time, but still die; and others in fact are persuaded not to starve themselves to death, but to eat and drink: however, the cavity no longer admits anything because the jejunum has grown together in that many days, and these people, too, die. It is also proven by the

---

[a] Cf. *Nature of the Child* 13 for a similar experiment.

---

[1] Del. Littré.

δὲ καὶ τῷδε τεκμήρασθαι· τὸ παιδίον ἑπτάμηνον
γόνιμον[1] γενόμενον, λόγῳ γεγένηται, καὶ ζῇ, καὶ
λόγον ἔχει τοιοῦτον καὶ ἀριθμὸν ἀτρεκέα ἐς τὰς
ἑβδομάδας· ὀκτάμηνον δὲ γενόμενον, οὐδὲν βιοῖ
πώποτε· ἐννέα δὲ μηνῶν καὶ δέκα ἡμερέων γόνος
γίνεται, καὶ ζῇ, καὶ ἔχει τὸν ἀριθμὸν ἀτρεκέα ἐς τὰς
ἑβδομάδας· τέσσαρες δεκάδες ἑβδομάδων ἡμέραι
εἰσὶ διηκόσιαι ὀγδοήκοντα·[2] ἐς δὲ τὴν δεκάδα τῶν
ἑβδομάδων ἑβδομήκοντα ἡμέραι. ἔχει δὲ καὶ τὸ
ἑπτάμηνον γενόμενον τρεῖς δεκάδας ἑβδομάδων, ἐς
δὲ τὴν δεκάδα ἑκάστην ἑβδομήκοντα ἡμέραι, τρεῖς
δεκάδες δὲ ἑβδομάδων αἱ σύμπασαι δέκα καὶ διηκόσιαι.

Καὶ αἱ νοῦσοι οὕτω τοῖς ἀνθρώποις ὀξύταται
γίνονται, ἡμερέων παρελθουσέων ἐν τῇσιν ἀνακρί-
νονται καὶ ἀπέθανον ἢ ὑγιέες ἐγένοντο, <τεσσάρων,
ἡμίσεος ἑβδομάδος· καὶ δευτεραῖαι ἐν μιῇ ἑβδο-
μάδι·>[3] καὶ τριταῖαι ἕνδεκα ἡμέρῃσιν, ἐν μιᾷ ἑβδομάδι καὶ ἡμίσει ἑβδομάδος· καὶ τεταρταῖαι ἐν δυσὶν
614   ἑβδομάσιν· καὶ αἱ πεμ|πταῖαι ἐν δυοῖν δεούσῃσιν
εἴκοσιν ἡμέρῃσι, δυοῖν τε ἑβδομάδοιν καὶ ἡμίσει
ἑβδομάδος. αἱ δὲ ἄλλαι νοῦσοι οὐκ ἔχουσι διαγνώ-
μην ἐν ὁκόσῳ ὑγιέες ἔσονται ἀποφαίνεσθαι. οὕτω
δὲ καὶ τὰ ἕλκεα τὰ μεγάλα τὰ ἐν τῇ κεφαλῇ καὶ τὰ
ἐν τῷ ἄλλῳ σώματι τεταρταῖα φλεγμαίνειν ἄρχεται,

[1] Potter: γόνον V: del. Ermerins.
[2] Froben: ἑβδομήκοντα V.
[3] Littré.

following. The child born viable at seven months is born in phase, and lives, and has the correct synchrony and precise numerical relationship to seven-day periods. A child born at eight months never survives. At nine months and ten days a child is born and lives, and has a precise numerical relationship to seven-day periods:

$4 \times 10 \times 7 = 280$ days,[a]
and $10 \times 7 = 70$ days.

The child born at seven months[b] counts exactly thirty seven-day periods:

$10 \times 7 = 70$ days,
and $3 \times 10 \times 7 = 210$ days.

Diseases are also most acute in patients according to this same pattern, the days in which they come to a crisis and either die or recover being <four (half a seven-day period), the second most acute occurring in a whole seven-day period,> the third most acute in eleven days (one and a half seven-day periods), the fourth most acute in two seven-day periods, and the fifth most acute in eighteen days (two and a half seven-day periods). Other diseases have no means by which to reveal in how many days patients will recover. Thus both large ulcers on the head and ulcers on the rest of the body begin to swell on

[a] For the Greeks, normal pregnancy in the healthy woman lasted exactly seven forty-day periods (280 days), and children born at that term were called "ten months'" (δεκάμηνοι). See W. H. Roscher, *Die Tessarakontaden und Tessarakontadenlehren der Griechen und anderer Völker*, Leipzig, 1909, p. 97f.

[b] Reckoning seven thirty-day months.

ἐν ἑπτὰ δὲ καθίστανται φλεγμήναντα καὶ ἐν τεσσα-
ρεσκαίδεκα <καὶ ἐν εἴκοσι>[1] δυοῖν δεούσῃσιν. ἢν
δέ τις ἀνακῶς θεραπεύῃ καὶ μὴ καταστῇ ἐν τούτῳ τῷ
χρόνῳ τὰ ἐν τῇσι κεφαλῇσι μεγάλα ἕλκεα, ἀποθνήσκουσιν οἱ ἄνθρωποι. θαυμάσειε δ᾽ ἄν τις καὶ τοῦτο
ὅστις ἄπειρος ᾖ, εἰ ἑπτάμηνον γίνεται παιδίον· ἐγὼ
μὲν οὖν αὐτὸς ὄπωπα καὶ συχνά· εἰ δέ τις βούλεται
καὶ τοῦτο ἐλέγξαι, ῥηΐδιον· πρὸς τὰς ἀκεστρίδας αἱ
πάρεισι τῇσι τικτούσῃσιν ἐλθὼν πυθέσθω. ἔστι δὲ
καὶ ἄλλο τεκμήριον· τοὺς ὀδόντας οἱ παῖδες ἑπτὰ
ἐτέων διελθόντων πληροῦσι· καὶ ἐν ἑπτὰ ἔτεσίν ἐστι
δὲ λόγῳ καὶ ἀριθμῷ ἀτρεκέως δεκάδες ἑβδομάδων
<ἓξ καὶ τριήκοντα καὶ ἥμισυ δεκάδος, ἑβδομάδες
πέντε καὶ>[2] ἑξήκοντα καὶ τριηκόσιαι. τῆς δὲ φύσιος
τὴν ἀνάγκην, διότι ἐν ἑπτὰ τούτων ἕκαστα διοικεῖ-
ται, ἐγὼ φράσω ἐν ἄλλοισιν.

[1] Zwinger in margin, from Calvus' *duovedeviginti*.
[2] Littré.

the fourth day, and the swelling subsides on the seventh, fourteenth and eighteenth days. If treatment is carefully carried out, but large ulcers of the head fail to heal up in this period, the patients die. (Someone who lacks experience might wonder whether a child can really be born in seven months, but I have seen it myself, and often. If anyone wishes proof, the matter is easy: let him go to the midwives that attend women who are giving birth and ask them.)[a] There is yet another proof. Children have acquired all their teeth by the time they reach seven years. Now there are in seven years by measure and number exactly thirty-six and a half tens of seven-day periods: 365 seven-day periods. The necessity of nature that compels each of these things to be ordered by seven, I shall explain elsewhere.

[a] Apparently this note is an after-thought to the paragraph above on seven-months' children.

PRORRHETIC I

INTRODUCTION

How the same recherché and presumably newly-coined title, *Prorrhetic*, came to be given to the two very different works we have before us here and below is unknown. Completely divergent in purpose, structure, and content, the most the *Prorrhetics* can be said to share is, as the name suggests, an interest in medical prediction.

Once assigned, however, the names persisted. In Erotian's census of Hippocratic writings we find among the semiotic: "*Prorrhetic I* and *II*—that it is not by Hippocrates, I shall demonstrate elsewhere."[1] Suggesting that the "it" in Erotian's parenthetical remark refers only to book 2 is the fact that over twenty glosses from book 1 appear in the *Glossary*, but none from book 2.[2] Six of these glosses to *Prorrhetic I* cite Bacchius of Tanagra,[3] evidence that the work was already in the Hippocratic Collection by the third century B.C.

Although Galen regards *Prorrhetic I* as a mixture of both genuine and spurious Hippocratic material drawn

[1] Erotian p. 9.
[2] See Nachmanson pp. 273–78.
[3] Erotian, s.v. A6, A8, E5, K6, Λ7, Σ1.

from many sources,[4] he valued the work enough to devote a lengthy commentary to it.[5] This commentary was written after his commentaries to *Epidemics I* and *II*, and before those to *Epidemics III* and *VI*.[6]

Caelius Aurelianus' Latin translation of Soranus' *Acute Diseases* cites *Prorrhetic I* 16 by name: "For in his book entitled *Prorrhetic* (*Praedictivo libro*) he says that phrenitics drink little, are agitated by every sound, and are affected by tremor."[7]

Prorrhetic I belongs to the genre of Hippocratic prognosis literature also represented by the *Aphorisms* and *Coan Prenotions*. It consists of a collection of 170 independent short prognostic or expectant chapters arranged by subject in the sections:

    1–38: Phrenitis, mania, mental derangement
   39–98: Bad or fatal signs
  99–124: Spasms and convulsions
125–152: Haemorrhages
153–170: Swellings beside the ears.

Interest is empirical and descriptive, with virtually no theoretic content. The conditions included are weighted towards the neurological (e.g. loss of speech, deafness,

[4] For a thorough discussion of Galen's position see V. Nutton, *Galen on Prognosis*, Corpus Medicorum Graecorum V 8,1, Berlin, 1978, pp. 175f.

[5] *Galeni In Hippocratis prorrheticum I commentarii*, ed. H. Diels, CMG V 9,2, Leipzig and Berlin, 1915, pp. 1-178.

[6] J. Ilberg, "Über die Schriftstellerei des Klaudios Galenos," *Rheinisches Museum* N.F. 44 (1889), pp. 229–38.

[7] Caelius pp. 376f.

delirium, coma, strabismus, headache, paralyses), although conditions of other kinds are also often included (e.g. ardent fever, diarrhoea, breathlessness). Most statements in the treatise have been worked through to the generalized form, but two frequent features suggest that the process of collecting and evaluating clinical data was still going on at the time of publication:

i) in twenty-seven chapters the author poses questions, (presumably to himself): e.g. (ch. 30) "Do patients that throb through their whole body end up by losing their speech?"

ii) in eleven cases specific patients are mentioned: e.g. (ch. 13) "In patients with phrenitis a white evacuation is bad, as it was for Archecrates."

The great interest this treatise commanded at a time when medicine's curative possibilities were limited is reflected in the many editions and translations listed by Littré (vol. 5, 509), and by the fact that in the eighteenth century it attracted two English translators:

> Francis Clifton, *Hippocrates upon . . . Prognosticks . . .*, London, 1734. (A large number of the chapters of *Prorrhetic I* are translated in pp. 250–389 of this book, but arranged according to the translator's rather than the treatise's order, and interspersed with copious other Hippocratic material.)
>
> John Moffat, *The Prognostics and Prorrhetics of Hippocrates translated . . . with Large Annotations . . .*, London, 1788. (pp. 55–292)

Besides appearing in the collected editions and translations down to Ermerins and Fuchs, *Prorrhetic I* was the subject of a 1954 Hamburg dissertation, published later as:

Hilde Polack, *Textkritische Untersuchungen zu der hippokratishen Schrift 'Prorrhetikos I'*, Hamburg, 1976.

Despite its modest title, this study contains a complete critical edition of the treatise followed by a very valuable philological commentary. My edition and translation depend almost entirely upon Dr. Polack's work.

ΠΡΟΡΡΗΤΙΚΟΝ Α

1. Οἱ κωματώδεες ἐν ἀρχῇσι γενόμενοι μετὰ κεφαλῆς, ὀσφύος, ὑποχονδρίου, τραχήλου ὀδύνης ἀγρυπνέοντες ἦρά γε φρενιτικοί εἰσιν; μυκτὴρ ἐν τούτοισιν ἀποστάζων ὀλέθριον, ἄλλως τε καὶ ἢν τεταρταίοισιν ἀρχομένοισιν.

2. Κοιλίης περίπλυσις ἐξέρυθρος κακὸν μὲν ἐν πᾶσιν, οὐχ ἥκιστα δὲ ἐπὶ τοῖσι προειρημένοισιν.

3. Αἱ δασεῖαι γλῶσσαι κατάξηροι φρενιτικαί.

4. Τὰ ἐπὶ ταραχώδεσιν ἀγρύπνοισιν οὖρα ἄχροα, μέλασιν ἐνῃωρημένα | παρακρουστικά· ἐφιδρῶντι[1] φρενιτικά.[2]

5. Ἐνύπνια τὰ ἐν φρενιτικοῖς ἐναργέα.

6. Ἀνάχρεμψις πυκνή γε, ἢν δή τι καὶ ἄλλο σημεῖον προσῇ, φρενιτικά.

7. Τὰ ἐγκαταλιμπανόμενα καύματα ἐν ὑποχονδρίῳ πυρετοῦ περιψυχθέντος κακά, ἄλλως τε καὶ ἢν ἐφιδρῶσιν.

8. Αἱ προαπαυδησάντων παραφροσύναι κάκισται, οἷον καὶ θρασύνοντι.

9. Τὰ φρενιτικὰ νεανικῶς τρομώδεα τελευτᾷ.

PRORRHETIC I

1. Do patients who have been comatose at the beginning, but later lie awake with pains in the head, loins, hypochondrium and neck develop phrenitis? A running nose in these is a fatal sign, especially if it begins on the fourth day.

2. A very red discharge from the cavity is bad in all cases, not least in the ones mentioned above.

3. A rough tongue that is very dry indicates phrenitis.

4. Colourless urines in persons with troubled sleeplessness, if they have dark material suspended in them, suggest derangement—in a person who perspires over his whole body, phrenitis.

5. The dreams of patients with phrenitis are vivid.

6. Frequent expectoration, if some other sign is present as well, indicates phrenitis.

7. If heat is left behind in the hypochondrium when a fever has grown cool, it is a bad sign, especially if the person perspires over his whole body.

8. Delirium in patients that are already failing is a very bad sign, as also in a person who becomes over-bold.

9. Phrenitis in the young ends with tremors.

---

[1] H² from Galen: -δρῶντα I.

[2] R from Galen: νεφρικά I.

10. Τὰ ἐν κεφαλαλγίῃσιν ἰώδεα ἐμέσματα μετὰ κωφώσιος ἀγρύπνῳ ταχὺ ἐκμαίνει.

11. Τὰ ἐν ὀξέσι κατὰ φάρυγγα ὀδυνώδεα σμι|κρά, πνιγώδεα, ὅτε χάνοι, μὴ[1] ῥηϊδίως συνάγοντι, ἰσχνῷ παρακρουστικὸν τὸ τοιοῦτον· φρενιτικοὶ ὀλέθριοι.

12. Ἐν τοῖσι φρενιτικοῖσιν ἐν ἀρχῇσι τὸ [μὴ][2] ἐπιεικές, πυκνὰ δὲ μεταπίπτειν, κακὸν τὸ τοιοῦτον. καὶ πτυελισμὸς κακόν.

13. Ἐν φρενιτικοῖσι λευκὴ διαχώρησις κακόν. ὡς καὶ τῷ Ἀρχεκράτει· ἦρά γε ἐπὶ τούτοισι καὶ νωθρότης ἐπιγίνεται; καὶ ῥῖγος ἐπὶ τούτοισι κάκιστον.

14. Τοῖσιν ἐξισταμένοισι μελαγχολικῶς, οἷσι τρόμοι ἐπιγίνονται, καὶ κακόηθες.

15. Οἱ ἐκστάντες ὀξέως ἐπιπυρέξαντες σὺν ἱδρῶτι φρενιτικοί.

16. Βραχυπόται, ψόφου καθαπτόμενοι τρομώδεες γίνονται.

17. †Τὰ ἐξ ἐμέτου ἀσώδεος, φωνὴ κλαγγώδης, ὄμματα ἐπίχνουν ἴσχοντα,[3]† μανικά. οἷον καὶ ἡ τοῦ Ἑρμοζύγου ἐκμανεῖσα ὀξέως ἄφωνος ἀπέθανεν.

18. Ἐν πυρετῷ καυσώδει ἤχων προγενομένων μετὰ δ' ἀμβλυωσμοῦ καὶ κατὰ τὰς ῥῖνας βάρεος προελθόντος ἐξίστανται μελαγχολικῶς.

---

[1] H[2] after Galen: μὲν I.
[2] Del. H[2] after Galen.
[3] H[2]: ἴσχνωντα I.

10. For a sleepless person who is deaf to vomit greenish material during a headache indicates a rapidly approaching mania.

11. Slight, suffocating pains in the throat, felt on opening the mouth by a person with an acute disease who cannot close it easily and who is feeble, announce mania. Those with phrenitis are doomed.

12. In phrenitis, for signs that are reasonable at the beginning to change frequently is bad[a]; salivation is also bad.

13. In patients with phrenitis a white evacuation is bad, as it was for Archecrates. Does torpor follow in these? Chills too are a very bad sign in these patients.

14. For trembling to come on in patients that are out of their wits with melancholy is a malignant sign.

15. For persons out of their wits to be suddenly attacked by an acute fever and sweating indicates phrenitis.

16. Persons who drink little and are over-sensitive to noise become tremulous.

17. † If subsequent to nausea and vomiting—the voice being shrill—the eyes take on a wool-like covering, † it announces mania, as in the daughter of Hermozygus who, while delirious, suddenly lost her voice and died.

18. If, during an ardent fever, ringing occurs in the ears together with dullness of vision and a heaviness in the region of the nose, these patients lose their wits in melancholy.

[a] Cf. chapter 28 below.

19. Αἱ παρακρούσιες σὺν φωνῇ κλαγγώδει γλώσ-
516 σης σπασμοὶ τρομώδεες γενόμε|νοι ἐξίστανται·
σκληρυσμὸς ταύτῃσιν ὀλέθριον.

20. Αἱ τρομώδεες γλῶσσαι σημεῖον οὐχ ἱδρυμέ-
νης γνώμης.

21. Ἐπὶ τοῖσι χολώδεσι διαχωρήμασι τὸ ἀφρῶ-
δες ἐπάνθισμα κακόν, ἄλλως τε καὶ ὀσφὺν προηλ-
γηκότι παρενεχθέντι.

22. Τὰ ἀραιὰ κατὰ πλευρὸν ἐν τούτοισιν ἀλγήματα παραφροσύνην σημαίνει.

23. Αἱ μετὰ λυγγὸς ἀφωνίαι κάκιστον.

24. Αἱ μετ' ἐκλύσιος ἀφωνίαι κάκιστον.

25. Ἐν ἀφωνίῃ πνεῦμα οἷον τοῖσι πνιγομένοισι
πρόχειρον πονηρόν· ἆρά γε καὶ παρακρουστικὸν τὸ
τοιοῦτον;

26. Αἱ ἐπ' ὀλίγον θρασέως[1] παρακρούσιες θηρι-
ώδεες.

27. Αἱ μετὰ καταψύξεως οὐκ ἀπύρῳ ἐφιδρῶοντι
τὰ ἄνω δυσφορίαι φρενιτικά, ὡς καὶ Ἀρισταγόρῃ,
καὶ μέντοι καὶ ὀλέθρια.

28. Τὰ ἐν φρενιτικοῖσι πυκνὰ μεταπίπτοντα
σπασμώδεα.

29. Τὰ οὐρούμενα μὴ ὑπομνησάντων ὀλέθρια·
518 ἦρα τούτοισιν οὐρεῖται | οἷον ἐπὴν ὑποστᾶσι ταρά-
ξειας;

_____
[1] Polack: θράσος I.

19. When, in delirium associated with shrillness of the voice, trembling spasms of the tongue occur, these patients lose their wits; hardening of the tongue in this condition is a fatal sign.

20. Trembling of the tongue is a sign that the mind is not settling down.

21. A frothy scum on bilious stools is a bad sign, especially in a person with diarrhoea and who had pain in his loins beforehand.

22. Intermittent pains in the side in such cases are a sign of derangement of the mind.

23. Loss of speech in conjunction with hiccups is a very bad sign.

24. Loss of speech in conjunction with faintness is a very bad sign.

25. In persons who have lost their speech perceptible breathing like that heard in suffocation bodes ill; does this also indicate oncoming mania?

26. Short periods of delirium characterized by over-boldness are malignant.[a]

27. Restlessness during a chill in a febrile patient who is perspiring over the upper part of his body indicates phrenitis, as in the case of Aristagora, and is a fatal sign.

28. Frequent changes in phrenitis announce convulsions.[b]

29. For a person to pass urine without intending to is a fatal sign. Is the urine passed in these cases like what you would stir up with a precipitate?

[a] See chapter 123 below.
[b] Cf. chapter 12 above.

30. Οἱ παλμώδεες δι᾽ ὅλου ἀρά γε ἄφωνοι τελευτῶσι;

31. Τὰ ἐν φρενιτικοῖσι μετὰ καταψύξιος πτυελίζοντα μέλανα ἐμεῖται.

32. Κώφωσις καὶ οὖρα ἐξέρυθρα, ἀκατάστατα ἐναιωρήματα παρακρουστικόν· τοῖσι τούτοισιν ἰκτεροῦσθαι κακόν. κακὸν δὲ καὶ ἐπὶ ἰκτέρῳ κώφωσις.[1] τούτους ἀφώνους, αἰσθανομένους δὲ συμβαίνει γίνεσθαι· τούτοισι δὲ καὶ κοιλίαι καταρρήγνυνται, οἷον Ἑρμίππῳ καὶ ἀπέθανε.

33. Κώφωσις ἐν ὀξέσι καὶ ταραχώδεσι παρακολουθοῦσα κακόν.

34. Αἱ τρομώδεες, ἀσαφέες, ψηλαφώδεες παρακρούσιες πάνυ φρενιτικαί, ὡς καὶ τῷ Διδυμάρχου ἐν Κῷ.

35. Αἱ ἐκ ῥίγεος νωθρότητες οὐ πάνυ παρ᾽ αὐτοῖσιν.

36. Οἱ περὶ ὀμφαλὸν πόνοι παλμώδεες ἔχουσι μέν τι καὶ γνώμης παράφορον, περὶ κρίσιν δὲ τούτοισι πνεῦμα ἁλὲς συχνὸν ξὺν πόνῳ διέρχεται. καὶ οἱ κατὰ γαστροκνημίην πόνοι ἐν τούτοισι γνώμης παράφοροι.

520   37. Τὰ κατὰ μη|ρὸν ἐν πυρετῷ ἀλγήματα ἔχει τι παρακρουστικόν, ἄλλως τε καὶ ἢν οὖρον ἐναιωρηθῇ λεῖον· καὶ ὁκόσα περὶ κύστιν ἴσχουσι τοιαῦτα ἅμα πυρετῷ· κοιλίη ταραχώδης τρόπον[2] χολερώδεα

---

[1] Potter: μώρωσις I.

30. Do patients that throb through their whole body end up by losing their speech?

31. What is salivated in cases of phrenitis with chills is vomited back up dark.

32. Deafness in conjunction with very red urines that contain suspended material but set down no deposit presages mania. For such a patient to become jaundiced is a bad sign. Deafness coming after jaundice is also a bad sign. These patients lose their speech but at the same time retain all their mental faculties. In such cases the cavity is also subject to violent discharges, as happened to Hermippus, and he died.

33. For deafness to come on in acute and disruptive diseases is a bad sign.

34. Delirium associated with trembling, confusion and groping with the hands points very definitely to phrenitis, as in the son of Didymarchus in Cos.

35. Patients who suffer torpor after chills are not altogether in their senses.

36. Pains about the navel accompanied by trembling may involve some disturbance of the mind, and at their crisis these patients pass a great quantity of wind and not without pain. Pains in the calf of the leg in such cases are also disturbing to the mind.

37. Pains in the thigh during fever point towards derangement of the mind, especially if the person's urine contains smooth, suspended material; so also do pains about the bladder during fever. When the cavity is moved and passes bilious stools, persons with comatose torpors

---

[2] Parisinus Graecus 2145: τρόπῳ I.

κωματώδεες νωθροὶ οὐ πάνυ παρὰ[1] αὐτοῖσιν.

38. Ἐπὶ κοιλίῃ ὑγρῇ, κοπώδει, κεφαλαλγικῷ, διψώδει, ὑπαγρύπνῳ, ἀσαφεῖ, ἀδυνάτῳ, οἷσι τὰ τοιαῦτα, ἐλπὶς ἐκστῆναι.

39. Οἱ ἐφιδρῶντες καὶ μάλιστα κεφαλὴν ἐν ὀξέσιν ὑποδύσφοροι, κακόν, ἄλλως τε καὶ ἐπ' οὔροισι μέλασι, καὶ τὸ θολερὸν πνεῦμα ἐν τούτοισι κακόν.

40. Αἱ παρὰ λόγον κενεαγγικῶς[2] ἀδυναμίαι οὐκ ἐούσης κενεαγγείης κακόν.

41. Κοιλίαι ἀπολελαμμέναι, σμικρὰ δὲ μέλανα σπυρα|θώδη κατ' ἀνάγκην χαλῶσαι μυκτήρ τε σὺν τούτοισιν ἐπιρρηγνύμενος κακόν.

42. Οἷσιν ὀσφύος ἄλγημα ἐπὶ πολὺ μετὰ καύματος ἀσώδεος, ἐφιδρῶντες οὗτοι κακόν· ἦρά γε τούτοισι τρομώδεα γίνεται; καὶ ἡ φωνὴ δὲ ὡς ἐν ῥίγει.

43. Ἄκραια δ' ἐπ' ἀμφότερα ταχὺ μεταπίπτοντα κακόν· καὶ δίψα δὲ ἡ τοιαύτη πονηρόν.

44. Ἐκ κοσμίου θρασεῖα ἀπόκρισις κακόν.

45. Οἷσιν φωνὴ ὀξείη,[3] ὑποχόνδρια τούτοισιν εἴσω εἴρυαται.

46. Ὄμματα ἀμαυρούμενα φαῦλον καὶ τὸ πεπηγὸς ἀχλυῶδες κακόν.

47. Ὀξυφωνίη κλαυθμώδης κακόν.

---

[1] Vaticanus Graecus 278 in margin: περὶ I.
[2] Polack: -γικῷ I.
[3] Polack: Φωνὴ ὀξείη οἷσιν I.

are not altogether in their senses.

38. In cases of diarrhoea, weariness, headache, thirst, mild sleeplessness, confusion and debility expect derangement.

39. To perspire in acute states, especially with a slight, unpleasant sweat over the head, is a bad sign, particularly in association with the passage of dark urines; laboured breathing in these patients is also a bad sign.

40. A weakness like the one that results from evacuant treatment, if occurring without a reason when no evacuation has taken place, is a bad sign.

41. If the cavity is blocked, but on being forced passes small, dark stools like those of sheep and goats, this, together with a haemorrhage from the nose, is a bad sign.

42. For patients suffering from pains in the loins in conjunction with nausea and fever to perspire over the body is a bad sign. Does trembling develop in them? Also, their voice changes the way it does in chills.

43. A rapid change in the extremities on both sides is a bad sign. A rapidly changing thirst also bodes ill.

44. An insolent reply from a polite person is a bad sign.

45. The voice becomes high by the hypochondrium being drawn tight inside.

46. For the eyes to lose their power of vision is an indifferent[a] sign, and for them to be fixed and clouded a bad one.

47. A high-pitched, broken voice is a bad sign.

[a] In this treatise, "indifferent", φαῦλος or φλαῦρος, is used euphemistically to mean "bad"; cf. *Oxford English Dictionary*, *s.v.* 7b.

48. Ὀδόντων πρίσιες ὀλέθριον οἷς μὴ σύνηθες καὶ ὑγιαίνουσι, πνιγμοὶ ἐν τούτοισι κακὸν πάνυ.

49. Προσώπου εὔχροια καὶ τὸ λίην σκυθρωπὸν πονηρόν.

50. Τὰ τελευτῶντα διαχωρήματα εἰς ἀφρώδεα ἄκρητα παροξυντικά.

51. Αἱ ἐκ καταψύξιος ἐν ὀξέσιν οὔρων ἀπολήψιες κάκισται.

524   52. Τὰ ὀλέθρια ἀσήμως ῥαστωνήσαντα θάνατον σημαίνει.

53. Ἐν ὀξέσι καὶ χολώδεσι ἔκλευκα ἀφρώδεα καὶ περίχολα διαχωρήματα κακόν, καὶ οὖρα τὰ τοιαῦτα κακόν· ἆρά γε τούτοισιν ἧπαρ ἐπώδυνον;

54. Αἱ ἐν πυρετοῖσιν ἀφωνίαι σπασμώδεα τρόπον ἐξίστανται· ὀλέθριον.

55. Αἱ ἐκ πόνου ἀφωνίαι δυσθάνατοι.

56. Οἱ ἐξ ὑποχονδρίων ἀλγήματος πυρετοὶ κακοήθεις.

57. Δίψα παραλόγως λυθεῖσα ἐν ὀξέσι κακόν.

58. Ἱδρὼς πολὺς ἐγγενόμενος ἅμα πυρετῷ ἐν ὀξέσι φαῦλον.

59. Καὶ οὖρα δὲ πέπονα πονηρὰ καὶ τὰ ἐρυθρὰ ἐκ
526   τούτων ἐπανθίσματα | ἰώδεα κατεχόμενα καὶ τὸ μικρὰ ἐπιφαίνεσθαι, στάξιες.

60. Καὶ ἔμετοι μετὰ ποικιλίης κακόν, ἄλλως τε καὶ ἐγγὺς ἀλλήλων ἰόντα.

61. Ὅσα ἐν κρισίμοισιν ἀλυσμοῖς ἐν ἱδρῶτι περιψύχεται κακόν· καὶ τὰ ἐπιρριγώσαντα ἐκ τούτων κακά.

48. To grind the teeth is a fatal sign in persons who do not habitually grind them when they are healthy, and difficulty in breathing in these is exceedingly bad.

49. A complexion of a good colour if associated with an over-sad countenance bodes ill.

50. Evacuations that at the end become frothy and unmixed indicate an exacerbation.

51. Stoppage of the urine subsequent to a chill in acute diseases is a very bad sign.

52. Fatal signs that relent without leaving any trace presage death.

53. In acute bilious diseases, very white frothy stools mixed with bile are a bad sign, as are like urines. Is the liver painful in these patients?

54. Loss of speech in fevers in which patients become deranged in a convulsive manner is a fatal sign.

55. Loss of speech arising from exertions brings a hard death.

56. Fevers arising from pains in the hypochondrium are malignant.

57. Thirst in acute diseases that resolves without a reason is a bad sign.

58. Copious sweating during fevers in acute diseases is an indifferent sign.

59. Mature urines, too, are an evil sign, and also a rusty red scum that separates out and covers them; also the appearance of little pieces—droplets—on the surface.

60. Attacks of vomiting in which material of different kinds comes up are a bad sign, especially if these occur closely one after another.

61. To be thoroughly cooled by a sweat during critical disturbances is a bad sign; an attack of shivering afterwards is also bad.

62. Ἐμέσματα ἄκρητα, ἀσώδεα πονηρά.

63. Τὸ καρῶδες ἆρά γε πανταχοῦ κακόν;

64. Μετὰ ῥίγεος ἄγνοια κακόν· κακὸν δὲ καὶ λήθη.

65. Αἱ ἐκ ῥίγεος καταψύξιες μὴ ἀναθερμαινόμεναι κακαί.

66. Οἱ ἐκ καταψύξιος ἱδρώδεες μὴ ἀναθερμαινόμενοι κακοί· ἐπὶ τούτοισιν ἐν πλευρῷ καῦμα καί τι ὀδυνῶδες καὶ τὸ ἐπιρριγῶσαι κακόν.

67. Τὰ κωματώδεα ῥίγεα ὑπό τι ὀλέθρια· καὶ τὸ φλογῶδες ἐν προσώπῳ μεθ᾽ ἱδρῶτα κακόν· ἐπὶ τούτοις ψύξις τῶν ὄπισθεν σπασμὸν ἐπικαλεῖται.

68. Οἱ ἐφιδρῶντες ἄγρυπνοι ἀναθερμαινόμενοι κακόν.

69. Ἐξ ὀσφύος ἀναδρομή, ὀφθαλμῶν ἴλλωσις κακόν.

70. Ὀδύνη ἐς τὸ στῆθος ἱδρυνθεῖσα σὺν νωθρότητι κακόν· ἐπιπυρετήναντες οὗτοι καυστι|κοὶ ὀξέως θνήσκουσιν.

71. Οἱ ἐπανεμεῦντες μέλανα ἀπόσιτοι παράφοροι καθ᾽ ἥβην μικρὰ κινδυνώδεες. οἷσιν ὄμμα θρασὺ ἢ κεκλιμένον, τούτους[1] μὴ φαρμακεύειν, ὀλέθριον γάρ. μηδὲ τοὺς ὑποιδέοντας, σκοτώδεας ἐν τῷ πλανᾶσθαι ἐκλιμπάνοντας ἀποσίτους ἀχρόους. μηδὲ τοὺς ἐν πυρετῷ καυματώδει κατακεκλασμένους, ὀλέθριον γάρ.

---

[1] Aldina after Galen: -τοισι I.

62. To vomit unmixed material in association with nausea bodes ill.

63. Drowsiness: is it a totally bad sign?

64. Loss of understanding in conjunction with chills is a bad sign; bad also is forgetfulness.

65. To be cooled subsequent to chills and not to be warmed again is bad.

66. To sweat subsequent to cooling and not be warmed again is a bad sign; if, in these patients, this is followed by burning heat in the side, any pain or subsequent chill is a bad sign.

67. Chills accompanied by coma are quite dangerous; fiery redness of the face after sweating is also a bad sign. In these patients a coldness of the back region announces a convulsion.

68. For sleepless persons who perspire over their body to be warmed up is a bad sign.

69. A pain shooting up from the loins, if accompanied by strabismus of the eyes, is a bad sign.

70. For a pain to become fixed in the chest at the same time there is torpor is a bad sign; if these patients later develop fever, they quickly succumb from the burning heat.

71. Persons in puberty that vomit up dark material, lose their appetite, and suffer derangement are in little danger. To persons who have one eye that protrudes or is slanted do not give purgative medications, since it would be fatal; nor to those with any degree of swelling, who are dizzy, who lose consciousness and have irregular movements, who lack appetite, or who have a bad colour; nor to those worn down by an ardent fever, for this would be lethal.

185

72. Καρδίης πόνος ἅμα ὑποχονδρίῳ ξυντόνῳ[1] καὶ κεφαλαλγίη κακόηθες καί τι ἀσθματῶδες ἐνίοτε· ἆρά γε οὗτοι ἐξαίφνης τελευτῶσιν, ὡς καὶ Λύσις ἐν Ὀδησσῷ; τούτῳ οὖρα ἐζυμωμένα κατέρυθρα ἐγένετο βιαίως.

73. Τραχήλου πόνος κακὸν μὲν ἐν παντὶ πυρετῷ· κάκιστον δὲ καὶ οἷσιν ἐκμανῆναι ἐλπίς.

74. Κοπιώδεες, κωματώδεες, ἄγρυπνοι, ἀχλυώδεες, ἐφιδρῶντες πυρετοὶ κακοήθεες.

530   75. Αἱ ἐκ νώτου φρίκαι πυκναὶ καὶ ὀξέως | μεταπίπτουσαι δύσφοροι. οὔρων ἀπόληψιν ἐπώδυνον σημαίνουσιν.

76. Οἱ ἀσώδεες ἀνημέτως παροξυνόμενοι κακόν.

77. Κατάψυξις μετὰ σκληρυσμοῦ σημεῖον ὀλέθριον.

78. Ἀπὸ κοιλίης λεπτὰ μὴ αἰσθανομένῳ διέναι ἐόντι παρ' ἑαυτῷ κακόν, οἷον τῷ ἡπατικῷ.

79. Τὰ μικρὰ ἐμέσματα χολώδεα κακόν, ἄλλως τε καὶ ἐὰν ἐπαγρυπνῶσι· μυκτὴρ ἐν τούτοισιν ἀποστάζων ὀλέθριον.

80. Αἷς ἐκ τόκων λευκά· ἐπιστάντων δὲ ἅμα πυρετῷ κώφωσις καὶ ἐς πλευρὸν ὀδύνη ὀξεῖα ἐνίσταται· ὀλέθριον.

81. [Οἱ][2] ἐν πυρετοῖσι καυσώδεσιν ὑποπεριψύχουσι, διαχωρήμασιν ὑδατοχόλοισι, συχνοῖσιν

---

[1] H[2] after Galen: -τομος I.
[2] Del. Littré.

72. Pain in the cardia, in conjunction with tension of the hypochondrium and headache, is malignant, and sometimes associated with a degree of breathlessness: do these patients suddenly die like Lysis in Odessus? He developed very effervescent, reddened urines.

73. A pain in the neck is a bad sign in every fever, but worst in patients in whom there is reason to expect delirium.

74. Fevers accompanied by weariness, coma, sleeplessness, dimness of vision, and sweating over the whole body are malignant.

75. Chills starting from the back that are frequent and change rapidly are hard to bear; they signal a painful blockage of the urine.

76. Exacerbations of nausea unaccompanied by vomiting are a bad sign.

77. Chills together with constipation are a fatal sign.

78. For thin evacuations of the cavity to be passed unawares by a person who is in his senses is a bad sign, just as in a patient with a liver complaint.

79. Scanty bilious vomitus is a bad sign, especially if the patients remain awake; in these a dripping nose is a fatal sign.

80. It is a fatal sign for women, who after giving birth have been subject to transient white fluxes which ended with a fever, to be attacked by deafness and a sharp pain in the side.

81. In ardent fevers accompanied by slight shivering and recurring watery, bilious evacuations, an eye looking

ὀφθαλμός [τε]¹ ἰλλαίνας² σημεῖον κακόν, ἄλλως τε
καὶ ἢν κάτοχοι γένωνται.

82. Τὰ ἐξαίφνης ἀποπληκτικὰ λελυμένως ἐπιπυ-
ρετήναντα χρόνῳ ὀλέθρια, οἷόν τι ἐπεπόνθει καὶ ὁ
Νουμηνίου υἱός.

83. Ἧισιν ἐξ ὀσφύος ἀλγήματος ἀναδρομὴ ἐς
καρδίην πυρετῶδες, φρικῶδες· ἀνεμέουσαι ὑδατώδεα,
532 λεπτά, πλέονα, | παρενεχθεῖσαι ἄφωνοι· ἐμέσασαι
μέλανα τελευτῶσιν.

84. Ὄμματος κατάκλεισις ἐν ὀξέσι κακόν.

85. Ἆρά γε τοῖσιν ἀσώδεσιν ἀνημέτοισιν,
ὀσφυαλγέσιν, ἢν ὀλίγα θρασέως παρακρούσωσιν,
ἐλπὶς μέλανα διελθεῖν;

86. Φάρυγξ ἐπώδυνος ἰσχνὴ μετὰ δυσφορίης
πνιγμώδης ὀλεθρίη ὀξέως.

87. Οἷς πνεῦμα ἀνέλκεται καὶ φωνὴ πνιγμώδης ὁ
σπόνδυλός τε ἐγκάθηται, τούτοισιν ἐπὶ τελευτῆς ὡς
συσπῶντός τινος πνεῦμα γίνεται.

88. Οἱ κεφαλαλγικοὶ κατόχως παρακρούοντες
κοιλίης ἀπολελαμμένης, ὄμμα θρασύνοντες, ἀνθη-
ροί, ὀπισθοτονώδεες γίνονται.

89. Ἐπὶ ὀμμάτων διαστροφῇ³ πυρετώδει, κοπιώδει ῥῖγος ὀλέθριον· [ὀξέως. οἷς πνεῦμα ἀνέλκεται·
καὶ φωνὴ ἀσαφής· πυρετώδει, κοπιώδει ῥῖγος ὀλέθ-

---

¹ Del. H² after Galen manuscripts.
² Polack: ἰλλύνας I.
³ H² from Galen: -στροφῆς I.

awry is a bad sign, especially if these patients become cataleptic.

82. Sudden paralytic conditions followed by mild fevers over an extended time are fatal, as for example the one Numenius' son suffered.

83. Women who experience a pain shooting up out of their loins towards the cardia and who are febrile and have chills vomit up copious, thin, watery material, become deranged and lose their speech, and they end[a] by vomiting dark vomitus.

84. The closing of an eye in acute diseases is a bad sign.

85. In patients who experience nausea but do not vomit, and who have pains in the loins, is it to be expected that, if they are delirious in a somewhat violent manner, they will evacuate dark stools?

86. A painful throat without swelling, if accompanied by a general distress, presages suffocation and a rapid death.

87. Patients in whom the breath is drawn short, the voice choked, and the spine depressed, breathe in the end like a person having a spasm.

88. Patients with headaches, who suffer a cataleptic delirium, whose cavity is intercepted, who protrude their eyes, and whose skin is of a good colour, become opisthotonic.

89. In a person who is febrile and wearied, strabismus of the eyes followed by a chill is a fatal sign [rapidly. Patients in whom the breath is drawn short, the voice unclear—in a person who is febrile and wearied, a chill is

[a] Alternative translation: "succumb".

189

ριον]¹ καὶ οἱ κωματώδεες δὲ ἐν τούτοισι πονηρόν.

90. Αἱ ἐν πυρετοῖσι πρὸς ὑποχόνδριον ὀδύναι ἀναύδως, ἱδρῶτι λυόμεναι, κακοήθεες, τούτοισιν ἐς ἰσχία ἀλγήματα ἅμα πυρετῷ καυσώδει· καὶ ἢν κοιλίη καταρραγῇ, ὀλέθριον.

91. Οἷσιν αἱ φωναὶ ἅμα πυρετῷ ἐκλείπουσι μετὰ 534 | ἀκρασίης τρομώδεες καὶ κωματώδεες τελευτῶσιν.

92. Οἷσι καυστικά, μεμωρωμένα, κάτοχα, ποικίλλοντα ὑποχόνδρια καὶ κοιλίην ἐπηρμένοι σίτων ἀπολελαμμένων ἐφιδροῦσιν· ἦρα τούτοισι τὸ θολερὸν πνεῦμα καὶ τὸ γονοειδὲς διελθὸν λύγγας σημαίνει; καὶ κοιλίη δὲ ἔπαφρα χολώδεα προσδιέρχεται; τὸ λαμπῶδες ἐν τούτοισιν οὐρηθὲν ἐπωφελέει· καὶ κοιλίαι δὲ τούτοισιν ἐπιταράσσονται.

93. Οἷσι κῶμα γίνεται ἐπ' ὀμμάτων διαστροφῆς, [ὀλέθριον ὀξέως· οἷσι πνεῦμα ἀνέλκεται, φωνὴ δὲ ἢ ἀσαφὴς]² ἐπάφρων προσδιελθόντων πυρετὸς παροξύνεται.

94. Καὶ ἐκ κεφαλαλγίης ἀφωνίαι ἅμα ἱδρῶτι πυρετώδεες λυγγώδεες· χαλῶντα ὑπὸ σφάς, ἐπαν 536 ιόντα χρονιώτερα· ἐπιρρι|γοῦν τούτοισι πονηρόν.

95. Χεῖρες τρομώδεες, κεφαλαλγέες, τραχήλου ὀδυνώδεες, ὑπόκωφοι, οὐρέοντες μέλανα δεδασυσ-

---

¹ Del. H² after Galen.
² Del. Polack after Galen.

ᵃ Repetitions from chapters 86 and 87 above, and from the beginning of this chapter.

a fatal sign],[a] and those of them with coma are also in an evil way.

90. In a person who loses his speech, pains to the hypochondrium that occur during fevers and are resolved with a sweat are malignant: these patients have pains that move to the hip-joints, and at the same time an ardent fever; if the cavity has a violent discharge, it is also a fatal sign.

91. Patients who lose their speech during a fever die incontinent, of tremor and coma.

92. In persons with burning fever, stupor, catalepsy and variation in the state of their hypochondrium, and who also swell up in the cavity, because of the obstruction of foods, and sweat over their whole body, does laboured breathing and the passage of material like seed indicate the onset of hiccups? Does their cavity also evacuate frothy, bilious stools? For these patients to pass urine with a scum on it is an advantageous sign. Their cavities are also disturbed.

93. In cases where coma comes on after strabismus of the eyes [it presages a rapid death. Patients in whom the breath is drawn short and the voice unclear],[b] if frothy evacuations occur the fever grows virulent.

94. Loss of speech subsequent to a headache, if accompanied by sweating, also indicates fever and hiccups[c]; evacuations during these are recurrent and quite longstanding. For these patients to have chills afterwards is an evil sign.

95. Trembling of the hands, headache, pains in throat, partial loss of hearing, dark, cloudy urines: in whomever

[b] A repetition from chapters 86 and 87 above; cf. also chapter 89.   [c] Cf. chapter 92 above.

μένα, οἶσι ταῦτα ᾖ, προσδέχεσθαι μέλανα ἥξειν,
ὀλέθριον.

96. Αἱ μετὰ ἐκλύσιος κατόχως ἀφωνίαι ὀλέθριοι.

97. Πλευροῦ ἀλγήματα ἐν πτύσει χολώδει ἀλό-
γως ἀφανισθέντα ἐξίσταται.

98. Ἐν τραχήλου ἀλγήματι κωματώδει, ἱδρώδει
538 κοιλίη φυσηθεῖσα, εἰ δέ τι πρὸς ἀνάγκην, ὑγρὰ
χαλῶσα, ὑποπεριπλυθεῖσα ἐκ τούτων ἄχολα, ἐξίσταται. τὰ τοιαῦτα διασωζόμενα μακροτέρως διανο-
σέει. ἦρά γε καί εἰσιν αἱ ἄχολοι περιπλύσιες εὐηθέ-
στεραι καὶ τὸ φυσῶδες ὄγκῳ προσωφελέει;

99. Κοιλίης περίτασις [πρὸς ἀνάγκην ὑγρὰ
χαλῶσα,][1] ταχὺ ὀγκυλωμένη, ἔχει τι σπασμῶδες,
οἷον καὶ τῷ Ἀσπασίου υἱῷ· τὸ ἐπιρριγοῦν τούτοισιν
ὀλέθριον· ἐκ τούτων σπασμώδης γενηθεὶς καὶ ἐμφυ-
σηθείς, μακροτέρως διανοσήσας, στόμα σαπεὶς
χλωρὸν ἀπεγένετο.

100. Τὰ κατ' ὀσφὺν καὶ λεπτὸν χρόνια ἀλγήματα
καί τι πρὸς ὑποχόνδρια γριφώμενα ἀποσιτικὰ ἅμα
540 πυρετῷ· | τούτῳ ἐς κεφαλὴν ἄλγημα σύντονον ἐλθὸν
κτείνει ὀξέως τρόπον σπασμώδεα.

101. Τὰ ἐπιρριγοῦντα καὶ ἐς νύκτα μᾶλλον παροξυνόμενα, ἄγρυπνα, φλεδονώδεα[2] ἐν τοῖς ὕπνοις,
ἔστιν ὅτε οὔρεα ἐφ' ἑαυτοὺς χαλῶντα ἐς σπασμοὺς
ἀποτελευτᾷ κωματώδεας.

---

[1] Del. Potter.
[2] Littré after Galen: φλεγμον- I.

these things occur expect dark material to come: a fatal sign.

96. Loss of speech in conjunction with a cataleptic weakness is a fatal sign.

97. When pains in the side associated with bilious expectoration disappear without any reason patients become delirious.

98. In the case of a pain in the neck accompanied by coma and sweating, if the cavity fills up with wind and if at all forced releases moist stools, and on being washed downwards non-bilious material comes out of these, the patient becomes deranged. Such cases, even when recovery occurs, remain ill for a longer period of time. Are these non-bilious washings really benign? Does the wind assist by means of its mass?

99. Distension of the cavity [, if forced, releases moist stools][a] that quickly swells up suggests a convulsion, as in the son of Aspasius. For chills to follow in these patients is a fatal sign. One patient who became convulsive and was puffed up with wind remained ill for quite a long time, and then, forming yellow-green pus in his mouth, died.

100. Chronic pains in the loins and flanks that lancinate to any degree towards the hypochondrium, provoke a distaste for food and bring on fever—in such a patient a violent pain moving to his head quickly kills him by convulsions.

101. Chills occurring during diseases that have their exacerbations more at night, if accompanied by sleeplessness, babbling in the sleep, and sometimes urinary incontinence, end in convulsions and coma.

[a] See chapter 98 above.

102. Οἱ ἐξ ἀρχῆς ἐφιδρῶντες οὔροισι πέποσι, καυστικοί, ἀκρίτως περιψύχοντες, διὰ ταχέων περικαέες, νωθροί, κωματώδεες, σπασμώδεες, ὀλέθριοι.

103. Τῇσι δὲ ἐπιφόροισι κεφαλαλγικὰ καρώδεα μετὰ βάρους γινόμενα φαῦλα, ἴσως δὲ ταύτῃσι καὶ σπασμῶδές τι παθεῖν ὀφείλει.

542   104. Καὶ τὰ ἐν | φάρυγγι ἰσχνῷ ἀλγήματα πνιγώδεα ἔχει τι σπασμῶδες, ἄλλως τε καὶ ἀπὸ κεφαλῆς ὁρμῶντα, οἷον καὶ τῇ Θρασύνοντος ἀνεψιῇ.

105. Τὰ τρομώδεα σπασμώδεα γινόμενα ἐφιδρῶσι φιλυπόστροφα· τούτοισι κρίσις ἐπιρριγώσασιν. οὗτοι ἐπιρριγέουσι περὶ κοιλίην καύματι προκληθέντες.

106. Ὀσφύος πόνος κεφαλαλγικῷ καὶ καρδιαλγικῷ μετὰ ἀναχρέμψιος βιαίης ἔχει τι σπασμῶδες.

107. †Τοῦτο ἄφονον, ἅμα κρίσις, ῥῖγος.†

544   108. Ἀπὸ κοιλίης | ὑποπέλια ταραχώδεα καὶ οὖρα λεπτά, ὑδατώδεα ὕποπτα.

109. Φάρυγξ τρηχυνθεὶς ἐπ᾽ ὀλίγον καὶ ἡ κοιλίη βορβορύζουσα κενῇσιν ἐξαναστάσεσι καὶ μετώπου ἄλγημα, ψηλαφώδεες, κοπιώδεες, ἐν στρώμασι καὶ ἱματίοισιν ὀδυνώδεες, τὰ ἐκ τούτων αὐξόμενα δύσκολα. ὕπνος ἐν τούτοισι πολύς, σπασμώδεες καὶ τὰ ἐς μέτωπον ἀλγήματα βαρέα καὶ οὔρησις δυσκολαίνουσα.

102. Patients that sweat over their whole bodies from the beginning, and have mature urines, that have burning heat, are cooled without a crisis and then quickly become burning hot again, and that suffer torpor, coma and a tendency to convulsions, are doomed.

103. In pregnant women, for stuporous headaches to arise in conjunction with a feeling of heaviness is an indifferent sign, and they are probably bound to suffer some sort of convulsion.

104. To have suffocating pains in the throat in the absence of any swelling gives an indication of convulsions, especially if the pains start out from the head, as for example in the female cousin of Thrasynon.

105. Trembling and convulsions that occur in conjunction with sweating over the whole body tend to relapse; in these patients, the crisis occurs when a later chill arrives; they become chilled after being warmed by burning heat about the cavity.

106. Pain in the loins in conjunction with headache, pain in the cardia, and violent expectoration gives an indication of convulsions.

107. † This loss of speech . . . at the same time crisis and chills. †

108. Mildly livid discharges from the cavity, and urines that are smooth and watery, are suspicious.

109. The throat becoming rough over a short period of time, the cavity rumbling with empty contractions, and pain in the forehead; persons who grope about with their hands, who are wearied, and who are pained by their blankets and clothes: what grow out of these things are troublesome. Sleep in these persons is excessive, they tend to have convulsions, the pains in their forehead are severe, and their micturition is difficult.

110. Καὶ οὔρου ἐπίστασις, οἷς ῥίγεα ἐπὶ τοῖσι σπασμώδεσι καὶ †ὡς αὐτὴ φρίξασα† ἐφιδρῶσιν.

546 111. Αἱ ἐς ἄκρητα τελευτῶσαι καθάρσιες | ἐν πᾶσι μὲν παροξυντικαί· τούτοισι δὲ καὶ πάνυ. ἐκ τοιούτων καὶ τὰ παρ' οὖς ἀνίστανται.

112. Καὶ αἱ ταραχώδεες θρασύταται ἐπεγέρσιες σπασμώδεες, ἄλλως τε καὶ μεθ' ἱδρῶτος.

113. Καὶ αἱ τοῦ τραχήλου καὶ μεταφρένου καταψύξιες δοκέουσαι[1] καὶ ὅλου τοῦ σώματος· ἐν τούτοισι δὲ καὶ πυώδεες οὐρήσιες [καὶ ἀφρώδεες][2] ἅμα ἀψυχίῃ καὶ ὄμματος ἀμαύρωσις σπασμὸν ἐγγὺς σημαίνει.

114. Πήχεων ἀλγήματα μετὰ τραχήλου σπασμώδεα, ἀπὸ προσώπου δὲ ταῦτα· καὶ κατὰ φάρυγγα ὠχροί, ἰσχνοί, πτυαλίζοντες, ἐν τούτοισιν 548 οἱ ἐν | ὕπνοισιν ἱδρῶτες [ἐν τούτοισιν][3] ἀγαθόν. ἆρά γε καὶ τῷ ἱδρῶτι κουφίζεσθαι τοῖσι πλείστοισιν οὐ πονηρόν; οἱ ἐς τὰ κάτω πόνοι τούτοισι εὔφοροι.

115. Οἱ ἐν πυρετοῖσιν ἐφιδρῶντες, κεφαλαλγέες κοιλίης ἀπολελαμμένης[4] σπασμώδεες.

116. Τὰ ὑποψάθυρα ὑγρὰ διαχωρήματα περιψύχοντα οὐκ ἀπύρως φλαῦρα, τὰ ἐπὶ τούτοισι ῥίγεα κύστιν καὶ κοιλίην ἐπιλαμβάνοντα ὀδυνώδεα. ἦρα τὸ κωματῶδες τούτοισιν ἔχει τι σπασμῶδες; οὐκ ἂν θαυμάσαιμι.

---

[1] R after Galen: ἐοῦσαι I.   [2] Del. Polack.
[3] Del. R after Galen.   [4] R after Galen: -μένοι I.

110. Stoppage of urine, too, occurs in persons who have chills subsequent to a convulsion, and † as it shudders † in persons who perspire over their whole body.

111. Cleanings that end in an unmixed state are in all cases associated with paroxysms, in the ones mentioned especially so. Subsequent to such things, the region beside the ear also swells up.

112. Waking up in a disturbed, over-bold state also indicates the onset of convulsions, especially if it is accompanied by sweating.

113. Also chills of the neck and back that seem to occupy the whole body; in these patients the passage of purulent [and frothy] urines, associated with fainting and dimness of the eye, indicates that a convulsion is near.

114. Pains of the forearms and the neck announce convulsions; these originate from the face. In patients that are yellow-green in the throat, have no swelling, and salivate, to sweat during sleep is a good sign. Is it not injurious, in most cases, to be relieved by the sweat? Pains moving to the lower parts in these patients are easy to bear.

115. If fever patients who sweat over their whole bodies and have headaches become obstructed in their cavity, they are liable to convulsions.

116. To pass loose, moist stools that chill and provoke fever is an indifferent sign; subsequent chills that affect the bladder and cavity are painful. Does the presence of coma in these patients give any indication of convulsions? I would not be surprised.

117. Τὰ ἐν ὀξέσιν ἐμετωδέως ἑλκόμενα φλαῦρα καὶ αἱ λευκαὶ διαχωρήσιες | δύσκολοι. περίγλισχρα ἐκ τοιούτων διελθόντα ἐξίσταται καύματι πολλῷ. ἆρα ἐκ τούτων κωματώδεες νωθροὶ ὑποσπασμώδεες [νωθροὶ][1] ἐπιγίνονται· ἐκ τοιούτων μακροτέρως ἐπινοσεῖ. ἠρά γε περὶ κρίσιν οὗτοι χολώδεες δύσπνοοι;

118. Τὰ ἐξ ὀσφύος ἐς τράχηλον καὶ κεφαλὴν ἀναδιδόντα παραλύσαντα παραπληκτικὸν τρόπον σπασμώδεα παρακρουστικά· ἆρα καὶ λύεται τὰ τοιαῦτα σπασμῷ; ἐκ τῶν τοιούτων ποικίλως διανοσέουσι διὰ τῶν αὐτῶν ἰόντες.

119. Οἱ ἐν ὑστερικοῖσιν ἀπύροις σπασμοὶ εὐχερέες οἷον καὶ Δορκάδι.

120. Κύστις ἀποληφθεῖσα ἄλλως τε καὶ μετὰ κεφαλαλγίης ἔχει τι σπασμῶδες. τὰ ναρκωδέως ἐν τοιούτοισιν ἐκλυόμενα δύσκολα, οὐ μὴν ὀλέθρια. ἆρά γε καὶ παρακρουστικὸν τὸ τοιοῦτον;

121. Ἠρά γε καὶ κατὰ | κρόταφον ὀστέων διακοπαὶ σπασμὸν ἐπικαλέονται· τὸν μεθύοντα πληγῆναι πολὺ ἐν ἀρχαῖσι τοῦτο ποιεῖ.

122. Τὰ σπασμώδεα ἐν ἱδρῶτι πτύελα παραρρέοντα πυρετώδει ἐόντι εὐήθεα. ἠρά γε τούτοισιν ἐπί τινας ἡμέρας κοιλίαι καθυγραίνονται; οἴομαι δὲ τούτοισιν ἐς ἄρθρον ἀπόστημα ἔσεσθαι.

123. Τὰ ἐπ᾽ ὀλίγον θρασέως παρακρούοντα μελαγχολικά· ἢν δὲ καὶ ἀπὸ γυναικείων ᾖ, θηριώ-

___
[1] Del. Polack.

117. In acute diseases, retching is an indifferent sign, and the passage of white motions troublesome. With sticky material passing after that, the person loses his senses and has great burning heat. Is this followed by a comatose torpor and a tendency to convulsions? From such conditions the illness goes on for a longer time. Do these patients suffer from bile and dyspnoea about the crisis?

118. Paralyses that move up from the loins to the neck and head, disabling in the manner of apoplexy, lead to convulsions and delirium. Are such conditions resolved by the convulsions? After them patients experience changing symptoms and go through the same stages.

119. Convulsions in afebrile hysterical disorders are easy to manage, as for example in the case of Dorcas.

120. Stoppage of the bladder, especially in association with headache, gives some indication of convulsions. For such patients to have numbness when they recover is troublesome, but not in the least dangerous. Does such a condition also incline towards delirium?

121. Do gashes to the bones at the temple also provoke a convulsion? For a drunken person to be struck hard does, right at the beginning.

122. Convulsions and excessive salivation during a sweat in a person with fever are benign. Does the cavity remain very moist over a number of days in these patients? I believe they will suffer an abscession to some joint.

123. Delirium characterized by over-boldness for short periods indicates melancholy[a]; if it arises from the

[a] Cf. chapter 26 above.

δεα· ἐπὶ πλέον δὲ ταῦτα ξυμπίπτει· ἦρά γε καὶ
σπασμώδεες αὗται; ἆρά γε καὶ μετὰ κάρου ἀφωνίαι
σπασμώδεες; οἷον τῇ τοῦ | σκυτέως θυγατρί· ἤρξατο
γυναικείων παρεόντων πυρέξαι.

124. Οἷσι σπασμώδεσι ὀφθαλμοὶ ἐκλάμπουσιν
ἀτενές, οὔτε παρὰ σφίσιν αὐτοῖς εἰσί, διανοσέουσί
τε μακροτέρως.

125. Τὰ ἀνάπαλιν αἱμορραγεῦντα κακὸν οἷον ἐπὶ
σπληνὶ [φλεγμαίνοντι]¹ μεγάλῳ ἐκ δεξιοῦ ῥυέντα
καὶ τὰ καθ᾽ ὑποχόνδριον ὡσαύτως. ἐφιδρῶντι δὲ
κάκιον.²

126. Τὰ ἐκ ῥινῶν μικροῖς ἱδρῶσι περιψυχόμενα
κακοήθεα.

127. Μεθ᾽ αἱμορραγίην μελάνων δίοδος κακόν,
πονηρὸν δὲ καὶ τὰ ἐξερυθρώδεα. ἦρά γε τεταρταῖα
αἱμορραγῆσαι; κωματώδεες ἐκ τοιούτων σπασμῷ
τελευτῶσιν· ἆρα μελάνων προδιελθόντων | καὶ κοι-
λίης ἐπαρθείσης;

128. Τὰ αἱμορραγέοντα ἐφιδροῦντα κακοήθεα·
οὗτοι συντόμως διαλεγόμενοι λαθραίως τελευτῶσι.

129. Μεθ᾽ αἱμορραγίην βραχείην καὶ μελάνων
διαχώρησιν ἐν ὀξέσι κώφωσις κακόν· αἵματος διαχώρησις τούτοισιν ὀλέθριον, κώφωσιν δὲ λύει.

130. Ἐν ὀσφύι ἐπωδύνῳ καρδιαλγικὰ προσιόντα
σημεῖα αἱμορροώδεα, οἶμαι δὲ καὶ προγενόμενα.

131. Τὰ τεταγμένοισι χρόνοισιν αἱμορραγεῦντα

¹ Del. Aldina.
² Aldina after Galen: κακόν I.

menses, it is malignant—these often coincide. Are these women subject to convulsions? Is loss of speech in conjunction with stupor also indicative of convulsions?—as for example in the shoemaker's daughter: her fever began during her menses.

124. Persons with convulsions whose eyes are bright and fixed, and who are out of their senses, remain ill for an extended period of time.

125. To lose blood on the opposite side is a bad sign, as for example to bleed on the right side in cases of splenomegaly or other such conditions about the hypochondrium; in a person who is perspiring over the whole body it is even worse.

126. To be cooled by a nose-bleed in association with minor sweating is a malignant sign.

127. To pass black stools after a haemorrhage is a bad sign, and very red evacuations bode ill too. Do these haemorrhages occur on the fourth day? Patients who become comatose after these signs end by having convulsions. Have dark stools passed through before; has the cavity been raised?

128. Haemorrhage in conjunction with sweating over the whole body is malignant; these patients speak only briefly, before dying without any obvious reason.

129. To become deaf in acute diseases after a brief haemorrhage and passing dark stools is a bad sign. For these patients to pass blood in their stools is a fatal sign, although it does relieve their deafness.

130. In a patient with lumbar pain the addition of a pain in the cardia presages a haemorrhage, and I think also indicates that a haemorrhage has occurred before.

131. Conditions that include haemorrhages at fixed

διψώδεα ἐκλυόμενα· μὴ αἱμορραγήσαντα ἐπιληπτικῶς τελευτᾷ.

558 132. Τὰ εὐθὺ τα|ραχώδεα καὶ ἄγρυπνα, ἀποστάξαντα, ἑκταῖα κουφισθέντα, νύκτα πονήσαντα, ἐς τὴν αὔριον ἐφιδρώσαντα, κατενεχθέντα, παρακρούσαντα αἱμορραγέει λάβρως.

133. Οἷσιν αἱμορραγίαι πλείους, παρεληλυθότος χρόνου κοιλίαι πονηρεύονται, ἢν μὴ τὰ οὖρα πεπανθῇ.[1] ἦρά γε τὸ ὑδατῶδες οὖρον τοιοῦτον σημαίνει;

134. Αἱ ἐν κρισίμοισι περιψύξιες τῶν αἱμορραγικῶν νεανικαὶ κάκισται.

135. Οἱ καρηβαρικοί, κατὰ βρέγμα ὀδυνώδεες ἄγρυπνοί τε αἱμορραγικοί, ἄλλως τε καὶ ἤν τι ἐς τράχηλον συντείνῃ.

136. Τὰ ἀγρυπνήσαντα ἐξαίφνης ἀλυσμῷ αἱμορραγικά, ἄλλως τε καὶ ἤν τι προερρυήκῃ· ἀρά γε καὶ φρίξαντα κάτοχα, κεφαλαλγικά;

137. Τραχήλου ὀδυνώδεα, βλέφαρα ὀδυνώδεα, ὄμματα ἐξέρυθρα ἐόντα, αἱμορραγικά.

560 138. Οἷσι κοιλίης ἐπιστάσης αἱμορραγέει, καὶ ἐπιρριγέουσιν· ἦρά γε τὸ αἱμορραγέειν τούτοισι κοιλίην λειεντεριώδη ποιέει καὶ ἐπίσκληρον ἢ ἀσκαρίδας ἢ καὶ ἀμφότερα;

139. Οἷσιν ἐξ ὀσφύος ἀναδρομὴ ἐς κεφαλὴν καὶ χεῖρα ναρκώδεες, καρδιαλγικοὶ ἢ χολώδεες αἱμορραγέουσι λάβρως καὶ κοιλίη καταρρήγνυται, τού-

[1] H[2] after Galen: πεπαίνῃ I.

times are resolved through thirst, but if no blood is passed, they end in epilepsy.

132. In a person who suddenly becomes troubled and sleepless, who has a nose-bleed which brings relief on the sixth day, but at night has a turn for the worse, and towards the morning sweats over his whole body, and who is weighed down and delirious, violent haemorrhages will occur.

133. In persons who have many haemorrhages, with the passage of time the cavity gets into a bad state, unless the urines are mature. Does the passage of watery urine point to this condition?

134. In patients with haemorrhages, violent chills on the critical days are a very bad sign.

135. Patients that have heaviness of the head and pain in the bregma suffer sleeplessness and haemorrhages, especially if there is tension in the neck.

136. To be sleepless and then suddenly restless points towards a haemorrhage, especially if there has been any haemorrhage before. Do chills in conjunction with catalepsy also give an indication of headache?

137. Pains in the neck, pains in the eyelids, and the eyes being very red indicate the onset of haemorrhage.

138. Patients who have haemorrhages in conjunction with a stoppage of the cavity will also have chills. Does their bleeding make the cavity lienteric and somewhat constipated? Or does it engender worms? Or does it do both?

139. Patients in whom a pain shoots up from the loins to the head, who become numb in a hand, who have pain about their cardia, and who are bilious bleed violently and have a violent discharge of the cavity: generally their

τοις γνῶμαι ταραχώδεες ὡς ἐπὶ τὸ πολύ.

140. Οἷσιν ἐφ᾽ αἱμορραγίη λάβρῃ πυκνὴ μελάνων συχνῶν διαχώρησις, ἐπιστάσης δὲ αἱμορραγέ-
ουσιν, οὗτοι κοιλίας | ὀδυνώδεες ἅμα δὲ τῇσι φύσῃ-
σιν εὔφοροι. ἦρα οἱ τοιοῦτοι ψυχροῖσιν ἐφιδροῦσι
πολλοῖσι; τὸ ἀνατεταραγμένον ἐν τούτοισιν οὖρον
οὐ πονηρὸν οὐδὲ τὸ ὑφιστάμενον γονοειδές. ἐπὶ
συχνὸν δὲ οὗτοι ὑδατώδεα οὐρέουσιν.

141. Οἷσιν ἀπὸ ῥινῶν ἐπὶ κωφώσει καὶ νωθρίῃ
μικρὰ ἀπόσταξις, ἔχει τι δύσκολον· ἔμετος τούτοισι
ξυμφέρει καὶ κοιλίης ταραχή.

142. Αἷς ἐκ ῥίγεος πυρετοὶ κοπιώδεες, γυναικεῖα
κατατρέχει· τράχηλος ἐν τούτοισιν ὀδυνώδης αἱμορ-
ραγικόν.

143. Τὰ σείοντα κεφαλῆς[1] καὶ τὰ ἠχώδεα αἱμορραγέει ἢ γυναικεῖα καταβιβάζει, ἄλλως τε καὶ ἢν
κατὰ ῥάχιν καῦμα παρακολουθῇ. ἴσως δὲ καὶ
δυσεντερικόν.

144. Οἱ κατὰ κοιλίην παλμοὶ ὑποχονδρίου ἐντά-
σει ὑπομάκρῳ | ὀγκώδει αἱμορραγικοί· φρικώδεες
οὗτοι.

145. Τὰ ἐκ ῥινῶν λάβρα[2] βίαια πολλὰ ῥυέντα
ἔστι δ᾽ οἷσι σπασμοὺς προάγεται· φλεβοτομίη λύει.

146. Αἱ πυκναὶ καὶ κατὰ μικρὰ ἐπαναστάσιες
ὑπόξανθοι, γλίσχρα ἔχουσαι μικρὰ κοπρώδεα μετὰ
ὑποχονδρίου ἀλγήματος καὶ πλευροῦ ἰκτεριώδεες.

---

[1] Potter: κεφαλὰς I.
[2] H[2] after Galen: λαῦρα I.

minds are disordered.

140. Persons who after a violent haemorrhage pass frequent, dark, copious stools, and haemorrhage again when these stop, have pain in the cavity, but bear the internal wind easily. Do these patients break out with copious, cold sweats over their whole body? If their urine contains material stirred up in it this does not bode ill, nor does the presence of a sediment resembling seed. They pass watery urine frequently.

141. For persons with deafness and torpor to have a weak haemorrhage from the nostrils indicates trouble. Vomiting is beneficial for them, as is also an evacuation of the cavity.

142. In women who have a debilitating fever after a chill, the menses run down; if their throat is sore, it suggests that a haemorrhage will follow.

143. Agitation of the head and ringing in the ears provokes a haemorrhage or brings down the menses, especially if followed closely by a burning heat along the spine. There is also some probability of dysentery.

144. A throbbing in the cavity together with persistent tension and swelling of the hypochondrium points to a haemorrhage; these patients are likely to shiver.

145. Copious, forceful, violent haemorrhages from the nostrils lead in some persons to convulsions; phlebotomy resolves the condition.

146. Yellowish stools with small, sticky pieces, passed frequently a little at a time in association with pains in the hypochondrium and the side, are a sign of jaundice. Are

ἠρά γε ἐπαναστάντων τούτων ἐκλύονται; οἶμαι δὲ
καὶ αἱμορραγεῖν τούτους, τὰ γὰρ ἐς ὀσφὺν ἀλγήματα ἐν τούτοισιν αἱμορραγικά.

147. Ὑποχονδρίου ἔντασις μετὰ καρηβαρίης καὶ
κωφώσιος καὶ τὰ πρὸς αὐγὰς[1] ὀχλέοντα αἱμορραγικά.

148. Αἱ ἑνδεκαταῖαι στάξιες δύσκολοι, ἄλλως τε
καὶ ἢν δὶς ἐπιστάξῃ.

566   149. Τὰ | ἐν φρίκῃσιν ἅμα ἱδρώσαντα κρισίμως·
ἐς δὲ τὴν αὔριον φρίξαντα, παραλόγως ἀγρυπνέοντα
αἱμορραγήσειν οἴομαι πεπαινομένων.

150. Οἶσιν ἐξ ἀρχῆς αἱμορραγίαι λάβροι ῥῖγος
ἵστησι.

151. Τὰ ἐξ αἱμορραγίης ῥίγεα πονηρά.

152. Οἶσι κεφαλαλγίαι καὶ τραχήλου πόνοι καὶ
ὅλου δέ τις ἀκράτεια τοῦ σώματος τρομώδης,[2]
αἱμορραγίαι λύουσιν. ἀτὰρ καὶ οὕτω χρόνῳ λύον-
ται.

153. Οὖρα τοῖσι παρὰ τὰ ὦτα ταχὺ καὶ ἐπ' ὀλίγον πεπαινόμενα φλαῦρα. καὶ τὸ καταψύχεσθαι ὧδε
πονηρόν.

154. Τὰ ὑποκαρώδεα καὶ ἰκτερώδεα οὐ πάνυ
[αἴτια][3] αἰσθανόμενα, οἷσι λύγγες, κοιλίαι καταρρήγνυνται, ἴσως δὲ καὶ ἐπίστασις,[4] οὗτοι ἀλλοιοῦν-
ται· ἦρα τούτοισι καὶ τὰ παρὰ τὰ ὦτα;

---

[1] Foes: αὐτὰς I.
[2] τις . . . τρομώδης Aldina from Galen: τοῦ σώματος τὰ
ἄκρα τρομώδεις I.

these patients delivered from the pains by going to stool? I think that they may bleed, too, since pains that move to the loins presage a haemorrhage for them.

147. Tension of the hypochondrium, in association with heaviness of the head, deafness, and impairment of vision, indicates the person will haemorrhage.

148. A nose-bleed on the eleventh day is troubling, especially if it occurs twice.

149. To sweat and at the same time shiver signals the arrival of the crisis; if on the following day there are shivering and unexplained sleeplessness, I think the person will haemorrhage mature blood.

150. In patients that have violent haemorrhages from the beginning, these are checked by a chill.

151. Chills subsequent to a haemorrhage bode ill.

152. In patients with headache, pains in the neck, and a degree of tremulous disability of the whole body, haemorrhages relieve the condition; but even thus, the relief comes only in time.

153. In patients with swellings beside their ears, urines that are passed early and that quickly become mature are an indifferent sign. To become chilly in this condition bodes ill.

154. If, in patients with lethargy and jaundice that are not very well in their senses, hiccups occur, then the cavity has a violent discharge and possibly also a stoppage; these patients are changeable. Does swelling beside the ears also occur in them?

---

³ Del. Aldina after Galen.
⁴ Aldina: -στάσης I.

155. Τὰ ἐπισχόμενα μετὰ ῥίγεος οὖρα πονηρά, ἄλλως τε καὶ προκαρωθέντι.[1] τὰ πρὸς οὖς ἦρα ἐπὶ τούτοισιν ἐλπίς;

156. Ἐκ στροφωδέων ὑπόστασις ἰλυώδης, ὑπο- 568 πέλιος, | κακή. ἆρά γε ἐκ τοιούτων ὑποχόνδριον ὀδυνᾶται; δοκέω δὲ δεξιόν. ἦρα καὶ ἀχλυώδεες οἱ τοιοῦτοι καὶ τὰ παρὰ <τὰ ὦτα>[2] τούτοισιν ὀδυνώδεα ἐπ᾽ ὀλίγον; κοιλίη καταρραγεῖσα τούτοισιν ὀλέθριον.

157. Ἐν τοῖσιν ἀσώδεσι τὰ παρ᾽ οὖς μάλιστα.

158. [Τὰ][3] ἐπὶ εἰλεοῖσι δυσώδεσι πυρετῷ ὀξεῖ, ὑποχονδρίῳ μετεώρῳ χρονιωτέρῳ, τὰ παρ᾽ οὖς ἐπαρθέντα κτείνει.

570 159. Ἐκ κωφώσιος ἐπιεικέως τὰ | παρ᾽ ὦτα, ἄλλως τε καὶ ἢν ἀσῶδές τι ἐπιγένηται, καὶ τοῖσι κωματώδεσιν. (160.) Ἐπὶ τούτοισι φλαῦρα τοῖς παραπληκτικοῖσι.

161. Τὰ σπασμώδεα τρόπον παροξυνόμενα κατόχως τὰ παρ᾽ οὖς ἀνίστησι.

162. Τὸ σπασμοτρομῶδες, ἀσῶδες κατόχως σμικρὰ πρὸς οὖς ἀνίστησιν.

163. Ἦρά γε, οἷσι τὰ παρ᾽ ὦτα, κεφαλαλγικοὶ οὗτοι; ἦρά γε ἐφιδροῦσι τὰ ἄνω, ἤν τι καὶ ἐπιρριγέωσιν; ἦρά γε καὶ κοιλίη καταρρήγνυται; καί τι καὶ κωματώδεες; ἀτὰρ καὶ τὰ ὑδατώδεα οὖρα ἐναιωρούμενα λευκοῖσι καὶ τὰ ποικίλως ἔκλευκα δυσώδεα

---

[1] Polack after Galen: -θέντα I.
[2] Polack: τὰ παρ᾽ ὦτα Galen.

155. For urines to be checked during a chill bodes ill, especially in a person already affected by drowsiness. Should you expect them to swell beside the ears?

156. A slimy, somewhat livid sediment in the urines appearing after a colic is a bad sign. Does pain in the hypochondrium develop out of this? On the right, I think. Do such patients suffer dimness and have pains beside <the ears> for a short time? A violent discharge of the cavity in these patients is a fatal sign.

157. Swelling beside the ear is most common in persons with nausea.

158. If, during an acute fever in a case of foul-smelling intestinal obstruction where the hypochondrium is raised for a longer time, the area beside the ear swells up, it kills the person.

159. After deafness, a swelling beside the ears is probable, especially if some kind of nausea is experienced, and in patients with coma. (160.) For deaf patients it (i.e. a swelling beside the ears) is an indifferent sign when paralysis is present.

161. Cataleptic crises that have a convulsive turn provoke swellings beside the ear.

162. Convulsions associated with tremor, nausea and catalepsy cause minor swellings beside the ear.

163. Do swellings beside the ears lead to headaches? Do these patients sweat over the upper part of their bodies, when they happen to have chills? Does their cavity have a violent downward discharge? Are they at all comatose? Do watery urines with white material suspended in them or variegated white, foul-smelling urines

---

[3] Del. Aldina after Galen.

ποιεῖ τὰ παρ' οὖς; ἠρά γε, οἷσι τὰ τοιαῦτα οὖρα, στάξιες πυκναί; ἠρά γε καὶ γλῶσσα τούτοισι λείη;

164. Οἷσι πνευματίῃσιν ἐοῦσιν ἴκτεροι καὶ πυρετοὶ ὀξέες ἐπιγίνονται, μεθ' ὑποχονδρίων σκληρῶν καταψυχθεῖσι τὰ παρὰ τὰ ὦτα μεγάλα ἀνίσταται.

165. Τὰ κωματώδεα, ἀσώδεα, ὑποχονδρίου ὀδυνώδεα [μικρὰ]¹ σμικρὰ ἐμετώδεα, ἐν τούτοισι τὰ 572 παρὰ τὸ οὖς ἐπανίσταται, πρόσθεν δὲ καὶ τὰ | περὶ τὸ πρόσωπον.

166. Κοιλίης μέλανα κοπρώδεα χολώδεα [κροκώδεα]² διείσης κῶμα ἐπιφανὲν τὰ παρ' οὖς ἀνίστησι.

167. Βήχια λεπτὰ μετὰ πτυελισμοῦ ἰόντα τὰ πρὸς οὖς λαπάσσει.³

168. Ἐκ κεφαλαλγίης κῶμα καὶ κώφωσις καὶ φωνῆς μώρωσις παρακολουθοῦντα παρ' οὖς τι ἐξερεύγεται.

169. Ὑποχονδρίου σύστασις μετὰ καύματος ἀσώδεος καὶ κεφαλαλγίης τὰ πρὸς οὖς ἐπαίρει.

170. Τὰ ἐπώδυνα παρ' οὖς ἀκρίτως⁴ καταμωλυνθέντα φλαῦρα.

¹ Del. Aldina.
² Del. Polack.
³ Parisinus Graecus 2145: λάπασι I.
⁴ Aldina: ἀκρήτως I.

provoke swellings beside the ears? In patients with urines of this kind, do frequent nose-bleeds occur? Is their tongue smooth?

164. Patients with difficult breathing in whom jaundice and acute fevers occur, and who have chills together with a hardness of the hypochondrium, swell up greatly beside the ears.

165. Patients with coma, nausea, [mild] pain in the hypochondrium, and a slight tendency to vomiting swell up beside the ear, and also anteriorly about the face.

166. If, with the cavity passing dark, fecal, bilious, [saffron-coloured] stools, a coma comes on, it leads to swellings beside the ear.

167. Mild coughs that produce sputum bring down swellings beside the ear.

168. Coma, deafness, and sluggishness of speech following closely upon a headache produce some kind of eruption beside the ear.

169. Contraction of the hypochondrium in conjunction with a nauseating burning-heat and headache provokes swellings beside the ear.

170. Painful swellings beside the ear that wither away without any crisis are an indifferent sign.

PRORRHETIC II

INTRODUCTION

Although he clearly knows of the treatise's existence, Erotian considers it spurious and does not include any word from it in his Glossary.[1] Galen, on the other hand, glosses six terms from "the greater *Prorrhetic* (which some call the second)."[2] In his treatise *Prognostic*, he also quotes the line "I, however, shall not prophesy anything like this" from ch. 2 of *Prorrhetic II*, taking the occasion to indicate that he agrees with "some physicians who hold *Prorrhetic II* not to be among the genuine Hippocratic books."[3]

The contents of *Prorrhetic II* can be summarized as follows:

---

[1] See Erotian p. 9, where he lists *Prorrhetic I* and *II* among the semiotic books, and Nachmanson p. 458.

[2] Galen vol. 19, 69 ἀγλίη; 19, 77 ἀμβλνωσμός; 19, 88 βδέλλῳ; 19, 109f. κατάστημα; 19, 120 μᾶσσον and 19, 149 ὑποξύρους. 19, 127 παράλαμψις also appears to be from *Prorrhetic II* (ch. 20), since this is the only occurrence of the word in the Hippocratic Collection.

[3] Galen vol. 14, 620 = CMG V 8,1, pp. 88–90.

The account is centred on the symptoms and course of the diseases described, their "differential diagnosis",[4] and their prognosis. No treatments are given, and no specific cases are presented.

The text of *Prorrhetic II*, besides appearing in all the collected editions, receives careful study in: Johannes Opsopoeus, *Hippocratis ... Prorrheticorum lib. II ... Graecus et Latinus contextus accurate renovatus ...*, Frankfurt, 1587. (= Opsopoeus) More recently, the work was the subject of a French thesis, which I have not been able to consult: B. Mondrain, *Édition critique, traduction et commentaire d'un traité hippocratique, le Prorrétique II*, Diss. Paris, 1984.

My edition is founded on a collation of the sole independent manuscript, I, made from microfilm.

[4] *"Differential diagnosis*, the determination of which one of two or more diseases or conditions a patient is suffering from, by systematically comparing and contrasting their symptoms." *Dorland's Illustrated Medical Dictionary*, s.v.

ΠΡΟΡΡΗΤΙΚΟΝ Β

1. Τῶν ἰητρῶν προρρήσιες ἀπαγγέλλονται
συχναί τε καὶ καλαὶ καὶ θαυμασταί, οἵας ἐγὼ μὲν
οὔτ᾽ αὐτὸς προεῖπον οὔτ᾽ ἄλλου του ἤκουσα προλέ-
γοντος. εἰσι δ᾽ αὐτῶν αἱ μὲν τοιαίδε· ἄνθρωπον
δοκέειν ὀλέθριον εἶναι καὶ τῷ ἰητρῷ τῷ μελεδαίνοντι
αὐτοῦ καὶ τοῖσιν ἄλλοισιν, ἐπεισιόντα δὲ ἰητρὸν ἕτερον εἰπεῖν ὅτι ὁ μὲν ἄνθρωπος οὐκ ἀπολεῖται,
ὀφθαλμῶν δὲ τυφλὸς ἔσται· καὶ παρ᾽ ἕτερον δοκέ-
οντα παγκάκως ἔχειν εἰσελθόντα προειπεῖν τὸν μὲν
ἄνθρωπον ἀναστήσεσθαι, χεῖρα δὲ χωλὴν ἕξειν καὶ
ἄλλῳ τῳ δοκέοντι οὐ περιέσεσθαι εἰπεῖν αὐτὸν μὲν
ὑγιέα ἔσεσθαι, τῶν δὲ ποδῶν τοὺς δακτύλους μελαν-
θέντας ἀποσαπήσεσθαι· καὶ τἆλλα τοιουτότροπα
προρρήματα λέγεται ἐν τοιούτῳ τῷ εἴδει. ἕτερος δὲ
τρόπος προρρήσιος, ὠνεομένοισί τε καὶ διαπρησσο-
μένοισι προειπεῖν τοῖσι μὲν θανάτους, τοῖσι δὲ
μανίας, τοῖσι δὲ ἄλλας νόσους, ἐπὶ πᾶσι τούτοισί τε
καὶ τοῖσι προτέροισι χρόνοισι προφητίζειν καὶ
πάντα ἀληθεύειν. ἄλλο δὲ σχῆμα προρρήσεων
τόδε[1] λέγεται· τοὺς ἀθλητὰς γινώσκειν καὶ τοὺς τῶν
νούσων εἵνεκα γυμναζομένους τε καὶ ταλαιπωρέον-

PRORRHETIC II

1. There are reports of physicians making frequent, true and marvellous predictions, predictions such as I have never made myself, nor ever personally heard anyone else make. Here are some examples. A person seems to be mortally ill both to the physician attending him and to others who see him, but a different physician comes in and says that the patient will not die, but go blind in both eyes. In another case where the person looked in a very poor state, the physician that came in foretold that he would recover, but be disabled in one arm; to another person who was apparently not going to survive, one said that he would recover, but that his toes would become black and gangrenous. Other predictions of this kind are reported in the same form. A different type of predicting is to foretell in merchants and adventurers death to some, madness to others, and other diseases to others; and in making revelations concerning present and past times to be correct in every detail. Another form of prediction is recounted as follows: to discover in athletes and in people that are carrying out exercises and exertions prescribed

---

[1] Foes: τάδε I.

τας, ἤν τι ἢ τοῦ σιτίου ἀπολίπωσιν, ἢ ἑτεροῖόν τι
φάγωσιν, ἢ ποτῷ πλέονι χρήσωνται, ἢ τοῦ περιπά-
του ἀπολίπωσιν, ἢ ἀφροδισίων τι πρήξωσι· τούτων
πάντων οὐδὲν λανθάνει, οὐδ' εἰ σμικρόν τι εἴη ἀπει-
8 θήσας ὤνθρωπος. | οὕτως ἐξηκριβῶσθαι οὗτοι πάν-
τες οἱ τρόποι λέγονται τῶν προρρησίων.

2. Ἐγὼ δὲ τοιαῦτα μὲν οὐ μαντεύσομαι, σημεῖα
δὲ γράφω οἷσι χρὴ τεκμαίρεσθαι τούς τε ὑγιέας
ἐσομένους τῶν ἀνθρώπων καὶ τοὺς ἀποθανουμένους,
τούς τε ἐν ὀλίγῳ[1] χρόνῳ ἢ ἐν πολλῷ ὑγιέας ἐσομένους ἢ ἀπολουμένους· γέγραπται δέ μοι καὶ περὶ
ἀποστασίων ὡς χρὴ ἐπισκέπτεσθαι ἑκάστας. [2.]
δοκέω δὲ καὶ τοὺς προειπόντας περί τε τῶν χωλω-
σίων καὶ τῶν ἄλλων τῶν τοιούτων ἤδη ἀποστηριζο-
μένου τοῦ νοσήματος προειπεῖν, καὶ δήλου ἐόντος
ὅτι οὐ παλινδρομήσει ἡ ἀπόστασις, εἴ περ νόον
εἶχον, πολὺ μᾶλλον ἢ πρὶν ἄρχεσθαι τὴν ἀπόστασιν γινομένην. ἐλπίζω δὲ καὶ τἄλλα προρρηθῆναι
ἀνθρωπινωτέρως ἢ ὡς ἐπαγγέλλεται· ἃ δὲ τοῖσιν
ὠνεομένοισί τε καὶ περναμένοισι λέγεται, προρρηθῆναι, θανάτους τε καὶ νοσήματα καὶ μανίας, ταῦτα δέ
μοι δοκέει τοιαῦτα γενέσθαι, καὶ οὐδέν τι χαλεπὰ
εἶναι προειπεῖν τῷ βουλομένῳ τὰ τοιάδε διαγωνίζε-
σθαι. πρῶτον μὲν γὰρ τοὺς ὑφύδρους[2] τε καὶ φθι-
νώδεας τίς ἂν οὐ γνοίη; ἔπειτα τοὺς παραφρονήσον-
τας ἐστὶ μὴ πολὺ λανθάνειν, εἴ τις εἰδείη οἷσι τὸ

[1] Aldina: ὀλιγίστῳ I.   [2] Linden: ἔφ- I.

because of illnesses whether they have failed to eat some of their meal, or have eaten something of a different kind, or have taken too much to drink, or omitted part of their walk, or practised venery; none of these things escapes their notice, not even if the person disobeys in but little. This is how precise all these kinds of predictions are reported to be.

2. I, however, shall not prophesy anything like this; rather I record the clinical signs from which one must deduce which persons will become well and which will die, and which will recover or die in a short or a long time. I have also written about the apostases,[a] and how one must meditate upon each of them. [2.][b] I believe, in fact, that those who make predictions about lameness and other conditions of that kind make their predictions, if they are sensible, only after the disease has become fixed and it is clear that the abscession will not revert, rather than before the abscession begins to occur. I also intend in the other kind of cases to make predictions more in line with human possibilities than what is reported. The reported predictions of such things as death, disease and madness in buyers and sellers, these things have, I think, come to pass somewhat as follows—indeed, it is not so difficult for a person who wants to have successes of this kind to make predictions. In the first place, who would not recognize patients with dropsy or consumption? Further, cases of impending mental derangement are not very likely to escape one's notice, if one knows in which

[a] See note on technical terms.

[b] Littré's chapter 2 begins here. I have followed Vander Linden in making a new chapter at the beginning of this paragraph, which seems a more appropriate break.

221

νόσημα τοῦτο ἢ ξυγγενές ἐστιν, ἢ πρόσθεν ποτ᾽
ἐμάνησαν· εἰ γὰρ οὗτοι οἱ ἄνθρωποι οἰνόφλυγες
εἶεν, ἢ κρεηφαγοῖεν, ἢ ἀγρυπνοῖεν, ἢ τῷ ψύχει ἢ τῷ
θάλπει ἀλογίστως ὁμιλοῖεν, πολλαὶ ἐλπίδες ἐκ τού-
10 των τῶν διαιτημάτων παραφρονῆσαι αὐτούς. | τούς
τε τὰς αἱμορροΐδας ἔχοντας, εἴ τις ὁρῴη τοῦ χειμῶνος πολυποτέοντάς τε καὶ εὐχρόους ἐόντας, ἔστι
προειπεῖν ἀμφὶ τούτων· ἐς γὰρ τὸ ἔαρ καταρραγῆναι
τὸ αἷμα πολλαὶ ἐλπίδες, ὥστε ἀχρόους τε καὶ ὑδαλέ-
ους ὑπὸ τὴν θερείην τούτους εἶναι.

Ἀλλὰ χρὴ προλέγειν καταμανθάνοντα πάντα
ταῦτα, ὅστις τῶν τοιούτων ἐπιθυμέει ἀγωνισμάτων·
ἔστι γὰρ ἐκ τῶν γεγραμμένων προειπεῖν καὶ θάνατον καὶ μανίην καὶ εὐεξίην. εἴποιμι δ᾽ ἂν καὶ ἄλλα
πάμπολλα τοιαῦτα, ἀλλὰ τὰ εὐγνωστότατα ἔδοξέ
μοι γράψαι· συμβουλεύω δὲ ὡς σωφρονεστάτους
εἶναι καὶ ἐν τῇ ἄλλῃ τέχνῃ καὶ ἐν τοῖσι τοιούτοισι
προρρήμασι, γνόντας ὅτι ἐπιτυχὼν μὲν ἄν τις τοῦ
προρρήματος θαυμασθείη ὑπὸ τοῦ ξυνεόντος ἀλγέ-
οντος, ἁμαρτὼν δ᾽ ἄν τις πρὸς τῷ μισεῖσθαι τάχ᾽ ἂν
καὶ μεμηνέναι δόξειεν. ὧν δὴ ἕνεκα κελεύω σωφρό-
νως τὰ προρρήματα ποιέεσθαι καὶ τἆλλα πάντα
[καὶ]¹ ταῦτα· καίτοι γε ἀκούω καὶ ὁρῶ οὔτε κρίνον-
τας ὀρθῶς τοὺς ἀνθρώπους τὰ λεγόμενά τε καὶ ποι-
εύμενα ἐν τῇ τέχνῃ οὔτ᾽ ἀπαγγέλλοντας.

3. Ἀμφὶ δὲ τῶν γυμναζομένων καὶ ταλαιπωρεόν-
των τὰς μὲν ἀτρεκείας τὰς λεγομένας ὡς λέγουσιν οἱ
λέγοντες οὔτε δοκέω εἶναι, οὔτ᾽ εἴ τις δοκέει κωλύω

persons this condition is hereditary, or which have ever been mad before; for if these persons give themselves over to drunkenness, or eat meat, or go without sleep, or thoughtlessly come into contact with cold or heat, then the likelihood is great that from these courses of action they will become deranged. Or if one sees sufferers from haemorrhoids who drink copiously and have a ruddy colour in winter, one can make a prediction about them: towards spring there is a great likelihood that a defluxion of blood will occur, so that they will be colourless and watery in the summer.

But anyone who desires to win such successes should make predictions only after learning about all these details; it is indeed possible from what is written to foretell death, or madness, or good health. I could mention a great many similar things, but have decided to record only what is most common. I advise you to be as cautious as possible not only in other areas of medicine, but also in making predictions of this kind, taking into account that when you are successful in making a prediction you will be admired by the patient you are attending, but when you go wrong you will not only be subject to hatred, but perhaps even be thought mad. For these reasons, then, I recommend that in making predictions and all other such practices you be cautious. And indeed I know, both by what I hear and by what I see, that people neither judge correctly what is said and done in medicine, nor report it accurately.

3. The alleged "precise knowledge" concerning exercises and exertions, as those who speak about it call it, I personally hold not to exist, although, if someone does

---

[1] Del. Littré.

δοκέειν· ὑπὸ σημείου μὲν γὰρ οὐδενὸς βλάπτεται τὰ
ὑπονοήματα οὔτε καλοῦ οὔτε κακοῦ, ᾧ χρὴ πιστεύ-
σαντα εἰδέναι εἴτε ὀρθῶς ἀπήγγελται εἴτ' οὔ· ἄλλως
δὲ ἐκποιέει τῷ βουλομένῳ πιστεύειν, οὐ γὰρ ἐμπο-
δὼν ἵσταμαι. δοκέω δὲ αὐτῶν εἴ τι ἀληθὲς λέγεται ἢ
τῶνδε τῶν περὶ τοὺς γυμναζομένους, ἢ ἐκείνων τῶν
πρότερον γεγραμμένων, πρῶτον μὲν τῶν σημείων
τεκμήρασθαι τοῦτο γνόντα, ἔπειτα ἐνδοιαστῶς τε
καὶ ἀνθρωπίνως προειπεῖν, ἅμα δὲ καὶ τοὺς ἀπαγγέλλοντας τερατωδεστέρως διηγεῖσθαι ἢ ὡς ἐγένετο.
ἐπεὶ οὐδ' ἐν τῇσι νούσοισιν εὐπετὲς γινώσκειν τὰ
ἁμαρτήματα· καί τοι κατάκεινταί γε οἱ ἄνθρωποι καὶ
12 διαιτήμασιν | ὀλιγοτρόφοισι χρῶνται, ὥστε μὴ
πάμπολλα δεῖ ὁρᾶσθαι ὑποσκεπτόμενον τὸν μελε-
δαίνοντα. οἱ μὲν γὰρ πίνουσι μόνον, οἱ δὲ πρὸς τῷ
πίνειν ἢ ῥύφημα ἢ σιτίον ὀλίγιστον ἐπιφέρονται·
ἀνάγκη οὖν ἐν τῷ τοιούτῳ τοὺς μὲν τῷ ποτῷ πλέονι
χρησαμένους δυσπνοωτέρους γίνεσθαι, καὶ οὐρέον-
τας πλέον φαίνεσθαι, τοὺς δὲ τῷ ῥοφήματι ἢ τῷ
σιτίῳ πλεονάσαντας διψᾶν τε μᾶλλον καὶ πυρεταί-
νειν· εἰ δέ τις ἀμφότερα, καὶ τῷ ποτῷ καὶ τοῖσι περὶ
τὰ σιτία ἀμέτρως χρήσαιτο, πρὸς τῷ πυρεταίνειν
καὶ δυσπνοεῖν καὶ τὴν γαστέρα περιτεταμένην ἂν
καὶ μείζω ἔχειν.

Ἔξεστι δὲ καὶ ταῦτα πάντα καταβασανίζειν κάλλιστα καὶ τἄλλα τοῖσι δοκιμίοισιν, οἷσιν ἔχομέν τε
καὶ χρεόμεθα εὖ πάντα. πρῶτον μὲν γὰρ τῇ γνώμῃ
τε καὶ τοῖσιν ὀφθαλμοῖσιν ἄνθρωπον κατακείμενον

believe in it, I will not oppose his belief. For such suppositions are not discredited by any sign, either good or bad, which you can trust in order to be certain whether or not the matter has been accurately reported. At all events, a person who wants to believe may do so, and I will not stand in his way. And I believe, if some of these cases—either the latter ones about exercises, or the former ones noted above—have in fact been accurately reported, first that they were recognized on the basis of signs, then that they were foretold tentatively as befits human knowledge, and furthermore that the reporters have related the tale more portentously than it really happened. And in truth even in diseases it is not easy to recognize deviations; for patients rest in bed and employ unnourishing diets, so that, of course, there is not very much for the treating physician to see when he makes his examination. Some patients only drink, while others take besides their drink a very little gruel or food; of necessity, then, in such an instance those who have taken more drink are seen to become more dyspnoic and to pass more urine, while those who have taken more gruel or food will rather be thirsty and have fever. If a patient makes both mistakes, taking excessive amounts of both drink and food, besides having fever and dyspnoea, he will have a distention of the belly and it will swell.

It is possible properly to examine all these conditions, as well as any others, by employing the methods of investigation that are available and effectively used. For, in the first place, it is easy to know by judgement and observation, in the case of a person lying ill in a fixed place and

ΠΡΟΡΡΗΤΙΚΟΝ Β

ἐν τῷ αὐτῷ καὶ ἀτρεκέως διαιτώμενον ῥᾷόν ἐστι
γνῶναι, ἤν τι ἀπειθήσῃ, ἢ περιοδοιπορέοντα καὶ
πάμπολλα ἐσθίοντα· ἔπειτα τῇσι χερσὶ ψαύσαντα
τῆς γαστρός τε καὶ τῶν φλεβῶν ἧσσόν ἐστιν ἐξαπατᾶσθαι ἢ μὴ ψαύσαντα. αἵ τε ῥῖνες ἐν μὲν τοῖσι
πυρεταίνουσι πολλά τε καὶ καλῶς σημαίνουσιν· αἱ
γὰρ ὀδμαὶ μέγα διαφέρουσιν· ἐν δὲ τοῖσιν ἰσχύουσί
τε καὶ ὀρθῶς διαιτωμένοισιν οὐκ οἶδα τί ἂν χρησαί-
μην, οὐδ᾽ ἐν τούτῳ τῷ δοκιμίῳ. ἔπειτα τοῖς ὠσὶ τῆς
φωνῆς ἀκούσαντα καὶ τοῦ πνεύματος, ἔστι διαγινώ-
σκειν, ἃ ἐν τοῖσιν ἰσχύουσιν οὐχ ὁμοίως ἐστὶ δῆλα.
ἀλλ᾽ ὅμως[1] πρόσθεν ἢ τὰ ἤθεα τῶν νοσημάτων τε
καὶ τῶν ἀλγεόντων ἐκμάθῃ ὁ ἰητρός, οὐ χρὴ προλέγειν οὐδέν· καὶ[2] γὰρ ἂν δυσπνούστερος ὤνθρωπος
14   γένοιτο, | ἔτι πλανωμένης τῆς νόσου, καὶ πυρετή-
νειεν ὀξυτέρῳ πυρί, καὶ ἡ γαστὴρ ἐπιταθείη· ὥστε
διὰ ταῦτα οὐκ ἀσφαλὲς προλέγειν πρόσθεν πρὶν ἂν
κατάστασιν λαβεῖν τὸ νόσημα· μετὰ δὲ τοῦτον τὸν
χρόνον ὅ τι ἂν παράλογον γένηται λέγειν χρή.
δῆλα δὲ τὰ διὰ τὴν ἀπειθίην γινόμενα κακά· αἵ τε
γὰρ δύσπνοιαι καὶ τἆλλα ταῦτα τῇ ὑστεραίῃ πεπαύ-
σεται, ἢν δι᾽ ἁμαρτάδα γένηται· ἢν οὖν τις ταύτην
τὴν κρίσιν προϊδὼν[3] λέγῃ, οὐχ ἁμαρτήσεται.

4. Ἐγὼ μὲν νῦν τόνδε τὸν τρόπον ἐσηγέομαι τῶν
ἐπισκεψίων, καὶ περὶ τῶν οἴκοι μενόντων, οἷα ἐξα-
μαρτάνουσιν, καὶ περὶ τῶν γυμναζομένων τε καὶ τῶν

[1] Littré: οἴως I.

charged with a strict regimen, whether he is disobeying in some matter, or whether he is taking his prescribed walks and eating a great deal. Furthermore, by using your hands to palpate his abdomen and the vessels, you are less likely to make a mistake than if you do not do this. Also, in patients with fevers your nose gives many true indications, for their odours differ greatly, although in persons that are well and following a proper diet I do not know what use I would find for even this criterion. And then, by listening with your ears to the voice and the breathing, it is possible to recognize things that are not equally clear in healthy persons. In any case, until a physician has learned the habits of particular diseases and patients, he ought not to make any prognostication, for while a person's disease is still unsettled, he may become quite short of breath, and suffer a very acute fever, and tightness of his belly; as a result, for this reason it is not safe to make an advance statement before the disease becomes settled. After that time, you should indicate whatever is abnormal; for evils that arise because of disobedience will be revealed as such, since the shortness of breath and the rest of the symptoms will cease on the following day, if they arose only because of a dietary mistake. So that if someone sees this crisis in advance and announces it, he will not go wrong.

4. Now I shall explain this type of investigation of the faults committed by those who remain at home, by those who do exercises, and by all the others. People who

---

ἄλλων πάντων· τὰς δ᾽ ἀκριβείας κείνας ἀκούω τε καὶ
καταγελῶ τῶν ἀπαγγελλόντων· σμικρὰ μὲν γὰρ
ἀπειθούντων τῶν ἀνθρώπων, οὐκ οἶδ᾽ ὅπως ἂν
ἐλέγξαιμι·[1] εἰ δ᾽ εἴη μείζονα τὰ ἁμαρτήματα, ὅντινα
τρόπον ὑποσκέπτεσθαι χρὴ γράφω. χρὴ δὲ πρῶτον
μὲν τὸν ἄνθρωπον ἐν ᾧ μέλλει τις γνώσεσθαι τὰ
ἀπειθεύμενα, ἅπασαν ἡμέρην ὁρᾶν ἐν τῷ αὐτῷ τε
χωρίῳ καὶ τὴν αὐτὴν ὥρην, μάλιστά τ᾽ ἦμος ἥλιος
νεωστὶ καταλάμπει· τοῦτον γὰρ τὸν χρόνον ὑποκεκε-
νωμένος ἂν εἴη, καὶ νῆστις ἂν ἔτι ἐστί, καὶ τεταλαι-
πωρηκὼς οὐδὲν πλὴν τῶν ὀρθρινῶν περιπάτων, ἐν
οἷς ἥκιστα ἀπειθεῖ, ἤν γε ἐπανεγερθεὶς ὤνθρωπος ἐς
τὴν περίοδον καταστῇ, ὥστε ἀνάγκη τὸν δικαίως
διαιτώμενον μάλιστα ταύτην τὴν ὥρην ὁμαλῶς ἔχειν
τὴν κατάστασιν τοῦ χρώματός τε καὶ τοῦ ξύμπαντος
σώματος, διὸ καὶ ὁ ἐπιμελόμενος ὀξύτατός τ᾽ ἂν εἴη
καὶ τὸν νόον καὶ τοὺς ὀφθαλμοὺς ὑπὸ τοῦτον τὸν
χρόνον. ἐνθυμέεσθαι δὲ χρὴ καὶ τοῦ ἀνθρώπου τῆς
τε γνώμης[2] τοὺς τρόπους, τοῦ τε σώματος τὴν δύνα-
μιν· ἄλλοι γὰρ ἄλλα ῥηιδίως ἐπιτελοῦσι τῶν προσ-
16 τασσομένων καὶ χαλεπῶς.

Πρῶτον μὲν οὖν ὁ λιμαγχεόμενος εἰ πλείονα
φάγῃ τε καὶ πίῃ, τούτοισι δῆλος ἔσται· καὶ ὀγκηρό-
τερον αὐτοῦ τὸ σῶμα φανεῖται, καὶ λιπαρώτερον καὶ
εὐχρούστερον ἔσται, ἢν μὴ κακῶς διακεχωρήκῃ τὰ
ἀπὸ τῆς γαστρὸς αὐτῷ· ἔσται δὲ καὶ εὐθυμότερος ἐν
τῇ ταλαιπωρίῃ. σκεπτέον δὲ καὶ ἤν τι ἐρυγγάνῃ ἢ
ὑπὸ φύσης ἔχηται· ταῦτα γὰρ προσήκει γίνεσθαι

report the kind of over-niceties mentioned above, I listen
to and just laugh at. For if the faults the people who are
disobedient make are slight, I do not know how I could
ever expose them; if on the other hand the faults are
greater, I shall describe now the way in which one should
investigate them. First you should inspect the person, in
whom you are intending to detect examples of disobedience, every day in the same location and at the same
time—best when the sun is just beginning to shine, since
at that hour the person will have been emptied downwards, will still not have eaten anything, and will not have
exerted himself except in his morning walks, in which
people rarely commit faults (if indeed on rising the person regularly takes a walk), since a person who follows a
well-balanced regimen must at that hour have the average
state of his facial colouring and of his whole body, and
also because at that time the examiner will have the
sharpest eyes and mind. One must also take into consid-
eration the inclinations of the person's mind and the
strength of his body. For different patients carry out
different instructions either easily or with difficulty.

First, then, if a person who is being reduced by a low
diet eats or drinks too much, he will be exposed by the
following: his body will appear more bulky, and his skin
will be moister and of a better colour, unless there has
been a pernicious excretion of the contents of his belly.
The person will also be in better spirits when he exerts
himself. You must also examine whether he has any eruc-
tations, or suffers from the passage of wind; for these

---

[1] Aldina: ἐλέξαιμι I.   [2] H: τήν τε γνώμην I.

τοῖσιν ὧδε διακειμένοισιν ἐπὶ ταύτῃ τῇ ἁμαρτάδι.
ἢν δὲ ἐσθίειν τε ἤδη ἀναγκαζόμενος συχνὰ καὶ
ταλαιπωρέειν ἰσχυρῶς, ἢ τὸ σιτίον μὴ κατα-
φάγῃ, ἢ θωρηχθῇ, ἢ μὴ περιέλθῃ ἀπὸ τοῦ δείπνου
συχνοῦ, ὧδ᾿ ὑποσκέπτεσθαι· τὸ μὲν δεῖπνον εἰ μὴ
καταφάγοι, περιπατῆσαι δὲ τὰ μεμαθηκότα, ἡδίων
τε ἂν προσιδεῖν, ὀξύτερός τε καὶ ἐργαστικώτερος ἐν
τοῖσι γυμνασίοισιν· ὁ δ᾿ ἀπόπατος σμικρότερός τε
καὶ ξυνεστηκὼς μάλιστ᾿ ἂν τούτῳ γένοιτο. ἢν δὲ τὸ
δεῖπνον καταφαγὼν μὴ περιπατήσῃ, ἐρυγγάνοι τ᾿ ἂν
καὶ φυσώδης εἴη, καὶ πλῆθος οὐκ ἐλάσσων φαίνοιτο, καὶ ἱδρῴη ἂν μᾶλλον ἢ πρόσθεν ἐν τῇ ταλαιπωρίῃ, καὶ δύσπνοος ἂν εἴη καὶ βαρύς· αἵ τε διέξ-
οδοι τῆς κοιλίης μέζονές τε καὶ ἧσσον γλίσχραι
τούτῳ γένοιντ᾿ ἄν. εἰ δὲ μήτε τὸ σιτίον καταναλώ-
σειε, μήτε περιπατήσειε, νωθρότερος ἂν εἴη καὶ
ὀγκωδέστερος. εἰ δὲ μεθυσθείη, ἱδρῴη τ᾿ ἂν μᾶλλον
ἢ πρόσθεν, καὶ δύσπνοος ἂν εἴη, καὶ βαρύτερος
αὐτὸς ἑωυτοῦ καὶ ὑγρότερος· εἴη δ᾿ ἂν καὶ εὐθυμότερος, ἢν μή τι αὐτῷ ἡ κεφαλὴ ἀνιῷτο. γυναικὶ δὲ
χρησάμενος ἅπαξ, ὀξύτερός τ᾿ ἂν εἴη καὶ λελυμένος
μᾶλλον· εἰ δὲ πλειστάκις διαπρήξαιτο, σκληρότερος
ἂν γένοιτο, καὶ αὐχμηρόν τι ἔχων, καὶ ἀχρούστερός
18   τε καὶ κοπιώδης | μᾶλλον. ἀποπάτους δὲ χρὴ
διαχωρέειν τοῖσι ταλαιπωρέουσιν, ἔστ᾿ ἂν ὀλιγοσι-
τέωσί τε καὶ ὀλιγοποτέωσι, σμικρούς τε καὶ σκλη-
ρούς, ἀνὰ δὲ πᾶσαν ἡμέρην· ἢν δὲ διὰ τρίτης, ἢ
τετάρτης, ἢ διὰ πλέονος χρόνου διαχωρέῃ, κίνδυνος

things are likely to happen in persons following this regimen if they commit these faults. If, again, a person who is being compelled to eat much and to work hard does not eat all his food, or becomes drunk, or does not go for his walk after his main meal, you must be on the lookout for the following. If he does not finish his dinner, but does take his accustomed walk, he will be more attractive in appearance, and sharper and more industrious in his exercises; this person's stools are normally less in amount and more compact. If he eats all his dinner but does not take his walk, he will suffer from belching and flatulence, in physical bulk he will not seem to be less, but he will sweat more than previously during his exertions, be short of breath, and feel heavy. The evacuations of the cavity in such a person will be greater in amount and less viscous. If he neither finishes his meal nor takes his walk, he will be more sluggish and more turgid. If he becomes drunk, he will sweat more than before, be short of breath, and heavier and moister than usual. He will also be more cheerful, unless his head should be distressed in some way. If a man avails himself of a woman once, he will be sharper and more relaxed. If however he does so repeatedly, he will become drier and somewhat parched, and be more pallid and weary. In persons that are subject to exertions, as long as they eat and drink but little, the stools should pass off every day in small amounts and be of firm consistency. If, however, they pass off every other day, every third day, or at even longer intervals, there

ἢ πυρετὸν ἢ διαρροίην ἐπιλαβεῖν. ὅσα δὲ ὑγρότερά
ἐστι τῶν διαχωρημάτων ἢ ὥστε ἐκτυποῦσθαι ἐν τῇ
διεξόδῳ, ταῦτα δὲ τοῖσι πάντα κακίω. τοῖσι δὲ
συχνὰ ἐσθίουσιν ἤδη καὶ πολλὰ ταλαιπωρέουσι τὴν
διέξοδον χρὴ μαλθακὴν ἐοῦσαν ξηρὴν εἶναι, πλῆθος
τῶν τε εἰσιόντων κατὰ λόγον <καὶ>[1] τῆς ταλαιπω-
ρίης· διαχωρέει δὲ ἀπὸ τῶν ἴσων σιτίων τοῖσι μὲν
ἐλάχιστα ταλαιπωρέουσι πλεῖστα, τοῖσι δὲ ταλαιπωρέουσι πλεῖστα σμικρόν, ἢν ὑγιαίνωσί τε καὶ
δικαίως διαιτῶνται· ἄλλα πρὸς ταῦτα συμβάλλε-
σθαι. αἱ δὲ ὑγρότεραι τῶν διαχωρήσεων καὶ ἄτερ
πυρετῶν γινόμεναι, καὶ ἑβδομαῖαι, καὶ θᾶσσον κρι-
νόμεναι, λυσιτελέες, ἐς ἅπαξ πᾶσαι γινόμεναι, καὶ
μὴ ὑποστρέφουσαι· εἰ δ' ἐπιπυρεταίνοιεν οἱ ἄνθρωποι, ἢ ὑποστρέφοιεν αἱ διάρροιαι, εἰ μακραὶ
γίνοιντο, πάντως πονηραί, εἴ τε χολώδεες εἴησαν, εἴ
τε φλεγματώδεες, εἴ τε ὠμαί, καὶ διαιτημάτων τε
ἰδίων προσδεόμεναι ἕκασται, καὶ φαρμακευσίων,
ἄλλαι ἄλλων.

Οὖρον δὲ χρὴ κατά τε τὸ τοῦ πινομένου πλῆθος
διουρέεσθαι, καὶ ἴσον αἰεὶ καὶ ἀθρόον ὡς μάλιστα,
καὶ ῥοπῇ ὀλίγῳ παχύτερον ἢ οἷον ἐπόθη. εἰ δὲ εἴη
ὑδατῶδές τε καὶ πλεῖον τοῦ προστασσομένου πίνε-
σθαι, σημαίνει μὴ πείθεσθαι τὸν ἄνθρωπον, ἀλλὰ
πλέονι ποτῷ χρῆσθαι, καὶ[2] οὐ δύνασθαι ἀνατραφῆ-
ναι, ἔστ' ἂν τὰ τοιαῦτα ποιέῃ τὸ οὖρον. εἰ δὲ κατ'
ὀλίγον στάζοι[3] τὸ οὖρον, σημαίνει ἢ φαρμακεύσιος
δεῖσθαι τὸν ἄνθρωπον, ἢ νόσημά τι τῶν περὶ κύστιν

is a danger that fever or diarrhoea will supervene. Stools that are too liquid to have any form when they are excreted are all very bad for patients. In persons that are already eating frequently and exerting themselves strenuously, the evacuations should be dry but soft, and in quantity proportional to what is ingested and to the person's exertions. From the same amount of food there pass off, in persons exerting themselves least, the greatest quantity, in those exerting themselves most, but a small amount, as long as they are healthy and following a normal regimen. Other cases are to be reckoned according to the same principle. Moister evacuations that occur in the absence of fevers and when the crisis is on the seventh day or sooner are advantageous if they all pass at once and do not recur. If, however, such patients later become febrile, or the diarrhoea returns and stays for a long time, such stools are absolutely malignant, whether they are bilious, or phlegmy, or raw (i.e. unconcocted), and each kind requires its own particular regimen, and different ones require different medications.

Urine should be excreted in an amount proportional to what is drunk and always evenly, and as far as possible all at once; it should be slightly thicker than what was drunk. But if the urine passed is watery and greater in quantity than what the person was ordered to drink, this is a sign that the person is not obeying, but that he is taking too much drink, and he cannot be built up as long as he passes urine like this. If the urine drips a little at a time, it shows either that the person needs a medication, or that

---

[1] Littré.   [2] Potter: ἢ I.
[3] Potter: τρίζοι I.

ΠΡΟΡΡΗΤΙΚΟΝ Β

ἔχειν. αἷμα δὲ οὐρῆσαι ὀλιγάκις μὲν καὶ ἄτερ πυρε-
τοῦ καὶ ὀδύνης οὐδὲν | κακὸν σημαίνει, ἀλλὰ κόπων
λύσις γίνεται· εἰ δὲ πολλάκις οὐρέοι, ἢ τούτων τι
προσγίνοιτο, δεινόν· ἀλλὰ προλέγειν, ἤν τε ξὺν
ὀδύνῃ οὐρέηται, ἤν τε ξὺν πυρετῷ, πύον ἐπιδιουρή-
σειν, καὶ οὕτω παύσεσθαι τῶν ἀλγεόντων. παχὺ δὲ
οὖρον λευκὴν ὑπόστασιν ἔχον σημαίνει ἢ περὶ
τὰ ἄρθρα τινὰ ὀδύνην καὶ ἔπαρσιν <ἢ περὶ τὰ
σπλάγχνα· χλωρὴν δέ, καθαρσίην τοῦ σώματος
δηλοῖ ἢ περὶ τὰ σπλάγχνα καὶ τούτων ὀδύνην τε καὶ
ἔπαρσιν.>[1] αἱ δ' ἄλλαι ὑποστάσεις αἱ ἐν τοῖσιν
οὔροισι τῶν γυμναζομένων πᾶσαι ἀπὸ τῶν νοσημάτων γίνονται τῶν περὶ τὴν κύστιν· δῆλον δὲ ποιήσουσι, ξὺν ὀδύνῃσί τε γὰρ ἔσονται καὶ δυσαπάλλακτοι.

Καὶ ταῦτα μὲν γράφω περὶ τούτων, καὶ λέγω τοιαῦτα ἕτερα. ὧν δὲ δι' ἀκρίβειαν κατηγορέονται τῶν
προρρήσεων, τοῖσι μὲν αὐτῶν αὐτὸς ξυνεγενόμην,
τῶν δὲ παισί τε καὶ μαθητῆσιν ἐλεσχηνευσάμην,
τῶν δὲ ξυγγράμματα ἔλαβον· ὥστε, εὖ εἰδὼς οἷα
ἕκαστος αὐτῶν ἐφρόνει, καὶ τὰς ἀκριβείας οὐδαμοῦ
εὑρών, ἐπεχείρησα τάδε γράφειν.

5. Περὶ δὲ ὑδρώπων τε καὶ φθισίων, καὶ τῶν
ποδαγρῶν, τῶν τε καὶ λαμβανομένων ὑπὸ τῆς ἱερῆς
νόσου καλεομένης,[2] τάδε λέγω, κατὰ μέν τι περὶ
πάντων τὸ αὐτό· ᾧ γὰρ ξυγγενές τι[3] τούτων τῶν

---

[1] H in margin.   [2] καλεομένης I above the line.

he has some disease associated with his bladder. To pass blood in the urine a few times without fever or pain is not a bad sign, but rather suggests the resolution of weariness. However, to pass bloody urine frequently or in conjunction with one of the other symptoms is ominous. If micturition is painful or associated with fever, predict that pus will be passed along with the urine, and that this will lead to an end of the suffering. Thick urine with a white deposit indicates there will be some pain and swelling about the joints <or about the inward parts. A yellow-greenish sediment indicates a cleaning of the body or about the inward parts, and pain and swelling in them>. Other sediments in the urines of persons who exert themselves all arise from diseases of the bladder; these conditions reveal themselves clearly, since they will be painful and hard to get rid of.

This is what I intend to write about such matters, keeping for oral presentation other similar points. Concerning those physicians whose predictions are vaunted for their nicety, with some of them I have conversed directly, others I know through having spoken with their children and students, and of others I have acquired the writings, so that, knowing full well what sort of thing each of them was thinking, but still not finding any of the reputed niceties anywhere, I have set to work to give a written account of the matter.

5. About dropsies, consumptions and gouty conditions as well as persons taken by what is called the sacred disease, I say that they all have something in common, namely that in whomever these diseases are to a degree

---

[3] Littré: οἱ γὰρ ξυγγενέσι I.

νουσημάτων, ἔστιν εἰδέναι δυσαπάλλακτον ἐόντα·
τὰ δὲ ἄλλα καθ' ἕκαστον γράψω.

22   6. Χρὴ δὲ τὸν ὑπὸ τοῦ ὕδρωπος ἐχόμενον καὶ
μέλλοντα περιέσεσθαι εὔσπλαγχνόν τε εἶναι, καὶ
ἀνατείνεσθαι κατὰ[1] φύσιν ἅμα πέπτεσθαί τε εὐπετέως, εὔπνοόν τ' ἐόντα ἀνώδυνον εἶναι, καὶ χλιαρὸν
ὁμαλῶς πᾶν τὸ σῶμα ἔχειν καὶ μὴ περιτετηκὸς περὶ
τὰ ἔσχατα· κρέσσον δὲ ἐπάρματα μᾶλλον ἔχειν ἐν
τοῖσιν ἀκρωτηρίοισιν, ἄριστον δὲ μηδὲ ἕτερον τούτων, ἀλλὰ μαλακά τε χρὴ καὶ ἰσχνὰ εἶναι τὰ ἀκρω-
τήρια· καὶ τὴν γαστέρα μαλθακὴν εἶναι ψαυομένην·
βῆχα δὲ μὴ προσεῖναι, μηδὲ δίψαν, μηδὲ τὴν γλῶσσαν ἐπιξηραίνεσθαι, ἔν τε τῷ ἄλλῳ χρόνῳ καὶ μετὰ
τοὺς ὕπνους, γίνεται δὲ ταῦτα κάρτα· τὰ δὲ σιτία
ἡδέως δέχεσθαι, καὶ ἐσθίοντα ἱκανὰ μὴ πονεῖσθαι·
τὴν δὲ κοιλίην πρὸς μὲν τὰ φάρμακα ὀξέην εἶναι,
τὸν δ' ἄλλον χρόνον διαχωρέειν μαλθακὸν ἐκτετυ-
πωμένον· τὸ δὲ οὖρον φαίνεσθαι περαιούμενον πρὸς
τὰ ἐπιτηδεύματα καὶ τῶν οἴνων τὰς μεταβολάς· τὴν
δὲ ταλαιπωρίην εὐπετῶς φέρειν, καὶ ἄκοπον εἶναι.
ἄριστον μὲν πάντα οὕτω διακεῖσθαι τὸν ἄνθρωπον,
καὶ ἀσφαλέστατ' ἂν γένοιτο ὑγιής· εἰ δὲ μή, ὡς
πλεῖστα τούτων ἐχέτω, ἐν ἐλπίδι γὰρ ἔσται περιγε-
νέσθαι· ὃς δ' ἂν μηδὲν τούτων ἔχῃ, ἀλλὰ τὰ ἐναντία,
ἀνέλπιστον ἐόντα εἰδέναι· ὃς δ' ἂν τούτων ὀλίγα
ἔχῃ, ἃ φημὶ χρηστὰ εἶναι τῷ ὑδρωπιῶντι προσόντα,
ὀλίγαι ἐλπίδες αὐτῷ.

Ὧι δ' ἂν αἱμορραγέῃ πολλὸν ἄνω καὶ κάτω, καὶ

hereditary, you can be sure that they are hard to get rid of. Their other features I shall describe disease by disease.

6. For a person to be attacked by dropsy and go on to survive he must have healthy inward parts, which are both expanded according to nature and at the same time digest favourably, draw his breath easily and without pain, be evenly warm over his whole body, and not be emaciated in the extremities—better than emaciation would be even to have swellings in the extremities, although it would be best to have neither, and that the extremities be soft and without swelling. The belly should be soft to the touch. Cough should not supervene, nor thirst, nor should the tongue become dry at any time, but especially after sleeping; these symptoms are frequent. Foods should be taken with pleasure, and after being eaten in sufficient quantity should not occasion pain. The cavity should be quick when stimulated by a medication, but otherwise pass soft, well-formed stools. The urine passed should appear appropriate to the person's regimen and changes of wine. The patient should bear exertions easily and remain unwearied. The dropsical patient will be in the best position if all these signs are as noted, and will recover with the greatest sureness. If not all, then let as many as possible of them be favourable, for he will have a good chance of recovering. A person with none of the favourable signs, but rather their opposites, you can be sure is hopeless; one with few of the features I list as favourable in dropsy has slight hope.

If a person has a copious haemorrhage both upwards

---

[1] Littré: καὶ I.

πυρετὸς ἐπιγένηται, ὕδατος ἐμπλησθῆναι πολλαὶ
ἐλπίδες τοῦτον, καὶ τῶν ὑδρώπων οὗτος ὀλιγοχρο-
νιώτατός τε καὶ ἐν τοῖσιν ἀφυκτοτάτοισιν· ἄλλῳ
δὲ προσημαίνειν περὶ τούτου. οἷσι δὲ οἰδήματα
μεγάλα γενόμενα καταμαραίνεται, καὶ αὖθις ἐπαίρε-
24   ται, | οὗτοι δὲ μᾶλλον περιγίνονται τῶν ἐκ τῶν αἱμά-
των τῆς ἀναρρήξιος ἐμπιπραμένων·[1] ἐξαπατέουσι δὲ
τοὺς ἀλγέοντας οὗτοι οἱ ὕδρωπες, ὥστε ποιέουσιν
αὐτοὺς ἀπιστέοντας[2] τοῖσιν ἰατροῖσιν ἀπόλλυσθαι.

7. Περὶ δὲ τῶν φθινόντων κατὰ μὲν τὸ πτύελον
καὶ τὴν βῆχα ταὐτὰ λέγω ἅπερ περὶ τῶν ἐμπύων
ἔγραφον. χρὴ γὰρ τὸ πτύελον τῷ μέλλοντι καλῶς
ἀπαλλάξειν εὐπετέως τε ἀναβήσεσθαι καὶ εἶναι
λευκόν, καὶ ὁμαλόν, καὶ ὁμόχροον, καὶ ἀφλέγμαν-
τον, τὸ δ' ἀπὸ τῆς κεφαλῆς καταρρέον ἐς τὰς ῥῖνας
τρέπεσθαι· πυρετὸν δὲ μὴ λαμβάνειν, <ἢ τοσοῦτον
λαμβάνειν,>[3] ὥστε τῶν δείπνων μὴ κωλύεσθαι, μηδὲ
διψῆν· ἡ δὲ γαστὴρ ὑποχωρείτω πᾶσαν ἡμέρην, καὶ
τὸ ὑποχωρέον ἔστω σκληρόν, πλῆθος δὲ κατὰ τὰ
εἰσιόντα· τὸν δὲ ἄνθρωπον ὡς ἥκιστα λεπτὸν εἶναι·
τὸ δὲ στῆθος ἐπαινεῖν χρὴ τετράγωνόν γε ἐὸν καὶ
λάσιον, καὶ ὁ χόνδρος αὐτοῦ μικρὸς ἔστω καὶ
σεσαρκωμένος ἰσχυρῶς. ὅστις μὲν γὰρ ταῦτα
πάντα ἔχει, περιεστικώτατος γίνεται· ὃς δ' ἂν μηδὲν
τούτων ἔχῃ, ὀλεθριώτατος.

Ὅσοι δ' ἂν ἔμπυοι γένωνται, νέοι ἐόντες, ἐξ ἀπο-

---

[1] I: -πιπλαμένων H.
[2] H²: ἀπιόντας I.   [3] H in margin.

and downwards, and fever supervenes, the chances are great that he will become filled with water, and this is the most short-lived of dropsies and among the most inescapable; foretell this to someone else who is present. Persons in whom great swellings form, and then dry up, and then swell up again, are more likely to survive than those who become febrile[a] from a haemorrhage upwards. The latter dropsies deceive patients, so that they lose faith[b] in their physicians, and thus perish.

7. In consumptive patients, with regard to the sputum and cough I have the same to say as I wrote with regard to internal suppuration: namely, that for those who are going to come off well it must be coughed up easily, and be white, uniform in consistency and colour, and without phlegm; and what flows down from the head should be directed to the nostrils.[c] No fever should be present, <or only so little as> not to hinder the patient from taking his meals; nor should there be thirst. The belly should evacuate downwards every day, and the stools should be firm and in proportion to the amount of food ingested. The person should be as little thin as possible: you must commend a chest that is square and covered with hair, and its cartilage (xiphisternum) should be small and well covered with flesh. Now whoever has all of these signs is most likely to survive, who has none of them, most likely to die.

Those who begin to suppurate internally while still

[a] Or, with the reading of H, "who become hydropic".

[b] For Foes and Grimm this loss of faith results in the patients deserting their physicians, while Littré and Fuchs ascribe fatal disobedience to the patients.

[c] Cf. *Glands* 13.

σκήψιος, ἢ σύριγγος, ἢ ἀπ' ἄλλου τινὸς τῶν τοιούτων, ἢ ἐκ παλινδρομίης ἀποστάσιος, οὐ περιγίνονται, ἢν μὴ πολλὰ κάρτα αὐτοῖσιν ἐπιγένηται τῶν
ἀγαθῶν σημείων. ἀπόλλυνται δὲ οἱ ἄνθρωποι οὗτοι
ἐς τὸ φθινόπωρον. (ἰσχυρῶς δὲ καὶ ἐκ τῶν ἄλλων
νοσημάτων μακρῶν ἐς τὴν ὥρην ταύτην τελευτῶσιν
οἱ πλεῖστοι.) τῶν δ' ἄλλων ἥκιστα περιγίνονται αἵ
τε παρθένοι καὶ αἱ γυναῖκες, ἧσιν ἀπολήψει ἐπιμη-
νίων ἡ φθίσις γένηται. εἰ δὲ μέλλοι τις περιέσεσθαι
τῶν παρθένων ἢ τῶν γυναικῶν, τῶν τε ἄλλων
σημείων τῶν [τε]¹ ἀγαθῶν δεῖ πολλὰ ἐπιγενέσθαι
καὶ τὰ ἐπιμήνια λαμπρῶς τε καὶ καθαρῶς ἐπιφαίνε-
σθαι, ἢ οὐδεμία ἐλπίς. οἱ δὲ ἐκ τῶν αἱμάτων τῆς |
26 ἀναρρήξιος ἔμπυοι γινόμενοι τῶν τε ἀνδρῶν καὶ τῶν
γυναικῶν καὶ τῶν παρθένων περιγίνονται μὲν οὐχ
ἧσσον, τὰ δὲ σημεῖα χρὴ πάντα ἀναλογισάμενον τά
γε περὶ τῶν ἐμπύων καὶ τῶν φθινόντων προλέγειν
τόν τε περιεσόμενον καὶ τὸν ἀπολούμενον. μάλιστα
δὲ περιγίνονται ἐκ τῶν τοῦ αἵματος ἀναρρήξεων
οἷσιν ἂν ἀλγήματα ὑπάρχῃ μελαγχολικὰ ἔν τε τῷ
νώτῳ καὶ ἐν τῷ στήθει, καὶ μετὰ τὴν ἀνάρρηξιν
ἀνωδυνώτεροι γένωνται· βῆχές τε γὰρ οὐ κάρτα ἐπιγίνονται, καὶ πυρετοὶ οὐ² πλεῖστοι διατελέουσιν ἐόντες, <καὶ>³ δίψαν εὐπετέως φέρουσιν·⁴ ὑποστροφαὶ
δὲ τῆς ἀναρρήξιος μάλιστα γίνονται τούτοισιν, εἰ
μὴ ἀποστάσιες ἐπιγένωνται· ἄρισται δὲ τῶν ἀποστασίων αἱ αἱματηρόταται. ὁκόσοισι δὲ ἐν τοῖσι
στήθεσιν ἀλγήματα ἔνεστι, καὶ διὰ χρόνου λεπτύ-

young—either from a determination of humours, or from a fistulous sore or anything else like that, or from a relapsing apostasis—do not survive unless a good many of the favourable signs appear in them; such persons perish towards fall. (From the other chronic diseases, too, it is mainly towards that season that most patients die.) Of the others, the least likely to live are the girls and women in whom the consumption arose along with a cessation of their menses; if any of the girls or women is going to survive, many of the other favourable signs must be present, and the menses must reappear vigorous and pure-flowing; otherwise, there is no hope. Those who suppurate internally as the result of a haemorrhage upwards—men, women and girls—are no less likely to survive: you should consider all the signs important in internal suppuration and consumption, to predict who will survive or succumb. When the condition has arisen from haemorrhaging upwards, the most likely to survive are persons who have melancholic pains in the back and chest and after the haemorrhage are freed from their pains. For they are not subject to excessive coughing, their fevers do not in most cases persist, and they bear their thirst easily. These patients usually have a recurrence of the haemorrhage, unless apostases take place; the best apostases are those with the most blood. Persons with pains in their chest,

---

[1] Del. Kühn.

[2] Littré: οἱ I.

[3] H.

[4] Recc.: -ρόντες I.

νονταί τε καὶ βήσσουσι, καὶ δύσπνοιαι γίνονται,
οὔτε πυρετῶν ἐπιλαμβανόντων, οὔτε ἐκπυημάτων
ἐπιγινομένων, τούτους ἐπανερέσθαι, ὁκόταν βήσ-
σωσί τε καὶ δύσπνοοι ἔωσιν, εἰ[1] ξυνεστραμμένον τι
καὶ μικρόν, ὀδμὴν ἔχον, ἐκβήσσουσιν.

8. Περὶ δὲ ποδαγρώντων τάδε· ὅσοι μὲν γέροντες
ἢ περὶ τοῖσιν ἄρθροισιν ἐπιπωρώματα ἔχουσιν, ἢ
τρόπον ἀταλαίπωρον[2] ζῶσι κοιλίας ξηρὰς ἔχοντες,
οὗτοι μὲν πάντες ἀδύνατοι ὑγιέες γίνεσθαι ἀνθρω-
πίνῃ τέχνῃ, ὅσον ἐγὼ οἶδα· ἰῶνται δὲ τούτους
28   ἄριστα | μὲν δυσεντερίαι, ἢν ἐπιγένωνται, ἀτὰρ καὶ
ἄλλαι ἐκτήξιες ὠφελέουσι κάρτα αἱ ἐς τὰ κάτω
χωρία ῥέπουσαι. ὅστις δὲ νέος ἐστὶ καὶ ἀμφὶ τοῖσιν
ἄρθροισιν οὔπω ἐπιπωρώματα ἔχει καὶ τὸν τρόπον
ἐστὶν ἐπιμελής τε καὶ φιλόπονος καὶ κοιλίας ἀγαθὰς
ἔχων ὑπακούειν πρὸς τὰ ἐπιτηδεύματα, οὗτος δὲ
ἰητροῦ γνώμην ἔχοντος ἐπιτυχὼν ὑγιὴς ἂν γένοιτο.

9. Τῶν δ' ὑπὸ τῆς ἱερῆς νόσου λαμβανομένων
χαλεπώτατοι μὲν ἐξίστασθαι, ὁκόσοισιν ἂν ἀπὸ
παιδὸς ξυμβήσηται καὶ συνανδρωθῇ τὸ νόσημα·
ἔπειτα δὲ ὅσοισιν ἂν γένηται ἐν ἀκμάζοντι τῷ
σώματι τῆς ἡλικίης (εἴη δ' ἂν ἀπὸ εἴκοσι καὶ πέντε
ἐτέων ἐς πέντε καὶ τεσσαράκοντα ἔτεα)· μετὰ δὲ τούτους, ὅσοις ἂν γένηται τὸ νούσημα μηδὲν προση-
μαῖνον, ὁκόθεν ἄρχεται τοῦ σώματος. οἷσι δὲ ἀπὸ
τῆς κεφαλῆς δοκέει ἄρχεσθαι, ἢ ἀπὸ τοῦ πλευροῦ, ἢ
ἀπὸ τῆς χειρός, ἢ τοῦ ποδός, εὐπετέστερα ἰᾶσθαι·
διαφέρει γὰρ καὶ ταῦτα· τὰ γὰρ ἀπὸ τῆς κεφαλῆς

who over time become thin and begin to cough, and who suffer from dyspnoea, but who have neither fevers nor suppurations, you must ask whether, when they are coughing and suffering from dyspnoea, they cough up little congealed clots that give off a smell.

8. As for patients with gout, those who are old, or who have concretions about the joints, or who live a life without exertions and are constipated, are all incurable by the human art, as far as I know, but are best cured by dysentery, if it should occur; also other attenuations which incline towards the lower parts are very helpful.[a] Anyone, on the other hand, who is young and does not yet have concretions about his joints, who is careful about his way of life and enjoys exertion, and whose bowels respond well to his regimen, if he chances upon a physician that has understanding, will recover.

9. Of patients seized by the sacred disease, the ones who have the most difficulty to escape are those whom the disease has accompanied from their childhood and has grown up along with them; next, those in whom the disease occurs as their body reached the fullness of age (that would be from twenty-five to forty-five years); after them, persons in whom the disease occurs without giving any indications from which part of the body it is originating. Those in whom the disease seems to begin from the person's head or side, or from his hand or foot, are more easily treated. But these cases differ too, for those where the disease begins from the head are the most difficult,

[a] Cf. ch. 9 below.

[1] Littré: καὶ I: ἢ H[2].
[2] Littré: ταλαί- I.

τούτων χαλεπώτατα· ἔπειτα τὰ ἀπὸ πλευροῦ· τὰ δὲ
ἀπὸ τῶν χειρῶν τε καὶ ποδῶν μάλιστα οἷά τε ἐξυγι-
αίνεσθαι. ἐπιχειρέειν δὲ χρὴ τούτοισι τὸν ἰητρόν,
εἰδότα τὸν τρόπον τῆς ἰήσιος, ἢν ἔωσιν οἱ ἄνθρωποι
νέοι τε καὶ φιλόπονοι· πλὴν ὅσον αἱ φρένες τι κακὸν
ἔχουσιν, ἢ εἴ τις ἀπόπληκτος γέγονεν· αἱ γὰρ
μελαγχολικαὶ αὗται ἐκστάσιες οὐ λυσιτελέες· αἱ δὲ
ἄλλαι αἱ εἰς τὰ κάτω τρεπόμεναι πᾶσαι ἀγαθαί· ἄρισται δὲ καὶ ἐνταῦθα πολλῷ αἱ αἱματηρόταται. ὁπόσοι δὲ γέροντες ἤρξαντο λαμβάνεσθαι, ἀποθνή-
σκουσί τε μάλιστα, καὶ ἢν μὴ ἀπόλωνται, τάχιστα
ἀπαλλάσσονται ὑπὸ τοῦ αὐτομάτου, ὑπὸ δὲ τῶν
ἰητρῶν ἥκιστα ὠφελέονται.

10. Οἷσι δὲ τῶν παιδίων ἐξαπίνης οἱ ὀφθαλμοὶ
διεστράφησαν, ἢ μεῖζόν τι κακὸν ἔπαθον, ἢ φύματα
ὑπὸ τὸν αὐχένα ἐφύη, ἢ ἰσχνο|φωνότεροι ἐγένοντο,
ἢ βῆχες ξηραὶ χρόνιοι προσέχουσιν, ἢ ἐς τὴν
γαστέρα μείζοσι γενομένοισιν ὀδύνη φοιτᾷ, καὶ οὐκ
ἐκταράσσεται, ἢ ἐν τοῖσι πλευροῖσι διαστρέμματα
ἔχουσιν ἢ φλέβας παχέας περὶ τὴν γαστέρα κιρσώ-
δεας, ἢ ἐπίπλοον καταβαίνει, ἢ ὄρχις μέγας γέγονεν,
ἢ χεὶρ λεπτὴ καὶ ἀκρατής, ἢ πούς, ἢ κνήμη ξύμ-
πασα ἐχωλώθη, ἄτερ προφάσιος ἄλλης, τούτοισι
πᾶσιν εἰδέναι ὅτι ἡ νοῦσος προεγένετο πρὸ τούτων
ἁπάντων, καὶ οἱ μὲν πλεῖστοι τῶν τρεφόντων τὰ παιδία ἐρωτώμενοι ὁμολογήσουσι, τοὺς δὲ καὶ λανθάνει
καὶ οὔ φασιν εἰδέναι τοιοῦτον οὐδὲν γενόμενον.

11. Τὸν δὲ περὶ τῶν ἑλκέων μέλλοντα γνώσε-

and then those where it begins from the side, whereas cases where it begins from both the hands and feet are the most possible to cure completely. The physician, if he knows the method of treatment, must take these cases on if the persons are young and active, unless their mind has some defect, or the patient is paralysed. Evacuations of dark bile in these patients are not advantageous, but all other movements that turn towards the lower parts are favourable; and best by far in that direction are ones that contain the most blood. Persons that have begun to be attacked by the disease in old age[a] generally succumb, but if they do not die, they recover spontaneously in a very short time, and are least benefited by physicians.

10. Children in whom the eyes suddenly look awry, or who suffer some serious accident, or in whom tubercles grow along the neck, or who become weaker in their speech, or in whom a chronic dry cough sets in, or who as they get older suffer a pain shooting to the belly unaccompanied by a motion of the bowels, or who have distortions (cramps?) in their sides, or wide varicose veins about the belly, or in whom the omentum descends or a testicle becomes large, or a hand becomes wasted and powerless, or a foot, or the whole lower leg becomes lame without any other cause—in all these instances, then, you can be sure that epilepsy was present before all of these events. Most of those who are caring for these children, if asked, will confirm this, but others it escapes, and they say they are unaware of any such thing having occurred.

11. A person who wants to know about ulcers, how

[a] Cf. ch. 8 above.

σθαι, ὅκως ἕκαστα τελευτήσει, πρῶτον μὲν χρὴ τὰ
εἴδεα τῶν ἀνθρώπων ἐξεπίστασθαι, τὰ δὲ ἀμείνω πρὸς
τὰ ἕλκεα καὶ τὰ κακίω· ἔπειτα τὰς ἡλικίας εἰδέναι,
ὁποίῃσιν ἕκαστα τῶν ἑλκέων δυσαπάλλακτα γίνεται·
τά τε χωρία ἐπεσκέφθαι τὰ ἐν τοῖσι σώμασιν, ὅσον
διαφέρει θάτερα θατέρων· τά τε ἄλλα ὁκοῖα ἐφ᾽
ἑκάστοισιν ἐπιγίνεται ἀγαθά τε καὶ κακὰ εἰδέναι.
εἰδὼς μὲν γὰρ ἄν τις ταῦτα πάντα εἰδῇ ἂν καὶ ὅπως
ἕκαστα ἀποβήσεται· μὴ εἰδὼς δὲ ταῦτα, οὐκ ἂν
εἰδῇ ὅπως αἱ τελευταὶ ἔσονται ἀμφὶ τῶν ἑλκέων.

Εἴδεα μὲν γὰρ ἀγαθά ἐστι τὰ τοιάδε· ἐλαφρὰ καὶ
ξύμμετρα, καὶ εὔσπλαγχνα, καὶ μήτε σαρκώδεα
ἰσχυρῶς μήτε σκληρά· κατὰ δὲ χρῶμα ἔστω λευκόν,
ἢ μέλαν, ἢ ἐρυθρόν· ταῦτα γὰρ πάντα ἀγαθὰ
ἄκρητα ἐόντα· εἰ δ᾽ εἴη μιξόχλωρον, ἢ χλωρόν, ἢ
πελιδνὸν τὸ χρῶμα, κάκιον γίνεται. Τὰ δὲ εἴδεα ὅσα
ἂν τοῖσι προγεγραμμένοισι τἀναντία πεφύκῃ, εἰδέναι χρὴ κακίω ἐόντα. Περὶ δὲ ἡλικιῶν, φύματα μὲν
32 ἔμπυα [γίνεται]¹ καὶ τὰ | χοιρώδεα, ταῦτα πλεῖστα
τὰ παιδία² ἴσχουσι, καὶ ῥᾷστα ἐξ αὐτῶν ἀπαλλάσ-
σει· τοῖσι δὲ γεραιτέροισί τε τῶν παιδίων καὶ νεηνί-
σκοισι φύεται μὲν ἐλάσσω, χαλεπώτερον δὲ ἐξ
αὐτῶν ἀπαλλάσσουσι. Τοῖσι δὲ ἀνδράσι τὰ μὲν
τοιαῦτα φύματα οὐ κάρτα ἐπιγίνεται· τὰ δὲ κηρία
δεινά, καὶ οἱ κρυπτοὶ καρκῖνοι οἱ ὑποβρύχιοι, καὶ οἱ
ἐκ τῶν ἐπινυκτίδων ἕρπητες, ἔστ᾽ ἂν ἑξήκοντα ἔτεα
συχνῷ ὑπερβάλλωσι. Τοῖσι δὲ γέρουσι τῶν μὲν
τοιουτοτρόπων φυμάτων οὐδὲν ἐπιγίνεται· οἱ δὲ

each will end, must first recognize clearly the different types of human beings, those better off with regard to ulcers, and those worse off. Then he must know in which ages each of the ulcers are difficult to cure, he must observe how the parts of the body differ from one another, and he must know which other evil and good things follow upon each. For if a person knows all these things, he will also know how each case will turn out, whereas if he does not, he will not know how the ulcers will end.

Now the following kinds of patients are better off: one who is nimble and well-proportioned, who has healthy inward parts, and who is neither too fleshy nor too thin; in colour he should be light, dark or ruddy—for all these are good as long as they are not mixed; but if he is yellowish, or yellow-green or livid in colour, it is rather bad. Kinds of patients who are the opposite of the ones just noted, you must know to be worse off. As far as the ages of life are concerned, it is mostly children who have pustular and scrofulous tubercles, and they get rid of them very easily. In older children and in young people these ulcers arise less, but patients also get rid of them with greater difficulty. In adults such tubercles are not very likely to occur, but rather malignant cysts, hidden deep-seated cancers, and shingles that arise from epinyctides[a]; this continues until they are well beyond the sixtieth year. In old persons, none of the former type of tubercles occur,

[a] A pustule rising in the night and disappearing in the morning; also, any pustule that is more painful at night.

---

[1] Del. Littré.

[2] Littré: τῶν παιδίων I.

καρκῖνοι οἱ κρυπτοὶ καὶ οἱ ἀκρόπαθοι γίνονται, καὶ
ξυναποθνήσκουσιν. Τῶν δὲ χωρίων μασχάλαι
δυσιητότεραι, καὶ κενεῶνές τε καὶ μηροί· ὑποστά-
σιές τε γὰρ ἐν αὐτοῖσι γίνονται καὶ ὑποστροφαί.
Τῶν δὲ περὶ ἄρθρα ἐπικινδυνότατοι οἱ μεγάλοι
δάκτυλοι, καὶ μᾶλλον οἱ τῶν ποδῶν. (οἷσι δὲ τῆς
γλώττης ἐν τῷ πλαγίῳ ἕλκος γίνεται πολυχρόνιον,
καταμαθεῖν τῶν ὀδόντων ἤν τις ὀξὺς τῶν κατ᾽ αὐτό.)

12. Τὰ δὲ τρώματα θανατωδέστερα μὲν τὰ ἐς τὰς
φλέβας τὰς παχείας τὰς ἐν τῷ τραχήλῳ τε καὶ τοῖς
βουβῶσιν, ἔπειτα εἰς τὸν ἐγκέφαλον καὶ ἐς τὸ ἧπαρ,
ἔπειτα τὰ ἐς ἔντερον καὶ τὰ ἐς κύστιν. ἔστι δὲ
πάντα ταῦτα, ὀλέθρια ἐόντα ἰσχυρῶς, οὐχ οὕτως
ἄφυκτα ὡς δοκέει· τά τε γὰρ χωρία ὀνόματα ἔχοντα
ταὐτὰ μέγα διαφέρει, καὶ οἱ αὐτοὶ τρόποι. Πολὺ δὲ
διαφέρει τοῦ αὐτοῦ ἀνθρώπου τοῦ σώματος ἡ παρα-
σκευή· ἔστι μὲν γὰρ ὅτε οὔτ᾽ ἂν πυρετήνειεν οὔτε
φλεγμήνειε τρωθείς· ἔστι δ᾽ ὅτε καὶ ἄνευ προφάσιος
34 ἐπυρέτηνεν | ἄν, καὶ φλεγμανθείη τι τοῦ σώματος
πάντως. ἀλλ᾽ ὅτε[1] ἕλκος ἔχων <μὴ>[2] παραφρονέει
εὐπετέως τε φέρει τὸ τρῶμα, ἐγχειρέειν χρὴ τῷ τρώ-
ματι ὡς ἀποβησομένῳ κατὰ λόγον τῆς ἰητρείης τε
καὶ τῶν ἐπιγινομένων. ἀποθνήσκουσι μὲν γὰρ οἱ
ἄνθρωποι ὑπὸ τρωμάτων παντοίων· πολλαὶ μὲν γὰρ
φλέβες εἰσὶ καὶ λεπταὶ καὶ παχεῖαι, αἵτινες αἱμορ-
ραγοῦσαι ἀποκτείνουσιν, ἢν αὐτῷ τύχωσιν ὀργῶσαι,
ἃς ἐν ἑτέρῳ καιρῷ διακόπτοντες ὠφελέουσι τὰ

[1] Littré after Cornarius᾽ at quando: ἀλλ᾽ ἢν ὅ τε I.

but hidden and superficial cancers arise, and remain with
them until they die. Of the parts of the body, the axilla is
rather difficult to treat, as are the flanks and the thighs,
for abscesses occur in them, and there are relapses. Of
the members affected with ulcers, the most dangerous
are the thumbs and the great toes, especially the latter.
(In cases where a chronic ulcer forms at the edge of the
tongue, find out whether one of the teeth on that side is
sharp.)

12. The more deadly wounds are those to the wide
vessels of the neck and the groins, then to the brain
and the liver, and then to the intestine and the bladder.
Still, all these are, while extremely dangerous, not so
inescapable as they might seem, for the particular parts,
though each bearing a common name,[a] differ greatly, and
so do the same kinds of wounds. There are considerable
differences, too, in the physical constitution of the body
of even the same person; for sometimes he will not have a
fever and inflammation on being wounded, while at
another time he has a fever without any obvious cause,
and some part of the body becomes completely inflamed.
Now when a person with an ulcer is not delirious and he
bears the wound easily, you must attempt to remove the
wound according to the principles of the medical art,
adjusted to the particular circumstances. People die as
the result of every kind of wound; for many are the vessels, both narrow and wide, which by haemorrhaging kill,
if in the particular person they happen to be swollen; if
the same vessels were incised at another time, it would

---

[a] Each brain, for example, is called "brain", suggesting that
all brains behave in the same way.

[2] Littré.

σώματα. πολλὰ δὲ τῶν τραυμάτων ἐν χωρίοισί τε
εἶναι εὐήθεσι καὶ οὐδέν τι δεινὰ φαινόμενα, οὕτως
ὠδύνησεν ἡ πληγὴ ὥστε μὴ δύνασθαι ἀναπνεῦσαι·
ἄλλοι δὲ ὑπὸ τῆς ὀδύνης τοῦ τρώματος οὐδὲν δή τι
δεινοῦ ἐόντος, τὸ μὲν πνεῦμα ἀνήνεγκαν, παραφρό-
νησαν δὲ καὶ πυρετήναντες ἀπέθανον· ὅσοι γὰρ ἂν ἢ
τὸ σῶμα πυρετῶδες ἔχωσιν ἢ τὰς γνώμας θορυβώ-
δεας, τὰ τοιαῦτα πάσχουσιν. ἀλλὰ χρὴ μήτε ταῦτα
θαυμάζειν, μήτε ὀρρωδέειν κεῖνα, εἰδότα ὅτι αἱ
ψυχαί τε καὶ τὰ σώματα πλεῖστον διαφέρουσιν αἱ
τῶν ἀνθρώπων, καὶ δύναμιν ἔχουσι μεγίστην. ὅσα
μὲν οὖν τῶν τραυμάτων καιροῦ ἔτυχεν, ἢ σώματός
τε καὶ γνώμης τοιαύτης, ἢ ὀργῶντος οὕτω τοῦ
σώματος, ἢ μέγεθος τοσαῦτα ἦν ὥστε μὴ δύνασθαι
καταστῆναι τὸν ἄνθρωπον †εἰς τὴν ἴησιν καταφρο-
νέοντα,† τοῖσι μὲν ἐξίστασθαι χρὴ ὁποῖα ἂν ᾖ,
πλὴν τῶν ἐφημέρων λειποθυμιῶν· τοῖσι δ' ἄλλοισι
πᾶσιν ἐπιχειρέειν, νεοτρώτοισιν ἐοῦσιν, ὡς ἂν τούς
τε πυρετοὺς διαφεύγωσιν οἱ ἄνθρωποι καὶ τὰς
αἱμορραγίας τε καὶ τὰς νομὰς περιεσόμενοι.[1] ἀτρε-
κέστατα δὲ καὶ ἐπὶ πλεῖστον χρόνον τὰς φυλακὰς
αἰεὶ τῶν δεινοτάτων ποιέεσθαι· καὶ γὰρ δίκαιον
οὕτως.

13. Αἱ δὲ νομαὶ θανατωδέσταται μὲν ὧν αἱ σηπε-
δόνες βαθύταται, καὶ μελάνταται, καὶ ξηρόταται·
πονηραὶ δὲ καὶ ἐπικίνδυνοι ὅσαι μέλανα ἰχῶρα ἀνα-
διδοῦσιν· αἱ δὲ λευκαὶ καὶ μυξώδεες τῶν σηπεδόνων
ἀποκτείνουσι μὲν ἧσσον, ὑποστρέφουσι δὲ μᾶλλον,

benefit the body. In many wounds that are apparently not
at all serious and in easily-treated places, the blow has
provoked so much pain that the person became unable to
breathe. Others, from the pain of a wound that was not at
all serious, had no difficulty in breathing, but became
delirious and died in fever; for all who have either a feverish body or a disturbed mind suffer such things. But you
should neither be surprised by the latter nor dread the
former, being aware that the minds and the bodies of
people differ very greatly, and that these differences have
great consequences. Now whenever wounds reach a danger point, when body and mind are in this condition, or if
the body is inflamed in the way mentioned, or if in magni-
tude the wound is such that the person cannot recover. . .,[1]
these you must abandon as they are, except for cases of
ephemeral fainting spells. All the others, though, you
must take on when they are newly wounded, so that these
persons escape from any fevers, haemorrhages, or spreading ulcers, and survive. Always let your measures against
the most dangerous conditions be most precise and most
extended, for this is as it should be.

13. Of spreading ulcers, most deadly are those whose
suppurations are deepest, darkest and driest; those which
produce a dark serum are troublesome and precarious,
whereas white and phlegmy suppurations kill less, but

---

[1] Ermerins after Opsopoeus in note: -σομένων I.

καὶ χρονιώτεραι γίνονται. οἱ δ' ἕρπητες ἀκινδυνότα-
τοι πάντων ἑλκέων ὅσα νέμεται, δυσαπάλλακτοι δὲ
μάλιστα, κατά γε τοὺς κρυπτοὺς καρκίνους. ἐπὶ
πᾶσι δὲ τοῖσι τοιούτοισι πυρετόν τε ἐπιγενέσθαι
ξυμφέρει μίην ἡμέρην καὶ πῦον ὡς λευκότατον καὶ
παχύτατον· λυσιτελεῖ δὲ καὶ σφακελισμὸς νεύρου, ἢ
ὀστέου, ἢ καὶ ἀμφοῖν, ἐπί γε τῇσι βαθείῃσι σηπε-
δόσι καὶ μελαίνῃσι· πῦον γὰρ ἐν τοῖσι σφακε-
λισμοῖσι ῥεῖ πολὺ καὶ λύει τὰς σηπεδόνας.

14. Τῶν δὲ ἐν τῇ κεφαλῇ τρωμάτων θανατωδέσ-
τατα μὲν τὰ ἐς τὸν ἐγκέφαλον, ὡς καὶ προγέγραπ-
ται· δεινὰ δὲ καὶ τὰ τοιαῦτα πάντα, ὀστέον ψιλὸν
μέγα, ὀστέον ἐμπεφλασμένον,[1] ὀστέον κατερρωγός·
εἰ | δὲ καὶ τὸ στόμα τοῦ ἕλκεος σμικρὸν εἴη, ἡ δὲ
ῥωγμὴ τοῦ ὀστέου ἐπὶ πολὺ παρατείνοι,[2] ἐπικινδυνό-
τερόν ἐστι· ταῦτα δὲ πάντα δεινότερα γίνεται καὶ
κατὰ ῥαφήν τε ὄντα, καὶ τῶν χωρίων αἰεὶ τὰ ἐν
τοῖσιν ἀνωτάτω τῆς κεφαλῆς.

Πυνθάνεσθαι δὲ χρὴ ἐπὶ πᾶσι τοῖσιν ἀξίοις
λόγου τρώμασιν, ἢν ἔτι νεότρωτοι αἱ πληγαὶ ἔωσιν,
εἰ βλήματα εἴη, ἢ κατέπεσεν ὥνθρωπος, ἢ εἰ
ἐκαρώθη· ἢν γάρ τι τούτων ᾖ γεγονός, φυλακῆς
πλείονος δεῖται, ὡς τοῦ ἐγκεφάλου ἐσακούσαντος
τοῦ τρώματος. εἰ δὲ μὴ νεότρωτος εἴη, ἐς τἆλλα
σημεῖα σκέπτεσθαι καὶ βουλεύεσθαι. ἄριστον μὲν
οὖν μήτε πυρετῆναι μηδαμᾶ τὸν τὸ ἕλκος ἔχοντα ἐν
τῇ κεφαλῇ, μήθ' αἷμα ἐπαναρραγῆναι αὐτῷ, μήτε
φλεγμονὴν μηδεμίην ἢ ὀδύνην ἐπιγενέσθαι· εἰ δέ τι

38

tend more to relapse, and become very chronic. Shingles are the least dangerous of all spreading lesions, but especially difficult to get rid of, like hidden cancers. It is advantageous in all such cases for fever to come on for a day, and for the pus to be very thick and white. Also of benefit is for a cord or a bone or both to form a sphacelus, at least in the case of deep suppurations and dark ones; for the pus in sphaceli flows copiously, and resolves such suppurations.

14. Of head wounds the most deadly are, as has been said above, those that reach the brain. But the following are all dangerous too: where the bone has a large bare area; where the bone is indented; and where the bone is comminuted; also it is very dangerous if the opening of the wound is small, but the fracture in the bone extends far. These lesions all become even more dangerous when they are over a suture, and whenever they occur in areas at the crown of the head.

In all cases of wounds worth mentioning you must find out whether the wounds are still fresh, whether shots were involved, whether the person fell down, and whether he became stupefied. For if one of these be the case, the person requires more watching, since the brain has felt the effect of the wound. If the wound is not fresh, you must look to other signs for your deliberations. Now best would be if the person with the wound in his head had no fever at all, nor any subsequent haemorrhage, nor any supervening inflammation or pain. If one of these

---

[1] Opsopoeus: -πεπλασ- I.
[2] Littré after Foes' *fissura longe pertingat*: -μείνοι I.

τούτων ἐπιφαίνοιτο, ἐν ἀρχῇσί τε γίνεσθαι ἀσφα-
λέστατον, καὶ ὀλίγον χρόνον παραμένειν. ξυμφέρει
δὲ ἐν τῇσιν ὀδύνῃσι καὶ τὰς φλεγμονὰς τὰς ἐπὶ τοῖσιν ἕλκεσιν ἐπιγίνεσθαι, τῇσι δ' αἱμορραγίῃσι πύον
ἐπὶ τῇσι φλεψὶ φαίνεσθαι.

Τοῖσι δὲ πυρετοῖσιν ἃ ἐν τοῖσιν ὀξέσι νοσήμασιν
ἔγραψα ξυμφέρειν ἐπὶ τούτοισι γενέσθαι, ταῦτα καὶ
ἐνθάδε λέγω ἀγαθὰ εἶναι, τὰ δ' ἐναντία κακά.
ἄρξασθαι δὲ πυρετὸν ἐπὶ κεφαλῆς τρώσει τεταρταίῳ,
ἢ ἑβδομαίῳ, ἢ ἑνδεκαταίῳ, θανατῶδες μάλα. κρίνεται δὲ τοῖσι πλείστοισιν, ἢν μὲν τεταρταίου ἐόντος
τοῦ ἕλκεος πυρετὸς ἄρξηται, ἐς τὴν ἑνδεκάτην· ἢν δ'
40 ἑβδομαῖος ἐὼν πυρετήνῃ ἐς τὴν | τεσσαρεσκαιδεκά-
την ἢ ἑπτακαιδεκάτην· ἢν δὲ τῇ ἑνδεκάτῃ ἄρξηται
πυρεταίνειν, ἐς τὴν εἰκοστήν, ὡς ἐν τοῖσι πυρετοῖσι
διαγέγραπται τοῖς ἄνευ προφάσεων ἐμφανέων γινο-
μένοισι. τῇσι δ' ἀρχῇσι τῶν πυρετῶν ἤν τε παρα-
φροσύνη ἐπιγένηται, ἤν τε ἀπόπληξις τῶν μελέων
τινός, εἰδέναι τὸν ἄνθρωπον ἀπολλύμενον, ἢν μὴ
παντάπασιν ἢ τῶν καλλίστων τι σημεῖον ἐπιγένηται, ἢ σώματος ἀρετὴ ὑπόκειται· [ἀλλ' ὑποσκε-
πτέσθω τὸν τρόπον τῷ ἀνθρώπῳ·]¹ ἔτι γὰρ αὕτη ἡ
ἐλπὶς γίνεται σωτηρίης· χωλὸν δὲ γενέσθαι τὸ
ἄρθρον ἐς ὃ ἀπεστήριξεν, ἀναγκαῖόν ἐστιν, ἢν ἄρα
καὶ περιγένηται ὁ ἄνθρωπος.

15. Τὰ δὲ τρώματα τὰ ἐν τοῖσιν ἄρθροισι μεγάλα
μὲν ὄντα καὶ τελέως ἀποκόπτοντα τὰ νεῦρα τὰ συν-
έχοντα, εὔδηλον ὅτι χωλοὺς ἀποδείξει. εἰ δὲ ἐνδοια-

signs does appear, it is least dangerous if this is at the beginning, and lasts only a short time. If pains are present, it is better for inflammations too to supervene in the lesions, and in haemorrhages for pus to appear in the vessels.

As for the signs that I have described as being beneficial when fevers occur in acute diseases, these I hold to be good here, too, and the opposite conditions bad. For a fever to begin in head wounds on the fourth, seventh, or eleventh day is especially deadly. The crisis occurs in most cases, if the fever began on the fourth day after the wound, towards the eleventh day; if the person became febrile on the seventh day, towards the fourteenth or seventeenth; if the person first had fever on the eleventh day, then towards the twentieth day, just as was described for fevers arising without visible causes. If derangement of the mind or paralysis of some limb is present at the beginning of the fevers, know that the person will die for certain unless either one of the most favourable signs appears or particular strength of body is present [but he must examine the particular form in the person][1], for there is still this hope of salvation. If the person does survive, it is inevitable for the joint in which the disease became fixed to become lame.

15. Wounds to the joints, if they are major and completely sever the cords that are holding the joint together, will clearly result in lameness. If there is doubt about the

---

[1] Del. Littré.

στὸν εἴη ἀμφὶ τῶν νεύρων, ὅπως ἔχοι, ὀξέος μὲν
ἐόντος τοῦ βέλεος τοῦ ποιήσαντος, ὀρθὸν τὸ τρῶμα
εἰδέναι[1] ἄμεινον ἐπικαρσίου· εἰ δ' εἴη βαρύ τε καὶ
ἀμβλὺ τὸ τρῶσαν, οὐδὲν διαφέρει· ἀλλ' ἐς τὸ βάθος
τε τῆς πληγῆς σκέπτεσθαι καὶ τἆλλα σημεῖα. ἔστι
δὲ τάδε· πύον ἢν ἐπιγένηται ἐπὶ τὸ ἄρθρον, σκληρότερον ἀνάγκη γενέσθαι· ἢν δὲ καὶ οἰδήματα συμ-
παραμείνῃ, σκληρὸν ἀνάγκη πολὺν χρόνον τοῦτο τὸ
χωρίον γίνεσθαι, καὶ τὸ οἴδημα, ὑγιέος ἐόντος τοῦ
ἕλκεος, παραμένειν· <καὶ βραδέως>[2] ἀνάγκη
συγκάμπτεσθαί τε καὶ ἐκτείνεσθαι ὁκόσα ἂν τοῦ
42   ἄρθρου καμπύλου ἐόντος θεραπεύηται. | οἷσι δ' ἂν
καὶ νεῦρον δοκέει ἐκπεσεῖσθαι, ἀσφαλεστέρως τὰ
περὶ τῆς χωλώσιος ἢ προλέγειν, ἄλλως τε καὶ ἢν
τῶν κάτωθεν νεύρων ᾖ τὸ ἐκλυόμενον· γνώσῃ δὲ
τοῖσι νεῦρον μέλλον ἐκπίπτειν, πύον λευκόν τε καὶ
παχὺ καὶ πολὺν χρόνον ὑπορρεῖ· ὀδύναι τε καὶ
φλεγμοναὶ γίνονται περὶ τὸ ἄρθρον ἐν ἀρχῇσι. (τὰ
δ' αὐτὰ ταῦτα γίνεται καὶ ὀστέου μέλλοντος ἐκπεσεῖσθαι.) τὰ δὲ ἐν τοῖσιν ἀγκῶσι διακόμματα ἐν
φλεγμονῇ μάλιστα ἐόντα ἐς διαπύησιν ἀφικνέεται
καὶ τομάς τε καὶ καύσιας.

16. Ὁ δὲ νωτιαῖος μυελὸς ἢν νοσέῃ ἤν τε ἐκ πτώματος, ἤν τε ἐξ ἄλλης τινὸς προφάσιος, ἤν τε ἀπὸ
αὐτομάτου, τῶν τε σκελέων ἀκρατὴς γίνεται ὁ
ἄνθρωπος, ὥστε μηδὲ θιγγανόμενος ἐπαΐειν, καὶ τῆς
γαστρὸς καὶ τῆς κύστιος, ὥστε τοὺς μὲν πρώτους
χρόνους μήτε κόπρον μήτε οὖρον διαχωρέειν, ἢν μὴ

state of the cords, know that if the projectile that made the wound was sharp a straight wound is more favourable than one at right angles; if, however, the cause of the wound was heavy and dull, the wound's axis makes no difference. You must inspect the depth of the blow, and also consider the other signs. They are as follows: if pus forms in the joint, it will inevitably become very stiff; also, if swellings remain there, the joint must stay stiff for a long time, and the swellings continue even when the lesion has healed. It is also inevitable that flexion and extension will return only slowly if the joint has been treated in the flexed position. In cases in which the cord seems to be about to mortify and separate, it is less difficult to predict lameness, especially if it is one of the cords of the lower limb that is separating. You will know that a cord is going to separate by the following: thick white pus flows down for a long time, and pains and inflammations arise about the joint at the beginning. (These same things also occur when a bone is going to separate.) Gashes in the elbows, generally being inflamed, go on to suppuration, incisions and cauterizations.

16. If the spinal marrow ails either as the result of a fall or some other manifest cause, or spontaneously, the person loses the power over his legs, so that on being touched he does not perceive it, and over his belly and bladder, so that in the early days of the disease he passes

---

[1] Littré: εἶναι I.

[2] Littré after Linden.

πρὸς ἀνάγκην. ὅταν δὲ παλαιότερον γένηται τὸ
νόσημα, οὐκ ἐπαΐοντι τῷ ἀνθρώπῳ ἥ τε κόπρος διαχωρέει καὶ τὸ οὖρον· ἀποθνήσκει δὲ μετὰ ταῦτα οὐ
πολλῷ ὕστερον χρόνῳ.

17. Ὧν δὲ ἐμπίπλαται αἵματος ἡ φάρυγξ, πολλάκις τῆς ἡμέρης τε καὶ τῆς νυκτὸς ἑκάστης, οὔτε
κεφαλὴν προαλγήσαντι, οὔτε βηχὸς ἐχούσης, οὔτε
ἐμέοντι, οὔτε πυρετοῦ λαμβάνοντος, οὔτε ὀδύνης
ἐχούσης οὔτε τοῦ στήθεος οὔτε τοῦ μεταφρένου, τού-
του κατιδεῖν ἐς | τὰς ῥῖνας καὶ τὴν φάρυγγα· ἢ γὰρ
ἕλκος τι ἔχων φανεῖται ἐν τῷ χωρίῳ τούτῳ, ἢ
βδέλλαν.

18. Ὀφθαλμοὶ δὲ λημῶντες ἄριστα ἐπαλλάττου-
σιν, ἢν τό τε δάκρυον καὶ ἡ λήμη καὶ τὸ οἴδημα
ἄρξηται ὁμοῦ γενόμενα. ἢν δὲ τὸ μὲν δάκρυον
τῇ λήμῃ μεμιγμένον ᾖ καὶ μὴ θερμὸν ἰσχυρῶς, ἡ
δὲ λήμη λευκή τε ᾖ καὶ μαλθακή, τό τε οἴδημα
ἐλαφρόν τε καὶ λελυμένον· εἰ γὰρ οὕτω ταῦτ' ἔχοι,
ξυμπλάσσοιτ' ἂν ὀφθαλμὸς ἐς τὰς νύκτας ὥστε
ἀνώδυνος εἶναι, καὶ ἀκινδυνότατον οὕτως ἂν εἴη καὶ
ὀλιγοχρονιώτατον. εἰ δὲ τὸ δάκρυον χωρέει πολὺ
καὶ θερμὸν ξὺν ὀλιγίστῃ λήμῃ καὶ σμικρῷ οἰδήματι,
εἰ μὲν ἐκ τοῦ ἑτέρου τῶν ὀφθαλμῶν, χρόνιον μὲν
κάρτα γίνεται, ἀκίνδυνον δέ· καὶ ἀνώδυνος οὗτος ὁ
τρόπος ἐν τοῖσι μάλιστα. τὴν δὲ κρίσιν ὑποσκέπτεσθαι, τὴν μὲν πρώτην, ἐς τὰς εἴκοσιν ἡμέρας· ἢν δ'
ὑπερβάλλῃ τοῦτον τὸν χρόνον, ἐς τὰς τεσσαράκοντα
προσδέχεσθαι· ἢν δὲ μηδ' ἐν ταύτῃσι παύηται, ἐν

neither stool nor urine, unless forced. As the disease becomes older, stool and urine pass without the person's perceiving it; he dies not very long after that.

17. Persons whose throat fills with blood many times each day and night, but who have neither suffered pain in the head beforehand, nor are subject to coughing, nor vomit, nor have fever, nor have pain in the chest or back: you must look into his nostrils and throat, for it will become obvious that he has either some lesion in that spot, or a leech.

18. Blear eyes (ophthalmia catarrhalis) heal best if the tears, rheum, and swelling all commence together. If the tears are mixed with the rheum and not excessively hot, if the rheum is white and soft, and if the swelling mild and loose, fine; for in this case the eye will grow together during the nights, and so be painless, and thus the condition will be least dangerous and chronic. But if the tears flow down copious and hot, with very little rheum and not much swelling, should this be from but one of the eyes, the condition becomes very chronic, but is not dangerous; this variety is especially painless. Look for the first crisis towards the twentieth day; if it exceeds this term, expect it about the fortieth; but if the condition does not cease

τῆσιν ἑξήκοντα κρίνεται. παρὰ πάντα δὲ τὸν χρόνον
τοῦτον ἐνθυμεῖσθαι τὴν λήμην, ἢν ἐν τῷ δακρύῳ τε
μίσγηται καὶ λευκή τε καὶ μαλθακὴ γίνηται,
μάλιστα δ' ὑπὸ τοὺς χρόνους τοὺς κρισίμους· ἢν
γὰρ μέλλῃ παύεσθαι, ταῦτα ποιήσει. εἰ δὲ οἱ
ὀφθαλμοὶ ἀμφότεροι ταῦτα πάθοιεν, ἐπικινδυνότεροι
γίνονται ἑλκωθῆναι· ἡ δὲ κρίσις ἐλάσσονος χρόνου
ἔσται. λῆμαι δὲ ξηραὶ ἐπώδυνοι κάρτα, κρίνονται δὲ
ταχέως, ἢν μὴ τρῶμα λάβῃ ὁ ὀφθαλμός. οἴδημα δὲ
ἢν μέγα ᾖ, ἀνώδυνόν τε καὶ ξηρὸν ἀκίνδυνον· εἰ δὲ
εἴη ξὺν ὀδύνῃ, κακὸν μὲν ξηρὸν ἐὸν καὶ ἐπικίνδυνον
ἑλκῶσαί τε τὸν ὀφθαλμὸν καὶ ξυμφῦσαι· δεινὸν δὲ
καὶ ξὺν δακρύῳ τε ἐὸν καὶ ὀδύνῃ· εἰ γὰρ δάκρυον
χωρέει θερμὸν καὶ ἁλμυρόν, κίνδυνος τῇ τε κόρῃ
46 ἑλκωθῆναι καὶ | τοῖσι βλεφάροισιν. εἰ δὲ τὸ μὲν
οἴδημα κατασταίη, δάκρυον δὲ πολὺ ἐπιχέεται
πολὺν χρόνον, καὶ λῆμαι εἰσί, τοῖσι μὲν ἀνδράσι
βλεφάρων ἐκτροπὴν προλέγειν, τῇσι δὲ γυναιξὶ καὶ
τοῖσι παιδίοισι ἕλκωσιν καὶ τῶν βλεφάρων ἐκτρο-
πήν. ἢν δὲ λῆμαι χλωραὶ ἢ πελιδναὶ ἔωσι, καὶ
δάκρυον πολὺ καὶ θερμόν, καὶ ἐν τῇ κεφαλῇ καῦμα
ᾖ, καὶ διὰ τοῦ κροτάφου ὀδύναι ἐς τὸν ὀφθαλμὸν
καταστηρίξωσι καὶ ἀγρυπνίη τούτοισιν ἐπιγένηται,
ἕλκος ἀνάγκη γενέσθαι ἐν τῷ ὀφθαλμῷ· ἐλπὶς δὲ καὶ
ῥαγῆναι τὸ τοιοῦτον. ὠφελέει δὲ καὶ πυρετὸς ἐπιγε-
νόμενος ἢ ὀδύνη ἐς τὴν ὀφρὺν[1] στηρίξασα. προλέγειν δὲ δεῖ τούτοισι τὰ ἐσόμενα, ἔς τε τὸν χρόνον
σκεπτόμενον ἔς τε τὰ ἐκ τοῦ ὀφθαλμοῦ ῥέοντα, ἐς

even in that number of days, it has its crisis in sixty days.
Throughout that whole time pay attention to the rheum,
to see whether it is mixed into the tears and becomes
white and soft, especially during the critical times; for if
the disease is going to end, the fluid will do these things.
If both eyes suffer these things, they will be in more danger of ulcerating, and the crisis will be in a shorter time.
Dry rheums are very painful, but reach their crisis
quickly, unless the eye ulcerates. A great swelling, if painless and dry, is without danger, but if the swelling involves
pain, it is bad for it to be dry, and there is a danger that
the eye will ulcerate and grow together. It is also dangerous if the swelling is accompanied with tears and pain; for
if the tears flow hot and salty there is a danger of ulceration of the pupil and eyelids. If the swelling subsides, but
there is a copious flux of tears over a long period, and
there are rheums, predict in the case of men that there
will be an eversion of the eyelids, and, in the cases of
women and children, ulceration and an eversion of the
eyelids. If the rheums are yellow-green or livid, the tears
are copious and hot, there is fever heat in the head, pains
move through the temple to settle in an eye, and these
persons become sleepless, then an ulcer of the eye must
occur, and there is a chance that such an eye will also rupture. It is advantageous either for fever to come on, or for
the pain to settle in the eyebrow. You must predict what
is going to happen to these persons, taking into considera-
tion the duration, the fluxes from the eye, the severe

---

[1] Foes in note: ὀσφὺν I.

τὰς περιωδυνίας τε καὶ ἀγρυπνίας.

19. Ἐπὴν δὲ καὶ τὸν ὀφθαλμὸν οἷόν τε ᾖ κατιδεῖν, ἢν μὲν εὑρεθῇ ἐρρωγώς τε καὶ διὰ τῆς ῥωγμῆς ὑπερέχουσα ἡ ὄψις, πονηρὸν καὶ χαλεπὸν καθιδρύσαι· εἰ δὲ καὶ σηπεδὼν ὑπῇ τῷ τοιούτῳ, τελέως ἄχρηστος ὁ ὀφθαλμὸς γίνεται. τοὺς δ' ἄλλους τρόπους τῶν ἑλκέων ἐς τὰ χωρία σκεπτόμενον¹ προλέγειν, καὶ τάς τε σηπεδόνας καὶ βαθυτῆτας· ἀναγκαῖον <γὰρ>² κατὰ τὴν ἰσχὺν τῶν ἑλκέων τὰς οὐλὰς γίνεσθαι. οἷσι μὲν οὖν ῥήγνυνται οἱ ὀφθαλμοί, καὶ μέγα ὑπερίσχουσιν ὥστε ἔξω τὴν ὄψιν τῆς χώρης εἶναι, ἀδύνατοι ὠφελέεσθαι καὶ χρόνῳ καὶ τέχνῃ εἰς τὸ βλέπειν· τὰ δὲ σμικρὰ μετακινήματα τῶν ὄψεων οἷά τε καθιδρύεσθαι, ἢν μήτε κακὸν ἐπιγένηται μηδέν, ὅ τε ἄνθρωπος νέος ᾖ.

20. Αἱ δὲ ἐκ τῶν ἑλκέων οὐλαί, οἷσιν ἂν μὴ κακόν τι ἄλλο προσῇ, πᾶσαι οἷαί τε ὠφελέεσθαι καὶ ὑπὸ τῶν χρόνων καὶ ὑπὸ τῆς τέχνης, μάλιστα δὲ αἱ νεώταταί τε καὶ ἐν τοῖσι νεωτάτοισι τῶν σωμάτων. |

48 τῶν δὲ χωρίων μάλιστα μὲν αἱ ὄψιες βλάπτονται ἑλκούμεναι, ἔπειτα τὸ ὑπεράνω τῶν ὀφρύων, ἔπειτα δὲ καὶ ὅ τι ἂν ἄγχιστα ᾖ τούτων τῶν τόπων. αἱ δὲ κόραι γλαυκούμεναι, ἢ ἀργυροειδέες γινόμεναι, ἢ κυάνεαι, οὐδὲν χρηστόν· τούτων δὲ ὀλίγῳ³ ἀμείνους, ὁκόσαι ἢ σμικρότεραι φαίνονται, ἢ εὐρύτεραι, ἢ γωνίας ἔχουσαι, εἴτ' ἐκ προφασίων τοιαῦται γενοίατο, εἴτ' αὐτόμαται. ἀχλύες, καὶ νεφέλαι, καὶ αἰγίδες ἐκλεαίνονταί τε καὶ ἀφανίζονται, ἢν μὴ τρῶμά τι

pains, and the sleeplessness.

19. When it is possible to examine the eye, if it is found to be ruptured, and because of the rupture the iris protrudes, this is serious and difficult to restore; if a putrefaction is also present, the eye becomes completely blind. About the other kinds of ulcers you must make predictions by examining them with regard to their locations, and their suppurations and depths, for it is, of necessity, according to the severity of the ulcerations that the scars will form. Now persons in whom the eyes rupture and greatly protrude, so that the iris is outside the socket, cannot be helped to see either by time or by skill, whereas small displacements of the iris can be restored, if no other evil is added, and if the person is young.

20. Scars from ulcers, in cases where no other evil is added, can all be helped both by time and by medical treatment, especially when they are freshest and in the youngest bodies. Of all the parts, the irises are most harmed by being ulcerated, then the part above the eyebrows, and then whatever is nearest to these places. Pupils that become grey, silvery, or blue do not function. Slightly better than these are pupils that appear either smaller or wider than normal, or to have angles, whether this happens as the result of a manifest cause or spontaneously. Mistinesses, clouds and specks become thinner and disappear, unless some other injury occurs in the

---

[1] Littré: -μένους I.

[2] Foes in note after Calvus' *nam*.

[3] Kühn: ὀλίγαι I.

ΠΡΟΡΡΗΤΙΚΟΝ Β

ἐπιγένηται ἐν τούτῳ τῷ χωρίῳ, ἢ πρόσθεν τύχῃ
οὐλὴν ἔχων ἐν τῷ χωρίῳ τούτῳ, ἢ πτερύγιον. ἢν δὲ
παράλαμψις¹ γένηται <καὶ>² ἀπολευκάνῃ τοῦ μέλανος μόριόν τι, εἰ πολὺν χρόνον παραμένῃ, καὶ τρη-
χείη τε καὶ παχείη <εἴη, οἵη τε>³ καὶ μνημόσυνον
ὑποκαταλιπεῖν.

21. Αἱ δὲ κρίσιες ὡς ἐν τοῖσι πυρετοῖσιν ἔγραψα,
οὕτως καὶ ἐνθάδε ἔχουσιν. ἀλλὰ χρὴ τὰ σημεῖα
ἐκμαθόντα προλέγειν τὰς μὲν διαφορὰς τῶν ὀφθαλ-
μιῶν·⁴ ὅταν τὰ κάκιστα τῶν σημείων ἐπιγένηται, τὰς
δὲ πολυχρονίους τῶν ὀφθαλμιῶν⁴ ὡς διαγέγραπται
ἐφ’ ἑκάστῃσι, τὰς | δὲ ὀλιγοχρονίους, ὅταν τὰ
σημεῖα προφαίνηται τὰ ἄριστα, τότε προλέγειν
ἑβδομαίας παύσασθαι, ἢ ἐγγὺς τούτων, καὶ ἄλλως
ἀσφαλῶς νομίζειν ἔχειν· τὰς δὲ ὑποστροφὰς προσδέχεσθαι, οἷσιν ἂν ῥᾴστωναι γένωνται, μήτ’ ἐν ἡμέρῃσι κρισίμοισι, μήτε σημείων ἀγαθῶν ἐπιφανέν-
των. πάντων δὲ χρὴ μάλιστα τὴν κατάστασιν τοῦ
οὔρου ἐν τοῖσι περὶ τοὺς ὀφθαλμοὺς ἐνθυμεῖσθαι· οἱ
γὰρ καιροὶ ὀξέες.

22. Αἱ δὲ δυσεντερίαι ξὺν πυρετῷ μὲν ἢν ἐπίω-
σιν, ἢ ποικίλοισί τε διαχωρήμασιν, ἢ ξὺν φλεγμονῇ
ἥπατος, ἢ ὑποχονδρίου, ἢ γαστρός, ἢ ὅσαι ἐπώδυνοι, ἢ ὅσαι τῶν σιτίων ἀπολαμβάνουσι δίψαν τε
παρέχουσιν, αὗται μὲν πᾶσαι πονηραί· καὶ ὃς ἂν

¹ Opsopoeus after Cornarius’ splendicans cicatrix: -ληψις I.
² Η². ³ Littré: εἴη Opsopoeus in note.
⁴ Aldina after Calvus’ oculorum malis: ὀφθαλμῶν I.

264

part, or the person happens to have an earlier scar in the part, or a pterygium.[a] If a shiny spot occurs on the cornea and turns a little spot of the black part of the eye white, and this persists for a long time and it is rough and thick, it can also leave a mark behind.

21. The crises, as I described them occurring in fevers,[b] apply in the same way here. You must predict the different kinds of ophthalmia by investigating the signs: when the worst signs occur, predict the chronic ophthalmias as they are described in each case; predict the shorter-lasting ones when the best signs appear; then predict that the latter will cease on the seventh day, or close to that, and at all events consider them to be without danger.[c] Expect relapses in patients who experience relief that does not come on critical days, or that is not accompanied by the appearance of favourable signs. Of all signs, you must pay most attention to the condition of the urine in cases involving the eyes, for the opportune moments are fleeting.

22. Dysenteries, if accompanied by fever, by variegated stools, or by inflammation of the liver, hypochondrium or belly, or if they are painful, or spoil the appetite or provoke thirst, are all troublesome, and whichever

[a] A patch of thickened conjunctiva extending over a part of the cornea; the membrane is usually fan shaped, with the apex towards the pupil and the base towards the inner canthus.

[b] See ch. 14 above.

[c] I have followed the manuscripts here. Littré's radical rearrangement of this sentence is certainly an improvement in clarity, but it is difficult to imagine how the version in the manuscripts could have grown out of it.

265

πλεῖστα ἔχῃ τούτων τῶν κακῶν, τάχιστα ἀπολεῖται·
ᾧ δ' ἂν ἐλάχιστα τῶν τοιούτων προσῇ, πλεῖστα
αὐτῷ ἐλπίδες. ἀποθνήσκουσι δὲ ὑπὸ ταύτης τῆς
νόσου μάλιστα παιδία τὰ πενταετέα, καὶ γεραίτερα
ἔς τε τὰ δεκαετέα· αἱ δ' ἄλλαι ἡλικίαι ἧσσον. ὅσαι
δὲ τῶν δυσεντεριῶν λυσιτελέες, τὰ μὲν κακὰ ταῦτα
οὐκ ἐμποιέουσιν, αἷμα δὲ καὶ ξύσματα διαχωρή-
σαντα ἐπαύσαντο ἑβδομαῖα, ἢ τεσσαρεσκαιδεκα-
ταῖα, ἢ εἰκοσταῖα, ἢ τεσσαρακονθήμερα, ἢ ἐντὸς
τούτων τῶν χρόνων. τὰ τοιαῦτα γὰρ διαχωρήματα
καὶ ὑπάρχοντα πρόσθεν ἐν τοῖσι σώμασι νοσήματα
ὑγιάζει, τὰ μὲν παλαιότερα ἐν[1] πλείονι χρόνῳ, τὰ δὲ
νεώτερα δύναται καὶ ἐν ὀλίγῃσιν ἡμέρῃσιν ἀπαλ-
λάσσειν. ἔτι τε αἱ[2] ἐν γαστρὶ ἔχουσαι καὶ αὗται
περιγίνονται, μᾶλλον ἐς τοὺς τόκους τε καὶ ἐκ τῶν
τόκων, καὶ τὰ ἔμβρυα διασῴζουσιν, αἵματός τε καὶ
ξύσματος διαχωρεόντων καὶ πολλοὺς μῆνας, ἢν μή
τις ὀδύνη αὐτῇσι προσῇ, ἢ[3] ἄλλο τι τῶν πονηρῶν ὧν
ἔγραψα σημείων εἶναι ἐν τῇσι δυσεντερίῃσιν ἐπι-
γένηται· εἰ δέ τι κείνων ἐπιφαίνοιτο, τῷ τε ἐμβρύῳ
52 ὄλεθρον σημαίνει καὶ τῇ | ἐχούσῃ κίνδυνον, ἢν μὴ
μετὰ τοῦ ἐμβρύου τὴν ἀπόφευξιν καὶ τοῦ ὑστέρου
τὴν ἀπόλυσιν ἡ δυσεντερίη παύσηται αὐθημερόν, ἢ
μετ' ὀλίγον χρόνον.

23. Αἱ δὲ λειεντερίαι ξυνεχέες μὲν καὶ πολυχρόνιοι καὶ πᾶσαν ὥρην ξὺν ψόφοισί τε καὶ ἄνευ ψόφων
ἐκταρασσόμεναι, καὶ ὁμοίως νυκτός τε καὶ ἡμέρης
ἐπικείμεναι, καὶ τοῦ διαχωρήματος ὑπιόντος ἢ ὠμοῦ

patient has most of these evils dies the soonest, whereas
the one who has least of such things has the greatest
chances of recovery. It is mostly children of five years
that die from this disease, and also older ones up to ten
years; other ages less. Dysenteries that are favourable do
not produce the evils mentioned above, but blood and
shreds of flesh passing off in the stools end them on the
seventh, fourteenth, twentieth or fortieth day, or within
that period; for such evacuations can also cure diseases
that have existed previously in the body, more long standing ones in a longer period,[1] whereas more recent ones can
end in even a few days. And besides, pregnant women
too survive the disease, more so towards the time of their
parturition, and afterwards, and they also save their
fetuses—with blood and shreds of flesh passing off below
sometimes for many months—if no pain befalls them, or
any other of the pernicious signs I described as accompa-
nying dysenteries. But if one of these signs should
appear, it signifies death to the fetus and danger to the
mother, unless,[2] after the exit of the fetus and the release
of the afterbirth, the dysentery stops on the same day or
in a short time.[3]

23. Lienteries that are continuous and of long duration, that are stirred up, with or without noises, at every
hour, that attack alike both by night and by day, and in
which the stool evacuated is either very undigested or

---

[1] Foes in note after Cornarius' *diutiore tempore*: ἦν I.

[2] Potter: ἐπεὶ καὶ I.

[3] προσῇ, ἢ Froben: προσῆν I.

ἰσχυρῶς, ἢ μέλανός τε καὶ λείου καὶ δυσώδεος,
αὗται μὲν πᾶσαι πονηραί. καὶ γὰρ δίψαν παρέχουσι, καὶ τὸ ποτὸν οὐκ ἐς τὴν κύστιν τρέπουσιν
ὥστε διουρέεσθαι, καὶ τὸ στόμα ἐξελκοῦσι, καὶ
ἔρευθος ἐξηρμένον ἐπὶ τῷ προσώπῳ ποιέουσι καὶ
ἐφήλιδας παντοῖα χρώματα ἐχούσας· ἅμα δὲ καὶ
τὰς γαστέρας ἀποζύμους τε καὶ ῥυπαρὰς ἀποδεικνύ-
ουσι καὶ ῥυτιδώδεας. ἐκ δὲ τῶν τοιούτων ἐσθίειν τε
ἀδύνατοι γίνονται οἱ ἄνθρωποι, καὶ τῇσι περιόδοισι
χρῆσθαι, καὶ τἆλλα τὰ πρασσόμενα ποιέειν. τὸ δὲ
νούσημα τοῦτο δεινότατον μὲν τοῖσι πρεσβυτέροι-
σιν, ἰσχυρὸν δὲ γίνεται καὶ τοῖσιν ἀνδράσι, τῇσι δὲ
ἄλλῃσιν ἡλικίῃσι πολλῷ ἧσσον. ὅστις δὲ μήτε ἐν
τῇσιν ἡλικίῃσίν ἐστι ταύτῃσιν ᾗσι φημι ὑπὸ τοῦ
νούσηματος τούτου κακῶς περιέπεσθαι, ἐλάχιστά τε
τῶν σημείων ἔχει τῶν πονηρῶν ἃ ἔγραψα εἶναι,
ἀσφαλέστατα διάκειται οὗτος. θεραπείης δὲ προσ-
δεῖται ἡ νοῦσος αὕτη, ἔστ᾽ ἂν τό τε οὖρον χωρέηται
τοῦ πινομένου κατὰ λόγον, καὶ τὸ σῶμα τῶν σιτίων
εἰσιόντων αὔξηται, καὶ τῶν χροιῶν τῶν πονηρῶν
ἀπαλλαχθῇ. αἱ δὲ ἄλλαι διάρροιαι ὅσαι ἄνευ πυρετῶν, [καὶ]¹ ὀλιγοχρόνιοί τε καὶ εὐήθεες· ἢ γὰρ κατα-
νιφθεῖσαι πεπαύσονται, ἢ ἀπὸ τοῦ αὐτομάτου. προ-
αγορεύειν² δὲ χρὴ παύεσθαι τὴν ὑπέξοδον, ὅταν τῇ |
54 τε χειρὶ ψαύοντι τῆς γαστρὸς μηδεμία κίνησις ὑπῇ
καὶ φῦσα διέλθῃ ἐπὶ τελευτῇ τοῦ διαχωρήματος.
ἕδραι δὲ ἐκτρέπονται, ἀνδράσι μὲν οὓς ἂν διάρροια
λάβοι ἔχοντας αἱμορροΐδας, παιδίοισι δὲ λιθιῶσί τε

dark, thin and evil-smelling, are all difficult. For they provoke thirst, they do not direct what is drunk to the bladder so that it can be passed through as urine, they make the mouth ulcerate, and they cause raised red patches on the face and many-coloured spots; at the same time they produce fermentation in the belly, and render its insides greasy and wrinkled. As a result of these symptoms, patients become unable to eat, to employ walks, or to carry out the other things they do. This disease is most dangerous in older people, also occurs forcefully in adults, but at other ages of life much less so. Anyone who is not in the age groups which I say are worst affected by the disease, and who has least of the signs I indicated as being pernicious, is safest. This disease requires treatment until the urine is passed in due proportion to what is drunk, until the body grows in proportion to the foods that are ingested, and until the person is freed from the skin disorders. Other diarrhoeas in which there is no fever are of short duration and mild; for they cease either on the administration of an enema, or spontaneously. You must predict that diarrhoea will cease when, on palpating the belly with your hand, there is no movement in it, and when wind is passed at the end of the evacuation. The anus prolapses in adults who get diarrhoea when they have haemorrhoids, in children who have urinary

---

[1] Del. Littré.

[2] Littré: προσαγ- I.

καὶ ἐν τῆσι δυσεντερίησι τῆσι μακρῆσί τε καὶ ἀκρί-
τοισι, πρεσβυτάτοισί τε οἷς ἂν προσπήγματα μύξῃς
ἐνῇ.

24. Τῶν δὲ γυναικῶν ὅσαι μᾶλλον καὶ ἧσσον ἐν
γαστρὶ λαμβάνειν πεφύκασιν, ὧδε ὑποσκέπτεσθαι·
πρῶτον μὲν τὰ εἴδεα· σμικραί τε γὰρ μειζόνων ἀμείνονες ξυλλαμβάνειν, λεπταὶ παχειῶν, λευκαὶ ἐρυ-
θρῶν, μέλαιναι πελιδνῶν, φλέβας ὅσαι ἐμφανέας
ἔχουσιν, ἀμείνονες ἢ ὅσῃσι μὴ καταφαίνονται·
σάρκα δὲ ἐπίθρεπτον ἔχειν πρεσβυτικῇ πονηρόν,
μαζοὺς δὲ ὀγκηρούς τε καὶ μεγάλους ἀγαθόν. ταῦτα
μὲν τῇ πρώτῃ[1] ὄψει δῆλά ἐστι. πυνθάνεσθαι <δὲ>[2]
χρὴ καὶ περὶ καταμηνίων, ἢν πάντας μῆνας φαίνωνται, καὶ ἢν πλῆθος ἱκανά, καὶ ἢν εὔχροά τε καὶ ἴσα
ἐν ἑκάστοισι τῶν χρόνων, καὶ ἢν τῆσιν αὐτῆσιν
ἡμέρῃσι τῶν μηνῶν· οὕτω γὰρ ταῦτα γίνεσθαι ἄριστον. τὸ δὲ χωρίον ἐν ᾧ ἡ σύλληψίς ἐστιν, ὃ δὴ
μήτρην ὀνομάζομεν, ὑγιές τε χρὴ εἶναι καὶ ξηρὸν
καὶ μαλθακόν, καὶ μήτ' ἀνεσπασμένον ἔστω μήτε
προπετές, μήτε τὸ στόμα αὐτοῦ ἀπεστράφθω μήτε
ξυμμεμυκέτω μήτ' ἐκπεπλίχθω·[3] ἀμήχανον γάρ, ὅ τι
ἂν ᾖ τῶν τοιούτων κωλυμάτων, σύλληψιν γενέσθαι.

Ὁκόσαι μὲν οὖν τῶν γυναικῶν μὴ δύνανται ἐν
γαστρὶ λαμβάνειν, φαίνονται δὲ χλωραί, μήτε πυρετοῦ μήτε τῶν σπλάγχνων αἰτίων ἐόντων, αὗται
φήσουσι κεφαλὴν ἀλγέειν, καὶ τὰ καταμήνια πονη-
ρῶς τε σφίσι καὶ ἀκρίτως γίνεσθαι καὶ ὀλίγως·
<ἢν>[4] δὲ | καὶ πολὺν χρόνον αὐτῇσιν οὕτως διακει-

56

stones or dysenteries that are long and do not come to a crisis, and in very old people in whom clots of phlegm are present.

24. Which women are by nature more, and which less, inclined to become pregnant you must investigate as follows. First, their physical appearance: small women conceive better than larger ones, thin better than obese, pale better than ruddy, dark better than livid, and those with visible vessels better than those in whom the vessels are not visible. For an old woman to have well-nourished flesh is pernicious, but to have large and bulky breasts good. These things are clear at first sight. You must also find out about the menses, whether they appear every month, whether they are adequate in amount, whether they flow freely and equally every time, and whether on the same days of the month; for it is best if this is the case. The part in which the conception takes place, which we call the "mother",[a] must be healthy, dry and soft, and neither retracted, nor prolapsed, nor have its mouth turned away, nor closed, nor gaping; for it is impossible, if any of such impediments be present, for conception to occur.

Women who are unable to become pregnant, but take on a yellow-greenish appearance, without there being any fever or reason in the inward parts for it, say that they have headaches, and that their menses are troublesome, irregular and scanty; if, in women who remain in this state

[a] A usage of "mother" now obsolete in English; cf. O.E.D. s.v. III.

---

[1] Aldina; $\pi\rho\grave{o}\varsigma$ $\tau\hat{\eta}$ I. [2] Linden.
[3] Littré: $-\pi\lambda\acute{\eta}\chi\theta\omega$ I. [4] Potter.

μένῃσιν ἀφανέα ᾖ, αἱ μῆτραι καθάρσιος ταύτῃσι
προσχρῄζουσιν. ὁκόσαι δὲ[1] εὔχροοί τέ εἰσι καὶ
σάρκα πολλήν τε καὶ πίειραν ἔχουσι, καὶ φλέβια
κεκρυμμένα, ἀνώδυνοί τέ εἰσι καὶ τὰ καταμήνια ταύτῃσιν ἢ παντάπασιν οὐ φαίνεται, ἢ ὀλίγα τε καὶ
ἀκρίτως γίνεται, τῶν τρόπων οὗτος ἐν τοῖσι χαλεπω-
τάτοισίν ἐστι καταναγκάσαι ὡς ἐν γαστρὶ λαμβάνειν. ἢν δέ, ἐπιφαινομένων τῶν καταμηνίων ἀπρο-
φασίστως, τό τε σῶμα ὧδε διάκειται ἡ γυνὴ καὶ μὴ
συλλαμβάνῃ, [τὸ χωρίον ἐν ᾧ] ἡ μήτρη αἴτιος,[2]
ὥστε μὴ δύνασθαι γίνεσθαι ἔκγονα· ἢ γὰρ ἀνεσπασμένον ἐστί, ἢ ἐκπεπλιγμένον· τὰ γὰρ ἄλλα
κακὰ γινόμενα ἐνταῦθα ξὺν ὀδύνῃσί τε γίνεται καὶ
ἀχροίῃσί τε καὶ τήξει.

Ἦισι δ' ἂν ἕλκος γένηται ἐν τῇσι μήτρῃσιν, εἴτε
ἐκ τόκου, εἴτε ἐκ φύματος, εἴτε ἐξ ἄλλης τινὸς προφάσιος, πυρετούς τε καὶ βουβῶνας ταύτῃσιν ἀνάγκη
ἐπιγίνεσθαι καὶ ὀδύνας ἐν τοῖσι χωρίοισι τούτοισιν.
εἰ δὲ καὶ τὰ λοχεῖα συναποληφθείη, ταῦτα ὑπάρχοντα κακὰ ταύτῃσιν[3] ἀκριτότερά τε καὶ χρονιώ-
τερα· καὶ πρὸς τούτοισιν ὑποχονδρίων τε καὶ κεφαλῆς ὀδύναι. ἕλκεος δὲ γενομένου καὶ ἐξυγιασθέντος,
τὸ χωρίον τοῦτο ἀνάγκη λειότερον καὶ σκληρότερον
γίνεσθαι, καὶ ἧσσον δύνασθαι ἐν γαστρὶ λαμβάνειν. εἰ δὲ μοῦνον ἐν τοῖσιν ἐπ' ἀριστερὰ γένοιτο
ἕλκος, ἡ δὲ γυνὴ ἐν γαστρὶ λάβοι, εἴτε τὸ ἕλκος ἔτι
ἔχουσα, εἴτε λοιπὸν ἤδη ὑγιὴς ἐοῦσα, ἄρσεν μᾶλλον
τεκεῖν αὐτὴν | ἐλπίς ἐστιν· εἰ δὲ ἐν τοῖσιν ἐπὶ δεξιὰ

58

for a long time, the menses disappear completely, the uterus requires a cleaning. Women who are of good colour, who are fleshy and plump with their vessels hidden, who are free of pains, but whose menses either do not appear at all or are scanty and irregular: this is one of the most difficult of types in which to force pregnancy to occur. If, on the other hand, with the menses appearing unhesitatingly, the body is in such a state and the woman does not conceive, the uterus is to blame that offspring cannot grow—for it must be either retracted or gaping, since the other evils that arise in it are accompanied by pains, paleness and melting of the tissue.

Women who have a lesion in the uterus, either as the result of parturition, of a tubercle, or of some other manifest cause, inevitably suffer fever, swellings in the groins, and pains in these parts. If in addition the lochia cease, these evils are in these women less likely to come to a crisis, but last longer, and besides pains occur in the hypochondria and head. When a lesion that was present has healed completely, the uterus inevitably becomes smoother and harder, and the woman less able to become pregnant. If the lesion involves only the left-handed parts, and the woman becomes pregnant—either still with the ulcer, or after it has already healed—there is a greater chance that she will bring forth a male child. But if the lesion involves the right-handed parts, and she

---

[1] Linden: τε I.

[2] Potter: τὸ χωρίον [ἐν ᾧ ἡ μήτρη] αἴτιον Littré. τὸ χωρίον ἐν ᾧ (sc. σύλληψις) is a gloss on μήτρη.

[3] Potter: πάντη ἂν I.

τὸ ἕλκος γένοιτο, ἡ δὲ γυνὴ ἐν γαστρὶ ἔχοι, θῆλυ
μᾶλλον τὸ ἔκγονον χρὴ δοκέειν ἔσεσθαι.

25. Ἢν δὲ πυρετοὶ γένωνται οὐ δυναμένῃ ἐν
γαστρὶ λαβεῖν, καὶ λεπτῆς τῆς γυναικὸς[1] ἐούσης,
πυνθάνεσθαι χρὴ μή τι αἱ μῆτραι ἕλκος ἔχουσιν, ἢ
ἄλλο τι τῶν πονηρῶν ὧν ἔγραψα· εἰ γὰρ ἐν τῷ
χωρίῳ τούτῳ μηδὲν ὑπεὸν κακὸν φαίνοιτο αἴτιον τῆς
λεπτύνσιός τε καὶ τοῦ μὴ συλλαμβάνειν δύνασθαι,
αἷμα ἐμέσαι τὴν γυναῖκα προσδόκιμον· τὰ δὲ καταμήνια τῇ τοιαύτῃ ἠφανίσθαι ἀνάγκῃ· ἢν δὲ ὁ πυρετὸς λυθῇ ὑπὸ τῆς ῥήξιος τοῦ αἵματος, καὶ τά τε
καταμήνια φανῇ, ἐν γαστρὶ λήψεται· ἢν δὲ τὰ τῆς
γαστρὸς πρὶν ἢ τὸ αἷμα ἀναρραγῆναι ὑγρὰ γένηται
πονηρὸν τρόπον, κίνδυνος ἀπολέσθαι τὴν γυναῖκα
ἔμπροσθεν ἢ τὸ αἷμα ἐμέσαι.

26. Ὁπόσαι δὲ ἐν γαστρὶ δοκέουσιν ἔχειν, οὐκ
ἔχουσαι, καὶ πολλοὺς μῆνας ἐξαπατῶνται, τῶν
καταμηνίων οὐ φαινομένων, καὶ τὰς γαστέρας ὁρῶ-
σιν αὐξανομένας τε καὶ κινεομένας, αὗται κεφαλήν
τε ἀλγέουσι καὶ τράχηλον καὶ ὑποχόνδρια· καὶ ἐν
τοῖσι τιτθοῖσι γάλα οὐκ ἐγγίνεται σφίσιν, εἰ μὴ
ὀλίγον τι καὶ ὑδαρές. ἐπὴν δὲ τὸ κύρτωμα τὸ τῆς
γαστρὸς ἀπολυθῇ, καὶ λαπαραὶ γένωνται, αὗται ἐν
γαστρὶ λήψονται, ἢν μή τι ἄλλο κώλυμα γένηται
σφίσιν· ἐπεὶ τὸ πάθος γε τοῦτο ἀγαθόν ἐστι μετα-
βολὴν ποιῆσαι ἐν τῇ ὑστέρῃ, ὥστε μετὰ τοῦτον τὸν
χρόνον ἐν γαστρὶ λαμβάνειν. τῇσι δ᾽ ἐχούσῃσιν ἐν
γαστρὶ τὰ ἀλγήματα ταῦτα οὐ γίνεται, ἢν μὴ ξυνή-

becomes pregnant, you must assume that the offspring is more likely to be female.

25. If fevers come on in a woman who is unable to become pregnant and who is thin, you must check whether the uterus does not have an ulceration in it, or any other of the disorders which I have described. For if no evil present in that part exists which appears to be to blame for the thinness and inability to become pregnant, expect that the woman will vomit up blood; in such a woman the menses must also disappear. If her fever is resolved by blood breaking out, and the menses reappear, the woman will become pregnant. But if the contents of the belly become loose in a pernicious manner before the blood breaks out upwards, there is a danger that the woman will die before she vomits the blood.

26. Women who think they are pregnant when they are not, who are deceived for many months when their menses do not appear, and who see their abdomens increase and quicken, suffer pains in their head, neck and hypochondrium, and no milk forms in their breasts, or at most a little, and that watery. When the swelling of their belly goes down and they become thin, these women do become pregnant, if no other impediment intervenes in them, since this affection is effective in bringing about such a change in their uterus that they later become pregnant. In truly pregnant women, such pains do not occur,

---

[1] τῆς γυναικὸς Littré after Cornarius' *attenuata sit muliercula*: βηχὸς I.

θεα ἔωσι [ταύτῃσι αἱ κεφαλαλγίαι][1] καὶ γάλα ἐν
τοῖσι τιτθοῖσιν ἐγγίνεται.

60   27. Τὰς δὲ ὑπὸ τῶν ῥόων τῶν πολυχρονίων ἐχο-
μένας ἐρωτᾶν, εἰ κεφαλὴν ἀλγέουσι καὶ ὀσφὺν καὶ
τὸ κάτω τῆς γαστρός· ἐρέσθαι δὲ καὶ περὶ αἱμωδίας,
καὶ ἀμβλυωσμοῦ, καὶ ἤχων.

28. Ὁκόσαι δὲ νήστιες ἐοῦσαι ὑπόχολα ἐμέουσι
πολλὰς ἡμέρας, μήτε ἐν γαστρὶ ἔχουσαι μήτε πυρε-
ταίνουσαι, πυνθάνεσθαι ἔλμινθας στρογγύλας εἰ
ξυνεμέουσιν· ἢν γὰρ μὴ ὁμολογέωσι, προλέγειν
αὐτῇσι τοῦτο ἔσεσθαι· γίνεται δὲ μάλιστα μὲν τῇσι
γυναιξὶ τὸ νόσημα τοῦτο, ἔπειτα δὲ καὶ παρθένοις,
τοῖσι δ' ἄλλοισιν ἀνθρώποισιν ἧσσον.

29. Ὅσαι[2] δὲ ἄνευ πυρετῶν ὀδύναι γίνονται,
θανάτους μὲν οὐκ ἐξεργάζονται, πολυχρόνιοι δὲ αἱ
πλείους εἰσὶ καὶ πολλὰς μεταστάσιας ἔχουσι καὶ
ὑποστροφάς.

30. Οἱ δὲ τρόποι πρῶτον μὲν τῶν περὶ τὴν κεφαλὴν ἀλγημάτων, τὰ μὲν εὐήθεα, τὰ δὲ χαλεπώτερα
πολλῷ. χρὴ δὲ ὑποσκέπτεσθαι ἑκάτερα αὐτῶν ὧδε·
ὁπόσοι δὲ αὐτῶν ἀμβλυώσσουσι καὶ ἔρευθός τι
ἔχουσιν ἐπὶ τῶν ὀφθαλμῶν [αὐτοῖσι γίνεται],[1] καὶ
κνησμὸς ἔχει τὸ μέτωπον, τούτοισιν ἀρήγει αἷμα
ῥυὲν ἀπὸ τοῦ αὐτομάτου καὶ ἐξ ἀνάγκης· ἁπλοῦς
οὗτος ὁ τρόπος. οἷσι δὲ ὀδύναι περὶ τὴν κεφαλὴν
καὶ τὸ μέτωπον ἔκ τε τῶν ἀνέμων τῶν μεγάλων
γίνονται καὶ ἐκ τῶν ψυχέων ὅταν θαλφθῶσιν ἰσχυρῶς, τούτους δὲ κόρυζαι μὲν τέλειαι μάλιστα ἀπαλ-

unless they have been habitually present before, and milk does form in their breasts.

27. Women who are suffering from chronic fluxes you must ask whether they have pains in their head, loins, and the lower part of their belly; also question them as to whether they feel their teeth set on edge, and about dimness of vision and ringing in their ears.

28. Women who in the fasting state vomit somewhat bilious material for many days, and are neither pregnant nor febrile: find out whether together with this they also vomit up round worms; if they do not answer in the affirmative, predict that they will do so. This disease occurs mainly in women, next in girls, but less in others.

29. Pains that arise without fevers do not cause deaths, but the greater part of them are of long duration and have frequent alterations and relapses.

30. First, types of headaches: the one kind is benign, while the other is much more difficult. You must distinguish between patients with headaches as follows: those who have dullness of vision and redness of the eyes, in conjunction with itching on the forehead, are helped by a haemorrhage, whether spontaneous or provoked: this type is straightforward. But persons in whom pains arise in the head and forehead as the result of violent winds, or from coldness when they have been strongly heated, are usually relieved by fully developed coryzas; also of benefit

---

[1] Del. Opsopoeus in note.

[2] Littré after Foes' *Dolores qui . . . invadunt*: Ὅσαις I.

λάσσουσιν, ὠφελέουσι δὲ καὶ πταρμοί, καὶ βλένναι
62 ἐν τῆσι ῥισὶ γινό|μεναι, μᾶλλον μὲν ἀπὸ τοῦ αὐτο-
μάτου, εἰ δὲ μή, ἐξ ἀνάγκης. <αἱ δὲ> κόρυζαι γενόμεναι[1] τελέως, ὥστε καὶ βῆχας ἐπιγενέσθαι, οἵ τε
πταρμοὶ ἐπιγενόμενοι, τὰς ὀδύνας ἢν μὴ παύωσι,
φύματα ἀνάγκη ἐπιγενέσθαι καὶ ἀχροίας τούτοισιν.

Ὁκόσοισι δὲ ὀδύναι ἄνευ προφάσεών τε γίνονται
καὶ πολυχρόνιοι καὶ ἐν πάσῃ τῇ κεφαλῇ ἰσχνοῖσί τε
ἐοῦσι καὶ ἀμενηνοῖσι, προορᾶσθαι τούτοισι τὸ
νόσημα πολλῷ χαλεπώτερον τοῦ πρόσθεν· ἢν δὲ καὶ
ἐς τὸν τράχηλόν τε καὶ ἐς τὸν νῶτον ἡ ὀδύνη κατα-
βαίνῃ τὴν κεφαλὴν ἀπολιποῦσα, καὶ αὖθις παλιν-
δρομέῃ ἐς τὴν κεφαλήν, καὶ ἔτι χαλεπώτερον γίνε-
ται· τούτων δὲ πάντων δεινότατον, εἰ ξυντείνοι ἐκ τῆς
κεφαλῆς ἐς τὸν τράχηλόν τε καὶ τὸν νῶτον. τὰς δὲ
ὠφελείας τούτοισι προσδέχεσθαι ἐξ ἀποστασίων
ἔσεσθαι, ἢ πῦον βήξασιν, ἢ αἱμορροΐδας ἔχουσιν, ἢ
ἐξανθήματα ἐν τοῖς σώμασι· λυσιτελέει δὲ καὶ πιτυ-
ρωθεῖσα ἡ κεφαλή. νάρκαι δὲ καὶ κνιδώσιες οἷσι
διὰ τῆς κεφαλῆς διαΐσσουσι, τοτὲ μὲν διὰ πάσης,
τοτὲ δὲ διὰ μέρους τινός, πολλάκις δὲ καὶ ψυχρόν τι
δοκέει αὐτοῖσι [εἰ] διαχωρέειν[2] διὰ τῆς κεφαλῆς,
τούτους ἐπανερέσθαι, εἰ καὶ ἐς τὴν γλῶσσαν ἄκρην
ἀφικνεῖται ἡ κνίδωσις· εἰ γὰρ τοῦτο ποιέοι, τέλεον
τὸ νόσημα γίνεται, καὶ χαλεπώτερον ἀπαλλάξαι,
εὐπετὲς δὲ ἄνευ τούτου. οἱ δὲ τρόποι τῶν ὠφελειῶν
ἐξ ἀποστασίων ὥσπερ προγέγραπται· ἧσσον μέντοι
ἐπιγίνονται ἀποστάσιες ἐπὶ τούτοισιν ἢ ἐκείνοισιν.

are sneezing, and for mucus to be formed in the nostrils, especially if this occurs spontaneously, but if not, then also if it is provoked. When the coryzas become fully developed so that coughs too supervene, and sneezing, if these do not bring the pains to an end, then tubercles inevitably follow in such patients, and paleness.

Persons in whom pains that arise without manifest causes are of long duration and occupy the whole head, while they themselves become thin and feeble, you must recognize to be suffering from a much more difficult disease than the one above. If the pain leaves the head and descends into the neck and back, and then returns to the head again, it becomes even more severe. The most dangerous of all these cases is if the pain extends simultaneously from the head into the neck and back. You may expect relief in these patients to be through an apostasis: they either cough up pus, or have haemorrhoids, or eruptions on their bodies; it is also beneficial if their head is affected with pityriasis. Patients in whom numbness and itching spread over the head—sometimes over the whole of it, sometimes over some particular part—and in whom some sort of coldness often seems to pass over the head, you should ask whether the itchiness has reached the tip of their tongue, for if it does this, the disease becomes full-blown, and is more difficult to escape, whereas if it does not, the case is favourable. The forms of relief are through apostases, as recorded above; however, apostases occur less in these latter cases than in the former ones.

---

[1] Littré: κόρυζαι γίνονται I.

[2] Opsopoeus in note: εἰ διαχωρέει I.

ὁκόσους δὲ ξὺν τῆσιν ὀδύνησιν σκοτόδινοι λαμβά-
64 νουσι, | δυσαπάλλακτον καὶ μανικόν· γέρουσι δὲ ὁ
τρόπος οὗτος μάλιστα γίνεται. αἱ δὲ ἄλλαι νόσοι αἱ
ἀμφὶ κεφαλὰς ἀνδράσι τε καὶ γυναιξὶν ἀσφαλῶς
ἰσχυρόταται καὶ πολυχρονιώτεραι· γίνονται δὲ καὶ
νεανίσκοισί τε καὶ παρθένοισι τῆσιν ἐν ἡλικίῃ, καὶ
μάλιστα τῶν καταμηνίων ἐς τὴν πρόοδον. τῆσι δὲ
γυναιξὶν ἐν τῆσι κεφαλαλγίῃσι τὰ μὲν ἄλλα πάντα
γίνεται ἃ καὶ τοῖσιν ἀνδράσιν· αἱ κνιδώσιες δὲ καὶ
τὰ μελαγχολικὰ ταύτῃσιν ἧσσον ἢ τοῖσιν ἀνδράσιν,
ἢν μὴ τὰ καταμήνια τελέως ἠφανισμένα ᾖ.

31. Οἷσι δὲ τὰ χρώματα νέοις ἐοῦσι πονηρά ἐστι
πολὺν χρόνον, ξυνεχέως δὲ μὴ ἰκτεριώδεα τρόπον,
οὗτοι καὶ τῶν ἀνδρῶν καὶ τῶν γυναικῶν κεφαλὴν
ἀλγέουσι, καὶ λίθους τε καὶ γῆν τρώγουσι, καὶ
αἱμορροΐδας ἔχουσιν. τὰ δὲ χλωρὰ χρώματα ὅσα
χρόνιά εἰσι, καὶ μὴ ἰσχυροὶ ἴκτεροί εἰσι, τὰ μὲν
ἄλλα τὰ αὐτὰ ποιέειν αὐτοῖσι ξυμβαίνει, ἀντὶ δὲ τῶν
λίθων τε καὶ τῆς γῆς τρώξιος αὐτούς[1] τὰ ὑποχόνδρια λυπέει μᾶλλον ἢ τοὺς ἑτέρους.

32. Ὁκόσοι δὲ πολὺν χρόνον ὠχροὶ φαίνονται,
καὶ τὰ πρόσωπα ἐπηρμένα ἔχοντες, εἰδέναι χρὴ
τούτους τὴν κεφαλὴν ὀδυνωμένους, ἢ περὶ τὰ
σπλάγχνα ἀλγήματα ἔχοντας, ἢ ἐν τῇ ἕδρῃ κακόν
τι φαινόμενον σφίσι.[2] τοῖσι δὲ πλείστοισι τῶν τοιούτων [φαινομένων][3] οὐχ ἕν τι τούτων τῶν κακῶν
φαίνεται, ἀλλ' ἔστιν ὅτε πολλὰ ἢ καὶ πάντα.

33. Οἱ δὲ τῆς νυκτὸς ὁρῶντες, οὓς δὴ νυκτάλωπας

People who become dizzy at the same time as the pains escape only with difficulty, and tend to madness; this sort of thing occurs most in the elderly. The other diseases of the head in men and women are invariably long and very severe; these also occur in youths and in girls at womanhood, especially towards the onset of the menses. In women with headaches everything else is the same as in men, except that they have less itchiness and melancholy than men, unless their menses are completely absent.

31. People who have poor colouring over a long period during their youth, but never of an icteric hue, these, both men and women, suffer headaches, eat stones and earth, and have haemorrhoids. A chronic yellow-greenish colouring, if without definite icterus, produces the same syndrome, except that instead of eating stones and earth these patients suffer more pain in the hypochondria than the others.

32. Persons who have a pale-yellow appearance over a long period, together with a swollen face, you must recognize as suffering pains in the head or about the inward parts, or some disorder arising in their seat. In the majority of these cases you find not just one of the evils, but sometimes many or even all of them.

33. Those who see better at night, whom we call nyc-

---

[1] Potter: πρὸς I: del. Froben.

[2] φαινόμενον σφίσι Littré: ἐν ἑωυτοῖσι I.

[3] Del. Littré.

καλέομεν, οὗτοι ἁλίσκονται ὑπὸ τοῦ νουσήματος
66 νέοι, ἢ παῖδες ἢ καὶ νεανίσκοι· καὶ | ἀπαλλάσσονται
ὑπὸ τοῦ αὐτομάτου, οἱ μὲν τεσσαρακονθήμεροι, οἱ δὲ
ἑπτάμηνοι. τισὶ δὲ καὶ ἐνιαυτὸν ὅλον παρέμεινε.
σημαίνεσθαι χρὴ περὶ τοῦ χρόνου ἔς τε τὴν ἰσχὺν
τοῦ νουσήματος ὁρῶντα ἔς τε τὴν ἡλικίην τοῦ νοσέοντος. αἱ δὲ ἀποστάσιες ὠφελέουσι μὲν τούτους
ἐπιφαινόμεναί τε καὶ ἐς τὰ κάτω ῥέπουσαι,[1] ἐπιγίνονται δὲ οὐ κάρτα διὰ τὴν νεότητα. αἱ δὲ γυναῖκες
οὐχ ἁλίσκονται ὑπὸ τοῦ νοσήματος τούτου, οὐδὲ
παρθένοι ᾗσι τὰ ἐπιμήνια φαίνονται.

34. Οἷσι δὲ ῥεύματα δακρύων πολυχρόνια ἢ
νυκτάλωπες γίνονται, τούτους ἐπανερωτᾶν, ἢν τὴν
κεφαλήν τι προηλγηκότες ἔωσι πρὸ τῶν ἀποστηριγ-
μάτων τούτων.

35. Ὁκόσοι δὲ μήτε πυρετήναντες μήτε ἄχροοι
ἐόντες ἀλγέουσι πολλάκις τήν τε κορυφὴν καὶ τοὺς
κροτάφους, ἢν μή τινα[2] ἄλλην φανερὴν ἔχωσιν ἀπόστασιν ἐν τῷ προσώπῳ, ἢ βαρὺ φθέγγωνται, ἢ
ὀδόντας ἀλγέωσι, τούτοισιν αἷμα ῥαγῆναι διὰ τῶν
ῥινῶν προσδέχεσθαι. οἷσι δὲ ἐκ τῶν ῥινῶν αἷμα
ῥεῖ, δοκέουσιν οἷδ᾿ ὑγιαίνειν τἄλλα, τούτους δὲ ἢ
σπλῆνα εὑρήσεις ἐπηρμένον ἔχοντας, ἢ τὴν κεφαλὴν ἀλγέοντάς τε καὶ μαρμαρυγῶδές τι πρὸ τῶν
ὀφθαλμῶν φαινόμενον σφίσι. τοῖσι δὲ πλείστοισι
τῶν τοιούτων ἅμα καὶ τὰ ἀπὸ τῆς κεφαλῆς οὕτως
ἔχοντα φαίνεται καὶ τὰ ἀπὸ τοῦ σπληνός.

36. Οὖλα δὲ πονηρὰ καὶ στόματα δυσώδεα οἷσι

talopes, are taken by the disease young, either as children
or also as young men; they escape from it spontaneously,
some on the fortieth day, others in seven months, and in
others it lasts as much as a whole year. One must estimate
the duration of the disease by observing both its severity
and the age of the sick person. Apostases are beneficial in
these cases, when they occur and incline downwards, but
they do not occur very often, because of the patients'
youth. Women are not subject to this disease, nor are
girls in whom the menses have appeared.

34. Ask those in whom chronic fluxes of tears or nyc-
talopia occur whether they have had any headache before
these determinations took place.

35. You must expect people who, without having
either fever or paleness, suffer frequent pains in the
crown of their head and their temples—if they have no
apparent apostasis on the face, deepening of the voice, or
toothaches—to haemorrhage through the nostrils. Those
in whom this epistaxis occurs seem to become healthy in
other respects, but you will discover either that their
spleen is swollen, or that they have headaches, and that
something like sparks appear before their eyes. In the
majority of such cases the symptoms of the head and
spleen appear simultaneously.

36. People with enlarged spleens have painful gums

---

[1] H[2]: ῥεόμεναι I.
[2] Mack: τὴν I.

σπλῆνες μεγάλοι. ὁκόσοι δὲ ἔχουσι σπλῆνας μεγά-
68 λους, μήτε αἱμορραγίαι γίνονται | μήτε στόμα
δυσῶδες, τούτων αἱ κνῆμαι ἕλκεα πονηρὰ ἴσχουσι
καὶ οὐλὰς μελαίνας. [ἢν ἄλλην φανερὴν ἔχωσιν
ἀπόστασιν ἐν τῷ προσώπῳ ἢ βαρὺ φθέγγωνται ἢ
ὀδόντας ἀλγέωσι, τούτοισιν αἱμορραγίην διὰ ῥινῶν
προσδέχεσθαι.]¹

37. Οἷσι δὲ τὰ ὑπὸ τοὺς ὀφθαλμοὺς ἐπαίρεται
ἰσχυρῶς, τούτους σπλῆνας μεγάλους εὑρήσεις ἔχον-
τας· εἰ δὲ καὶ ἐν τοῖσι ποσὶν οἰδήματα προσγίνον-
ται, καὶ ὕδωρ φανήσονται ἔχοντες, ἀλλὰ τὴν
γαστέρα καὶ τὴν ὀσφὺν ἐπικατιδεῖν.

38. Τὰ δὲ ἐν τοῖσι προσώποισι παραστρέμματα
ἢν μηδενὶ ἄλλῳ τοῦ σώματος ἐπικοινωνέῃ, ταχέως
παύεται, καὶ αὐτόματα καὶ πρὸς ἀνάγκας· οἱ δ'
ἄλλοι ἀπόπληκτοι.

39. Οἷσι μὲν τῷ μὴ δύνασθαι κινέειν λεπτύνεται
τὸ νενοσηκὸς τοῦ σώματος, οὗτοι ἀδύνατοι εἰς τωὐτὸ
καθίστασθαι· οἷσι δὲ ξυντήξιες μὴ ἐπιγίνονται,
οὗτοι δ' ἔσονται ὑγιέες. περὶ δὲ τοῦ χρόνου ὁπότε
ἔσονται, προλέγειν ἔς τε τὴν ἰσχὺν τοῦ νοσήματος
ὁρῶντα, καὶ ἐς τὸν χρόνον, καὶ ἐς τὴν ἡλικίην τοῦ
ἀνθρώπου, καὶ εἰς τὴν ὥρην, εἰδὼς ὅτι τὰ παλαιό-
τατα τῶν νουσημάτων καὶ τὰ κάκιστα καὶ κυλινδού-
μενα βαρύτατα ὑπακούει, καὶ τὰ ἐν τοῖσι γεραιτά-
τοισι τῶν σωμάτων· ἔστι δὲ καὶ τὸ φθινόπωρόν τε
καὶ ὁ χειμὼν τοῦ ἦρός τε καὶ τοῦ θέρεος ἀνεπιτη-
δειότερα ταῦτα τὰ νοσήματα ἀφιέναι.

and foul-smelling mouths. But splenomegaly without haemorrhages or a foul-smelling mouth is associated with painful ulcerations on the lower legs and dark scars. [If they have any apparent apostasis on the face, deepening of the voice, or toothaches, expect them to haemorrhage through the nostrils.][1]

37. People in whom the areas beneath the eyes swell up greatly you will discover to have large spleens; if in addition their feet swell, they will turn out to have dropsy, too; you should also examine their belly and loins.

38. Distortions of the face, if they coincide with no other disorder of the body, quickly cease, either spontaneously or as the result of treatment. Otherwise there is paralysis.

39. People in whom a diseased part of the body becomes thin, because they are unable to move it, cannot be returned to normal, whereas those in whom no wasting supervenes will recover. As to the time when they will recover, make your prediction by examining the force of the disease, and its duration, the age of the patient, and the season, knowing that the most long-standing and severest diseases, as well as those which are cyclical, yield the most reluctantly, and those in the oldest bodies; also the fall and winter are less favourable than the spring and summer for getting rid of these diseases.

---

[1] Del. Cornarius; cf. ch. 35 above.

40. Αἱ δὲ ἐν τοῖσιν ὤμοισι γενόμεναι ὀδύναι, ὁκόσαι μὲν ἐς τὰς χεῖρας ἐπικαταβαίνουσαι νάρκας τε καὶ ὀδύνας παρέχουσι, ταύτῃσιν ἀποστάσιες μὲν οὐκ ἐπιγίνονται, ὑγιάζονται δὲ μελαίνας χολὰς ἐμέοντες· ὁκόσαι δὲ αὐτοῦ μένουσιν ἐν τοῖσιν ὤμοισιν, ἢ καὶ ἐς τὸν νῶτον ἀφικνέονται, ταύτας πῦον ἐμέσαντας ἐκφυγγάνουσιν, ἢ μέλαιναν | χολήν. καταμανθάνειν δὲ περὶ τούτων ὧδε· ἢν μὲν γὰρ εὔπνοοι ἔωσι καὶ ἰσχνοί, μέλαιναν χολὴν αὐτοὺς μᾶλλον ἐλπὶς ἐμέσαι· εἰ δ᾽ αὖ[1] δυσπνοώτεροι, καὶ ἐπὶ τοῦ προσώπου ἐπιτρέχει τι αὐτοῖσι χρῶμα, ὃ πρόσθεν οὐκ ἐπεγένετο, ὑπέρυθρον, εἴτε μέλαν, τούτους πῦον ἐλπὶς μᾶλλον πτύσειν. σκέπτεσθαι δὲ πρὸς τούτοισι καὶ εἰ ἐν τοῖσι ποσὶν οἰδήματα ἔνεστι· καὶ γὰρ τοῦτο τὸ σημεῖον τούτοισιν ὁμολογέον ἐστίν. τὸ δὲ νόσημα τοῦτο τοῖσιν ἀνδράσι προσγίνεται ἰσχυρότατον τοῖσι ἀπὸ τεσσαράκοντα ἐτέων ἐς τὰ ἑξήκοντα.

41. Τὴν ἡλικίην δὲ ταύτην μάλιστα ἰσχιάδες βιάζονται. σκέπτεσθαι δεῖ ὧδε περὶ ἰσχιάδων· ὁκόσοισι γὰρ τῶν γεραιτέρων αἵ τε νάρκαι ἰσχυρόταται καὶ καταψύξιες τῆς ὀσφύος τε καὶ τῶν σκελέων, καὶ τὸ αἰδοῖον ἐπαίρειν ἀδυνατέουσι, καὶ ἡ γαστὴρ οὐ διαχωρέει, εἰ μὴ πρὸς ἀνάγκην, καὶ κοπρώδης μύξα πολλὴ διεξέρχεται, τούτοισι χρονιώτατον τὸ νούσημα ἔσται, καὶ προλέγειν ἐνιαυτὸν τὸ ἐλάχιστον, ἀφ᾽ οὗ χρόνου ἤρξατο τὸ νούσημα γίνεσθαι, καὶ τὰς ὠφελείας ἐς τὸ ἔαρ τε καὶ τὸ θέρος προσ-

40. Pains in the shoulders that spread downwards into the arms, causing numbness and pains there, are not followed by apostases, but rather such patients recover by vomiting dark bile. Pains that remain fixed in the shoulders, or even move to the back, patients escape by vomiting pus or dark bile. Decide between these two possibilities as follows: if patients breathe easily and have no swelling, there is more chance that they will vomit dark bile: if, on the other hand, they have marked difficulty in breathing, and some coloration that was not there before—either reddish or dark—spreads over their face, they are more likely to vomit pus. Also look for any swelling present in their feet, for this sign reinforces the others. The disease occurs most violently in men between forty and sixty years.

41. In this same age sciaticas too press most forcefully[a]; you must investigate them as follows. In old persons, in whom the attacks of numbness are very severe and there are chills in the loins and the legs, who are unable to erect their penis, whose belly does not evacuate unless it is forced, and then passes copious mucous stools, the disease will be of very long duration, and you should predict that it will go on at least a year from the time when it began; expect the remissions towards spring and

---

[a] Chapter 40 ends here in Littré, but I have reverted to Vander Linden's division (vol. 1, 519).

[1] Littré: ἂν I.

δέχεσθαι. τοῖσι δὲ νεανίσκοισιν ἐπώδυνοι μὲν οὐχ
ἧσσον αἱ ἰσχιάδες, βραχύτεραι δέ· καὶ γὰρ τεσσαρακονθήμεροι ἀπαλλάσσονται· ἀλλ' οὐδὲ αἱ νάρκαι
ἐπιγίνονται ἰσχυραί, οὔτε αἱ καταψύξιες τῶν σκελέων τε καὶ τῆς ὀσφύος. οἷσι δὲ τὸ νούσημα τοῦτό
ἐστι μὲν ἐν τῇ ὀσφύι καὶ τῷ σκέλει, βιάζεται δὲ οὐχ
οὕτως ὥστε κατακέεσθαι, ξυστρέμματα σκέπτεσθαι
μὲν εἴ που ἐν τῷ ἰσχίῳ, καὶ ἐπανερέσθαι εἰ εἰς τὸν
βουβῶνα ἡ ὀδύνη ἀφικνεῖται· ἢν γὰρ ταῦτ' ἔχῃ
ἄμφω, χρόνιον τὸ νούσημα γίνεται· ἐπανερέσθαι δὲ
72 καὶ εἰ ἐν τῷ μηρῷ νάρκαι ἐγγίνονται, | καὶ ἐς τὴν
ἰγνύην ἀφικνοῦνται· καὶ ἢν φῇ, αὖθις ἐρέεσθαι, καὶ
ἢν διὰ τῆς κνήμης, ἐπὶ τὸν ταρσὸν τοῦ ποδός. ὁπόσοι δ' ἂν τούτων τὰ πλεῖστα ὁμολογέωσι, εἰπεῖν
αὐτοῖσιν ὅτι τὸ σκέλος σφὶν τοτὲ μὲν θερμὸν γίνεται, τοτὲ δὲ ψυχρόν. ἡ δὲ νοῦσος αὕτη ὁκόσοισι μὲν
[ἀνὰ]¹ τὴν ὀσφὺν ἐκλείπουσα ἐς τὰ κάτω τρέπεται,
θαρσύνει. ὁκόσοισι δὲ τά τε ἰσχία καὶ τὴν ὀσφὺν
μὴ ἐκλείπουσα εἰς τὰ ἄνω τρέπεται, προλέγειν δεινὰ
εἶναι.

42. Οἷσι δὲ περὶ τὰ ἄρθρα ὀδύναι τε γίνονται καὶ
ἐπάρσιες καὶ καταπαύονται, οὐκ ἐν τῷ ποδαγρικῷ
τρόπῳ, εὑρήσεις τά τε σπλάγχνα μεγάλα καὶ ἐν τῷ
οὔρῳ λευκὴν ὑπόστασιν· καὶ τοὺς κροτάφους, ἢν
ἐπέρῃ, φήσει πολλάκις ἀλγέειν· φήσει δὲ καὶ ἱδρῶ-
τας αὐτῷ γίνεσθαι νυκτερινούς. ἢν δὲ μήτε ὑπὸ τῷ
οὔρῳ ὑφίσταται ἡ ὑπόστασις αὕτη, μήτε οἱ ἱδρῶτες
γίνονται, κίνδυνος ἢ χωλωθῆναι τὰ ἄρθρα, ἢ ὃ δὴ

summer. In young men sciaticas are not less painful, but they are shorter, for they end on the fortieth day, and neither the attacks of numbness are severe nor the chills of the legs and loins. In people in whom this disease occupies the loins and leg, but does not press them so forcefully that they are bedridden, look to see whether there are tumours anywhere in the hip region, and ask whether the pain extends to the groin, for if both these signs are present, the disease becomes chronic. Also ask whether there is numbness in the thigh, extending to the loin; if the patient says there is, ask again whether it also extends down the lower leg to the flat of the foot. To those who answer in the affirmative on most of these points, say that their leg will at one time become hot, at another cold. In cases where this disease departs from the loins, and turns downwards, there is reason for confidence, but in cases where it does not leave the hips and loins, but moves upward, you must predict danger.

42. In people in whom pains arise about the joints together with swellings, and then cease—but not in the manner of gout—you will discover that the inward parts are enlarged, and that there is a white precipitate in the urine; if you ask the person, he will say that his temples often have pains, and he will also say that he has night-sweats. If, however, neither this precipitate comes down in the urine nor the sweats set in, there is a danger either that the joints will become lame, or that what they call

---

[1] Del. Opsopoeus in note.

μελικηρίδα καλέουσι γίνεσθαι ὑπ᾽ αὐτοῖσι. γίνεται
δὲ τὸ νόσημα τοῦτο οἷσιν ἐν τῇ παιδίῃ τε καὶ νεό-
τητι ξύνηθες ἐὸν αἷμα ῥεῖν ἐκ τῶν ῥινῶν πέπαυται.
ἐπανερέσθαι οὖν περὶ τῆς τοῦ αἵματος ῥήξιος, εἰ
ἐγένετο ἐν τῇ νεότητι· καὶ αἱ κνιδώσιες ἔν τε τῷ
στήθει καὶ τῷ μεταφρένῳ εἰ ἔνεισι· καὶ ὁκόσοις αἱ
κοιλίαι ἰσχυρὰς ὀδύνας παρέχουσιν ἄνευ ἐκταρα-
ξίων· καὶ ὁκόσοισιν αἱμορροΐδες γίνονται· αὕτη γὰρ
ἡ ἀρχὴ τῶν νουσημάτων τούτων. ἢν δὲ κακόχροοι
οἱ ἄνθρωποι οὗτοι φαίνωνται, ἐπανερέσθαι καὶ
κεφαλήν, εἰ ὀδυνῶνται· φήσουσι γάρ. τούτων δὲ
ὁκόσοισιν αἱ κοιλίαι ἐπώδυνοι ἔν γε τοῖς δεξιοῖς
εἶεν, τὰ ἀλγήματα ἰσχυρότερα γίνεται, καὶ μάλιστα
ὅταν πρὸς τῷ ὑποχονδρίῳ κατὰ τὸ ἧπαρ τὸ ὑπό-
λειμμα τῆς ὀδύνης ᾖ. ὠφελέει δὲ ταύτας τὰς ὀδύνας
74   τὸ παραυτίκα ψόφος ἐν τῇ γαστρὶ γενόμενος· | ὁκό-
ταν δὲ ἡ ὀδύνη παύσηται, τὸ οὖρον παχὺ καὶ χλωρὸν οὐρέουσιν. ἔστι δὲ θανατώδης μὲν οὐδαμῶς ὁ
τρόπος οὗτος, χρόνιος δὲ κάρτα· ὁκόταν δὲ παλαιὸν
ἤδη ᾖ τὸ νούσημα, ἀμβλυώσσουσιν οἱ ἄνθρωποι ὑπ᾽
αὐτοῦ. ἀλλ᾽ ἐπανερέσθαι περὶ τοῦ αἵματος, εἰ νέῳ
ἐόντι ἔρρει, καὶ περὶ τοῦ ἀμβλυωσμοῦ, καὶ περὶ τοῦ
ὀτῆς παχύνσιος[1] <καὶ τῆς>[2] χλωρότητος, καὶ ἀμφὶ
τῶν ψόφων εἰ ἐγγίνονταί τε καὶ ὠφελέουσιν ἐπιγινό-
μενοι· φήσουσι γὰρ πάντα ταῦτα. [εἴτε ἐρυθρόν τε
εἴτε μέλαν, τούτοισι πῦον ἐλπίζειν μᾶλλον ἢ πτύειν·
σκέπτεσθαι δὲ τούτοισι καὶ ἐν τοῖσι ποσὶν οἰδή-
ματα· καὶ γὰρ τοῦτο τὸ σημεῖον τούτοισιν ὁμολογόν
ἐστιν.][3]

meliceris[a] will arise in them. This disease arises in people
in whom an epistaxis that was common in their childhood
and youth has stopped. Thus you must ask whether such
an epistaxis occurred when the person was young; also
whether there is itching of the chest and back, whether
the cavities are causing violent pains without any apparent disturbance, and whether haemorrhoids have developed—for this is how these diseases begin. If these
people appear to have a poor colour, ask whether they
have pains in the head; they will say they do. In those
whose cavities are painful on the right side, the pains are
more violent, and especially when what remains of the
pain is at the hypochondrium in the region of the liver.
These pains are relieved at once by rumblings in the
belly; when the pain stops, thick yellow-green urine is
passed. This form of the complaint, though not at all
mortal, is especially chronic. When the disease is already
long-standing, patients become dim of vision from it. You
must ask about the blood—whether the patient haemor-
rhaged when he was young—about the dimness of vision,
about the thickening and greenness of the urine, and
about the rumblings—whether they occurred and
whether they gave any relief when they did occur;
patients will say that all these things were so. [Either reddish or dark, expect pus rather than expectoration. Also
look for swellings in their feet, for this sign reinforces the
others.]

---

[a] A cyst filled with a honey-like substance.

---

43. Λειχῆνες δὲ καὶ λέπραι καὶ λεῦκαι, οἷσι μὲν νέοισιν ἢ παισὶν ἐοῦσιν ἐγένετό τι τούτων, ἢ κατὰ μικρὸν φανὲν αὔξεται ἐν πολλῷ χρόνῳ, τούτοισι μὲν οὐ χρὴ ἀπόστασιν νομίζειν τὸ ἐξάνθημα, ἀλλὰ νόσημα· οἷσι δὲ ἐγένετο τούτων τι πολύ τε καὶ ἐξαπίνης, τοῦτο ἂν εἴη ἀπόστασις. γίνονται δὲ λεῦκαι μὲν ἐκ τῶν θανατωδεστάτων νοσημάτων, οἷον καὶ ἡ νοῦσος ἡ Φοινικίη καλεομένη. αἱ δὲ λέπραι καὶ οἱ λειχῆνες ἐκ τῶν μελαγχολικῶν. ἰῆσθαι δὲ τούτων εὐπετέστερά ἐστιν, ὅσα νεωτάτοισί τε γίνεται καὶ νεώτατά ἐστι, καὶ τοῦ σώματος ἐν τοῖσι μαλθακωτάτοισί τε καὶ σαρκωδεστάτοισι φύεται.

43. Lichen, lepra and *leuce*: when one of these has arisen in a young person or a child, or when it appears it grows but little over a long time, you must consider the eruption not to be an apostasis, but a disease; whereas when one of them appears suddenly and over a large area, it is an apostasis. *Leuce* is among the most mortal of diseases, like the so-called Phoenician disease.[a] Lepra and lichen are melancholic. These diseases are easier to heal when they occur in the youngest patients, are of the most recent origin, and grow in the softest and fleshiest parts of the body.

[a] Littré follows Galen vol. 19, 153, in suggesting elephantiasis.

PHYSICIAN

INTRODUCTION[1]

Very little is known about the position of *Physician* in the history of medicine. "Cet opuscule," says Littré,[2] "n'est mentionné par aucun des anciens critiques." And later on; "Dans le silence des anciens commentateurs il n'est pas possible de se faire une idée sur l'origine de l'opuscule du Médecin."[3] Littré does, however, point out passages that are parallel to parts of *Physician* in *Surgery*, *Ancient Medicine* and several other Hippocratic works, concluding that these similarities of content assure the treatise a legitimate place in the Hippocratic collection.[4]

After the first chapter, which outlines the attributes desirable in a physician, the piece goes on to discuss the arrangement of the surgery, the preparation of bandages, instruments, and so forth. Then follows a short discussion of tumours and sores, and the book finishes with a recommendation to a student to attach himself to mercenary troops in order to have practice in surgery—a fairly

---

[1] Jones edited and translated Ch. 1 of *Physician* in vol. 2 of the Loeb *Hippocrates*, pp. 303–313. The introduction here printed is to a large part taken directly from Jones'.

[2] Littré vol. 1, 412.

[3] Littré vol. 1, 414.

[4] Littré vol. 9, 198–201.

sure indication of a date later than 400 B.C.[5]

J. F. Bensel holds that *Physician* is closely connected with the treatises *Precepts* and *Decorum*. It is most important to come to some conclusion as to whether there is a real connection, or whether there are merely resemblances.

Bensel's monograph (it is really an edition of *Physician*) is very instructive. The author sees that all three books are intended for young beginners; he points out that the artifices we associate with the style of Isocrates are to be seen in *Physician*, and in particular that in some cases there are verbal parallels. These tend to indicate that the date of *Physician* is 350–300 B.C.

Up to this point it is easy to agree with Bensel. But when he goes on to assert that *Physician* is contemporary with *Precepts* and *Decorum*, and that the last shows Epicurean tendencies, it is difficult to follow his argument. *Physician* is comparatively simple, and the Greek is rarely strange or obscure. There are none of the signs of late date. *Precepts* and *Decorum*, on the other hand, are not only strange but even fantastic. No extant Greek prose shows such peculiar vagaries in diction. The signs of late date are many and insistent. Finally, the supposed Epicureanism of *Decorum* cannot possibly be reconciled with the assertion made in that work that physicians give way before the gods, and know that their art is under the direction of a higher power. Surely this is Stoic rather than Epicurean doctrine. The truth seems to be that what Bensel takes to be Epicureanism is really the

[5] Fleischer (p. 56f.) argues on the basis of examples of Hellenistic word usage for a date in the third century.

received ethical teaching of later Hellenistic times, which is in part common to both schools of thought.

The likeness, then, between *Physician* and the other two works is a similarity of subject. All are addresses to young men at the beginning of their medical course, and lay down the rules of conduct and practice that such students must follow. In the face of the evidence it is illegitimate to go further, and to assert that all were written at the same time. On the contrary, there is every reason to think that *Physician* is considerably earlier than the other two.

Physician has attracted considerable interest among Hippocratic scholars over the centuries.[6] Noteworthy recent studies of the treatise include:

J. F. Bensel, "Hippocratis qui fertur De medico libellus ad codicum fidem restitutus," *Philologus* 78 (1922), p. 88–131. (= Bensel)

U. Fleischer, *Untersuchungen zu den pseudohippokratischen Schriften* ΠΑΡΑΓΓΕΛΙΑΙ, ΠΕΡΙ ΙΗΤΡΟΥ und ΠΕΡΙ ΕΥΣΧΗΜΟΣΤΥΝΗΣ, Berlin, 1939. (= Fleischer)

M. Moisan, *Edition critique et commentée, avec traduction française, de deux oeuvres de la collection hippocratique*: *Préceptes et Médecin*, Diss. Paris, 1993. (= Moisan)

The text of *Physician* printed here is based on the text and apparatus of Heiberg.

[6] See Littré vol. 9, 203.

ΠΕΡΙ ΙΗΤΡΟΥ

1. Ἰητροῦ μὲν εἶναι προστασίην ὁρῆν <ὡς>[1] εὔχρως τε καὶ εὔσαρκος ἔσται πρὸς τὴν ὑπάρχουσαν αὐτῷ φύσιν· ἀξιοῦνται γὰρ ὑπὸ τῶν πολλῶν οἱ μὴ εὖ διακείμενοι τὸ σῶμα οὕτως ὡς οὐδ᾽ ἂν ἑτέρων ἐπιμεληθῆναι καλῶς. ἔπειτα <τὰ>[2] περὶ αὐτὸν καθαρείως ἔχειν ἐσθῆτι χρηστῇ καὶ χρίσμασιν εὐόδμοις ὀδμὴν ἔχουσιν ἀνυπόπτως· πρὸς ἅπαντα ταῦτα γὰρ ἡδέως ἔχειν ξυμβαίνει τοὺς νοσέοντας, δεῖ δὲ τοῦτο σκοπέειν. τὰ δὲ περὶ τὴν ψυχὴν σώφρονα, μὴ μόνον τὸ σιγᾶν, ἀλλὰ καὶ περὶ τὸν βίον πάνυ εὔτακτον· μέγιστα γὰρ ἔχει πρὸς δόξαν ἀγαθά. τὸ δὲ ἦθος εἶναι καλὸν καὶ ἀγαθόν, τοιοῦτον δ᾽ ὄντα πᾶσι καὶ σεμνὸν καὶ φιλάνθρωπον· τὸ γὰρ προπετὲς καὶ τὸ πρόχειρον καταφρονεῖται, κἂν πάνυ χρήσιμον ᾖ. σκοπὸν δὲ ἐπὶ τῆς ἐξουσίης· τὰ γὰρ |

αὐτὰ παρὰ τοῖς αὐτοῖς σπανίως ἔχουσιν ἀγαπᾶται. σχήμασι δὲ ἀπὸ μὲν προσώπου σύννουν μὴ πικρῶς· αὐθάδης γὰρ δοκέει εἶναι καὶ μισάνθρωπος· ὁ δὲ εἰς γέλωτα ἀνέμενος καὶ λίην ἱλαρὸς φορτικὸς ὑπολαμβάνεται, φυλακτέον δὲ τὸ τοιοῦτον οὐχ ἥκιστα.

---

[1] Zwinger in margin. [2] Littré.

PHYSICIAN

1. The dignity of a physician requires that he should look healthy, and as plump as nature intended him to be; for the common crowd consider those who are not of this excellent bodily condition to be unable to take care of others. Then he must be clean in person, well dressed, and anointed with sweet-smelling unguents that are beyond suspicion. For all these things are pleasing to people who are ill, and he must pay attention to this. In matters of the mind, let him be prudent, not only with regard to silence, but also in having a great regularity of life, since this is very important in respect of reputation; he must be a gentleman in character, and being this he must be grave and kind to all. For an over-forward obtrusiveness is despised, even though it may be very useful. Let him look to the liberty of action that is his; for the same things, if done but seldom to the same patients, are appreciated.[a] In appearance, let him be of a serious but not harsh countenance; for harshness is taken to mean arrogance and unkindness, while a man of uncontrolled laughter and excessive gaiety is considered vulgar, and vulgarity especially must be avoided. In every social rela-

[a] The only thing scholars agree upon about this passage is its difficulty.

301

δίκαιον δὲ πρὸς πᾶσαν ὁμιλίην εἶναι· χρὴ γὰρ
πολλὰ ἐπικουρέειν δικαιοσύνην, πρὸς δὲ ἰητρὸν οὐ
μικρὰ συναλλάγματα τοῖσι νοσοῦσίν ἐστι· καὶ γὰρ
αὑτοὺς ὑποχειρίους ποιέουσι τοῖς ἰητροῖς, καὶ
πᾶσαν ὥρην ἐντυγχάνουσι γυναιξί, παρθένοις καὶ
τοῖς ἀξίοις πλείστου κτήμασιν· ἐγκρατέως οὖν δεῖ
πρὸς ἅπαντα ἔχειν ταῦτα. τὴν μὲν οὖν ψυχὴν καὶ τὸ
σῶμα οὕτω διακεῖσθαι.

2. Τὰ δὲ εἰς τὴν ἰητρικὴν τέχνην παραγγέλματα,
δι' ὧν ἔστιν εἶναι τεχνικόν, ἀπ' ἀρχῆς συνοπτέον,
ἀφ' ὧν καὶ μανθάνειν ἄνθρωπος ἄρξαιτο· τὰ τοίνυν
ἐν ἰητρείῳ θεραπευόμενα σχεδὸν μανθανόντων
ἐστίν. δεῖ δὲ πρῶτον μὲν τόπον ἔχειν οἰκεῖον,[1] ἔσται
δὲ τοῦτο ἐὰν μήτε πνεῦμα εἰς αὐτὸν παραγινόμενον
ἐνοχλῇ μήθ' ἥλιος ἢ αὐγὴ λυπέῃ. φῶς δὲ τηλαυγὲς
μὲν τοῖς θεραπεύουσιν, ἄλυπον δὲ τοῖς θεραπευομένοις ὑπάρχειν·[2] πάντως μὲν οὖν τοιαύτην τὴν αὐγὴν
μάλιστα, διὸ συμβαίνει τοὺς ὀφθαλμοὺς νοσέειν· τὸ
μὲν οὖν φῶς τοιοῦτον εἶναι παρήγγελται· τοῦτο δέ,
ὅπως μηδαμῶς ἐναντίως ἕξει τῷ προσώπῳ τὰς
αὐγάς· προσενοχλεῖ γὰρ τὴν ὄψιν ἀσθενέως ἔχου-
σαν· πᾶσα δ' ἱκανὴ πρόφασις ἀσθενέοντας ὀφθαλμοὺς ἐπιταράξαι· τῷ μὲν οὖν φωτὶ τοῦτον τὸν τρόπον
χρηστέον ἐστίν. τοὺς δὲ δίφρους ὁμαλοὺς εἶναι τοῖς
208   ὕψεσιν ὅτι μάλιστα, ὅπως | κατ' αὐτοὺς ὦσιν. χαλ-
κώματι δὲ πλὴν τῶν ὀργάνων μηδενὶ χρήσθω· καλ-
λωπισμὸς γάρ τις εἶναί μοι δοκεῖ φορτικὸς σκεύεσι
τοιούτοισι χρῆσθαι. τὸ δ' ὕδωρ παρέχειν δεῖ πότι-

tion he will be fair, for fairness must be of great service. The intimacy also between physician and patient is close. Patients in fact put themselves into the hands of their physician, and at every moment he meets women, maidens, and possessions very precious indeed. So towards all these self-control must be used. Such then should the physician be, in body and in soul.

2. As for the precepts of the medical art that make proficiency possible, you must first consider which ones a person should take as the starting point of his instruction. Now the things treated in the surgery are perhaps what the beginner should learn. First you must have a suitable location: it will be thus, if neither wind blows in, causing annoyance, nor sunshine or brightness does harm; light must be bright for those who are treating, but not do any harm to those who are being treated. In every instance, then, the brightness must be such, for it does happen that the eyes become ill. It is a rule, then, that the light should be like this; also that the patient should under no circumstances directly face the brightness; for light disturbs even more a vision that is weak, and any factor is sufficient to trouble ailing eyes. Light, then, must be arranged in this way. The couches should be level in their heights as far as possible, so that they will be convenient.[a] The surgeon should avoid the use of bronze, except for his instruments, for the use of such gear seems to me to be nothing but vulgar ostentation. You must provide

[a] Littré understands "afin que le médecin et le patient soient de niveau" and Fuchs "damit sie der Körperform der Patienten entsprechen"; the passage is unclear.

[1] Littré: τῆς οἰκίης V.   [2] Fleischer: -ἀρχει V.

μον τοῖς θεραπευομένοις καὶ καθαρόν. τοῖς δὲ ἀπο-
μάγμασιν καθαροῖς καὶ μαλθακοῖς χρῆσθαι, πρὸς
μὲν τοὺς ὀφθαλμοὺς ὀθονίοις, πρὸς δὲ τὰ τραύματα
σπόγγοις· αὐτόματα γὰρ ταῦτα βοηθεῖν δοκεῖ
καλῶς. τὰ δ' ὄργανα πάντα εὐήρη πρὸς τὴν χρείαν
ὑπάρχειν δεῖ τῷ μεγέθει καὶ βάρει καὶ λεπτότητι.

3. Τὰ δὲ προσφερόμενα ἅπαντα μὲν χρὴ συνορῆν ὅπως συνοίσει·[1] μάλιστα δὲ πλεῖστον, εἰ ὁμι-
λεῖν μέλλει τῷ νοσοῦντι μέρει· ταῦτα δέ ἐστιν ἐπι-
δέσματα καὶ φάρμακα καὶ τὰ περὶ τὸ ἕλκος ὀθόνια
καὶ τὰ καταπλάσματα· πλεῖστον γὰρ χρόνον ταῦτα
περὶ τοὺς νοσέοντάς ἐστι τόπους. ἡ δὲ μετὰ ταῦτα
ἀφαίρεσις τούτων, ἀνάψυξίς τε καὶ περικάθαρσις,
καὶ τῶν ὑδάτων κατάντλησις, ὀλίγου τινός ἐστι χρό-
νου· καί τι ποιῆσαι ὅκου χρή, μᾶλλόν τε καὶ ἧσσον
ἐσκέφθαι δεῖ· τούτων γὰρ ἀμφοτέρων ἡ χρῆσις
εὔκαιρός τε καὶ μὴ γενομένων μεγάλην ἔχει δια-
φορήν.

4. Ἔστι δὲ οἰκείη ἐπίδεσις τῆς ἰητρικῆς, ἀφ' ἧς
ὠφελεῖσθαι τὸν θεραπευόμενον·[2] μέγιστα δὲ ὠφελεῖ
δύο ταῦτα, οἷς ἐστι χρηστέον, πιέσαι ὅκου δεῖ καὶ
210   ἀνειμένως ἐπιδῆσαι· πρὸς δὲ τοὺς χρόνους | τῆς
ὥρης, πότε δεῖ σκεπαστικῶς καὶ μή, συνορῆν, ὅκως
μηδὲ ἀσθενῆ λεληθὸς ποτέρῳ τούτων ἐνιαχοῦ χρη-
στέον· εὐρύθμους δὲ ἐπιδέσιας καὶ θεητρικὰς μηδὲν
ὠφελούσας ἀπογινώσκειν· φορτικὸν γὰρ τὸ τοιοῦτον
καὶ παντελῶς ἀλαζονικόν, πολλάκις τε βλάβην
οἶσον τῷ θεραπευομένῳ· ζητεῖται ὁ νοσέων οὐ καλ-
λωπισμόν, ἀλλὰ τὸ συμφέρον.

potable and clean water for the patients. Use swabs that are clean and soft, linen ones for the eyes, sponges for wounds; for these seem to serve well just by themselves. All the instruments must be well fitted for their use in size, weight, and fineness.

3. You must take care that all the things you apply will be benign, most especially if they are to be in contact with the ailing part, i.e. bandages, medications, linen cloths about the lesion, and cataplasms. For these remain about the ailing places for a very long time, whereas afterwards their removal, and the cooling, cleaning off and bathing of the lesion in water lasts only a short time. And where something must be done, you must consider the degree; for the use of more and less must be well-timed, and when they are not it makes a great difference.

4. There is a proper way of bandaging in medicine, by which the person being treated is benefited. The following two types help most: ones you must employ to exert pressure where it is required, and ones to bind loosely. Consider when the bandage should be covered, and when not, in relation to the times of the season, so that the patient too may know which covering should be employed under the various circumstances. Reject graceful and showy bandages as doing no good, for this sort of thing is vulgar and purely a matter of display, and will often bring harm to the person being treated: the ill person is not looking for what is decorative, but for what is beneficial.

---

[1] Chartier: -οισι V.
[2] Foes in note: -πεύοντα V.

5. Ἐπὶ δὲ τῶν χειρουργιῶν, ὅσαι διὰ τομῆς εἰσιν
ἢ καύσιος, τὸ ταχέως ἢ βραδέως ὁμοίως ἐπαινεῖται·
χρῆσις γάρ ἐστιν ἀμφοτέρων αὐτῶν. ἐν οἷς μὲν γάρ
ἐστι διὰ μιῆς τομῆς ἡ χειρουργία, χρὴ ποιέεσθαι
ταχεῖαν τὴν διαίρεσιν· ἐπεὶ γὰρ συμβαίνει τοὺς
τεμνομένους πονέειν, τὸ μὲν[1] λυπέον ὡς ἐλάχιστον
χρόνον δεῖ παρεῖναι· τοῦτο δὲ ἔσται ταχείης τῆς
τομῆς γενομένης. ὅπου δὲ πολλὰς ἀναγκαῖον γενέσθαι τὰς τομάς, βραδείη χρηστέον τῇ χειρουργίᾳ· ὁ
μὲν γὰρ ταχὺς ξυνεχῆ ποιέει τὸν πόνον καὶ πολύν·
τὰ δὲ διαλείποντα[2] ἀνάπαυσιν ἔχει τινὰ τοῦ τόνου
τοῖς θεραπευομένοις.

6. Τὸ δ' αὐτὸ ἐπὶ τῶν ὀργάνων λέγοιτ' ἄν· τοῖς
δὲ μαχαιρίοις ὀξέσι δὲ χρῆσθαι καὶ πλατέσιν οὐκ
ἐπὶ πάντων ὁμοίως παραγγέλλο|μεν· μέρη γάρ τινά
ἐστι τοῦ σώματος, ἃ ἐν τάχει μὲν ἔχει τὴν ῥύμην
τοῦ αἵματος, καὶ κατασχεῖν ἐστιν οὐ ῥηΐδιον· ταῦτα
δέ ἐστιν οἵ τε κίρσοι καί τινες ἄλλαι φλέβες· τὰς
μὲν τομὰς χρὴ εἶναι τῶν τοιούτων στενάς· οὐ γὰρ
οἷόν τε[3] τὴν ῥύσιν γενέσθαι κατακορῆ· ξυμφέρει δέ
ποτε ἀπὸ τῶν τοιούτων αἵματος ἀφαίρεσιν ποιέεσθαι. πρὸς δὲ τοὺς ἀκινδύνους τόπους καὶ περὶ οὓς
μὴ λεπτόν ἐστι τὸ αἷμα, πλατυτέροις χρῆσθαι τοῖς
μαχαιρίοισι· τὸ γὰρ αἷμα πορεύοιτ' ἄν, ἄλλως δὲ
οὐδαμῶς· πάνυ δ' ἐστὶν αἰσχρῶς μὴ ξυμβαίνειν ἀπὸ
τῆς χειρουργίης ὅ τι θέλει.

7. Σικυῶν δὲ δύο τρόπους εἶναι χρησίμους· ὅτε
μὲν γὰρ ῥεῦμα ξυνεστηκὸς πόρρω τῆς ἐπιφαινομέ-

5. In surgical operations that consist in incising or cautery, speed or slowness are commended alike, for each has its value. In cases where the surgery is performed by a single incision, you must make it a quick one; for since the person being cut generally suffers pain, this suffering should last for the least time possible, and that will be achieved if the incision is made quickly. However, when many incisions are necessary, you must employ a slow surgery, for a surgeon that was fast would make the pain sustained and great, whereas intervals provide a break in its intensity for the patients.

6. The same principle might be enunciated on the use of instruments. We recommend the use of both pointed scalpels and broad ones, but not indifferently in every case. For there are some parts of the body from which the flow of blood comes rapidly, and is not very easy to hold back, for example from varices and certain other vessels. Incisions in these must be narrow, for thus it is impossible for the haemorrhage to become too intense; it does sometimes convey a benefit to draw blood from these. In undangerous locations and those where the blood is not thin, employ broader scalpels: for in this way the blood will be successfully brought to flow, although otherwise not at all, and it is very disgraceful not to achieve with an operation what you wish to.

7. Of cupping instruments two types are useful. When the flux has collected far from the surface tissue,

---

νης σαρκός, τὸν μὲν κύκλον αὐτῆς εἶναι δεῖ[1] βρα-
χύν, αὐτὴν δὲ μὴ γαστρώδη, προμήκη τὸ πρὸς τὴν
χεῖρα μέρος, μὴ βαρεῖαν· τοιαύτην γὰρ οὖσαν
ἕλκειν ἐς ἰθὺ ξυμβαίνει, καὶ τοὺς ἀφεστῶτας ἰχῶρας
καλῶς ἀνεσπᾶσθαι πρὸς τὴν σάρκα· τοῦ δὲ πόνου
πλείονος κατεσκεδασμένου τῆς σαρκός, τὰ μὲν ἄλλα
παραπλησίην, τὸν δὲ κύκλον μέγαν· οὕτω γὰρ ἐκ
πλείστων μερῶν εὑρήσεις ἄγουσαν ἐς ὃν δεῖ τὸ
λυποῦν τόπον· οὐ γὰρ οἷόν τε[2] μέγαν εἶναι τὸν
κύκλον, μὴ συναγομένης τῆς σαρκὸς ἐκ πλείονος
τόπου. βαρεῖα δ᾽ οὖσα ῥέπει καὶ ἐς τοὺς ἄνω τόπους·
κάτω δὲ μᾶλλον τὴν ἀφαίρεσιν, καὶ πολλάκις ὑπολείπεσθαι τὰς νούσους. τοῖσι μὲν οὖν ἐφεστῶσι ῥεύ-
214 μασι καὶ μα|κρὰν ἀπέχουσιν ἀπὸ τῶν ἄνω τόπων οἱ
πλατεῖς κύκλοι πολλὰ ξυνεπισπῶνται παρὰ τῆς
ἄλλης σαρκός· ἐπιπροσθεῖν οὖν ξυμβαίνει τὴν
ἐντεῦθεν ἑλκομένην νοτίδα τῷ ξυναγομένῳ κάτωθεν
ἰχῶρι, καὶ τὰ μὲν ἐνοχλεῦντα ὑπολείπεσθαι, τὰ δ᾽
οὐδὲν λυπέοντα ἀφαιρεῖσθαι. μέγεθος δὲ σικύης τί
χρήσιμον στοχάζεσθαι χρὴ πρὸς τὰ μέρη τοῦ
σώματος, οἷς ἂν δέη προσβάλλειν. ὅταν δὲ κατα-
κρούῃ, κάτωθεν δέχεσθαι· τὸ γὰρ αἷμα φανερὸν
εἶναι δεῖ τῶν χειρουργουμένων τόπων· ἄλλως δὲ
οὐδὲ τὸν κύκλον τὸν ἑλκυσθέντα χρὴ κατακρούειν·
εὐτονωτέρη γάρ ἐστιν ἡ σὰρξ τοῦ πονήσαντος·
μαχαιρίοις δὲ τοῖς καμπύλοις ἐξ ἄκρου μὴ λίην στε-

---

[1] Aldina: διά- V.   [2] Foes after Gorraeus: οἴονται V.

the circle at the mouth of the instrument must be short,
the instrument itself not too bellied, and the handle long
and not heavy. For if it has this shape, it will attract in a
straight line and draw the dispersed sera up nicely
towards the tissue. But if the pain is more spread out over
the tissue, the cupping instrument should be otherwise
the same, but its circle large; for if it is like this, you will
find that it draws the disease material from as many parts
as possible to the place where it should be; for it is impossible for the circle to be large without the tissue attracting
from a wider area. If the instrument is heavy, it presses
on the upper parts as well, whereas the removal of fluid
ought rather to take place lower down, and often the diseases are left behind. Now when fluxes become blocked
and are held far away from the upper parts, instruments
with wide circles attract together material from the rest of
the tissue, so it turns out that the moisture drawn from
there (i.e. from the superficial tissue) stands in the way of
the serum collected from below, so that what is troublesome is left behind while that which was doing no harm is
removed. What size of cupping instrument will be useful
you must estimate according to the bodily part to which it
is to be applied. When you make scarifications, the blood
must be taken from below—for the blood of the incised
parts must be visible. Otherwise you should not scarify
even the circle that is drawn up by the cupping instrument, since the tissue at the surface is more distended
than the area which is ill.[a] Use curved scalpels that are not

---

[a] See Daremberg p. 70, note 23. If you scarify when the surface tissue is compacter than the ailing region below it, the blood
will come from the surface, and do no benefit.

νοῖς· ἐνίοτε γὰρ ἰχῶρες ἔρχονται γλίσχροι καὶ
παχεῖς· κίνδυνος οὖν ἐστιν ὑποστῆναι τῆσι τομῆσιν,
ὅταν στεναὶ τμηθέωσιν.

8. Τὰς δ' ἐπὶ τῶν βραχιόνων φλέβας τῆσι κατα-
λήψεσι χρὴ φυλάσσειν· ἡ γὰρ καλύπτουσα σὰρξ
πολλοῖς οὐ καλῶς συνήρμοσται τῇ φλεβί· τῆς γὰρ
σαρκὸς ὀλισθηρῆς οὔσης, οὐ καθ' ἑαυτὰς ξυμβαίνει
τὰς τομὰς ἀμφοτέρων γίνεσθαι· τὴν γὰρ φλέβα
ἐκφυσᾶσθαι ξυμβαίνει καλυφθεῖσαν, καὶ τὴν ῥύσιν
τοῦ αἵματος κωλύεσθαι, πολλοῖσι δὲ καὶ πῦος διὰ
τοῦτο ξυνίστασθαι· δοκεῖ δὴ[1] δύο βλάβας φέρειν ἡ
τοιαύτη χειρουργίη, τῷ μὲν τμηθέντι πόνον, τῷ δὲ
τέμνοντι πολλὴν ἀδοξίην· τὸ δ' αὐτὸ κατὰ πασῶν
παρήγγελται γίνεσθαι.

9. Τὰ μὲν οὖν κατ' ἰητρεῖον ἀναγκαῖα ὄργανα,
καὶ περὶ ἃ δεῖ | τεχνικὸν εἶναι τὸν μανθάνοντα, ταῦτ'
ἐστίν· ὀδοντάγρησι γὰρ καὶ σταφυλάγρησι χρῆσθαι τὸν τυχόντα ἐστίν· ἁπλῆ γὰρ ἡ χρῆσις αὐτῶν
εἶναι δοκεῖ.

10. Περὶ δὲ φυμάτων καὶ ἑλκέων, ὁκόσα μειζόνων
ἐστὶ νοσημάτων, τὰ μὲν φύματα τεχνικώτατον ὑπει-
ληφέναι δεῖ δύνασθαι διαλύειν, καὶ τὰς συστάσεις
αὐτῶν κωλύειν· ἐχόμενον δὲ τούτων, στέλλειν εἰς τὸν
ἐπιφανῆ τόπον ὡς εἰς βραχύτατον, καὶ τὴν σύστα-
σιν ὁμαλῶς διὰ παντὸς ποιεῖσθαι τοῦ φύματος·
ἀνωμάλως γὰρ ἔχοντος αὐτοῦ, ῥαγῆναί τε καὶ
δυσθεράπευτον τὸ ἕλκος κίνδυνός ἐστι γενέσθαι·
ἐξομαλίζειν τε χρὴ πέσσοντα[2] πανομοίως, καὶ μήτε

216

too narrow at the point, for sometimes the sera that come are sticky and thick; there is therefore a danger that they will become blocked in the incisions if these are cut too narrow.

8. The vessels in the arm you must secure with ligatures, for the tissue that covers them is, in many persons, not closely attached to the vessel. So, since the tissue is slippery, the cuts in them both (i.e. vessel and tissue) do not occur at the same place. The vessel then comes to puff out, since it is covered, and the flow of blood is impeded; in many persons pus as well forms for this reason. Indeed, such an operation seems to bear a double harm: pain to the person incised, and to the one incising great ill-repute. The same thing (i.e. use of ligatures) is recommended for all vessels.

9. These then are the instruments necessary in the surgery, and with which the learner must be proficient. The tooth-forceps and the uvula-forceps anyone can employ, since their use appears to be straightforward.

10. Growths and lesions, which are among the major disorders: growths, it must be understood, it is most proficient to be able to resolve, and to prevent their formation; but when one exists, to restrict it to a surface location as narrow as possible, and to make the consistency homogeneous through the whole growth. For if it is not homogeneous, there is a danger that the lesion will rupture and be difficult to treat. You must allow maturation to take place evenly all through the growth, and neither cut it

---

[1] Zwinger: δὴ δοκεῖ V.

[2] Froben: πεσόντα V.

διαιρεῖν πρότερον μήτε αὐτόματον ἐὴν ῥαγῆναι· τὰ
δὲ ἐκπέψαι[1] δυνάμενα ὁμαλῶς ἐν ἑτέροις εἴρηται.

11. Τὰ δ' ἕλκεα δοκεῖ πορείας ἔχειν τέσσαρας,
μίαν μὲν ἐς βάθος· ταῦτα δ' ἐστὶ τὰ συριγγώδη καὶ
ὅσα ὕπουλά ἐστι, καὶ ἔντοσθεν κεκοιλασμένα· ἡ δ'
ἑτέρη εἰς ὕψος, τὰ ὑπερσαρκεῦντα· τρίτη δέ ἐστιν εἰς
πλάτος· ταῦτα δ' ἐστὶ τὰ καλεόμενα ἑρπηστικά·
τετάρτη ὁδός ἐστιν, αὕτη δὲ μόνη κατὰ φύσιν εἶναι
δοκέει κίνησις. αὗται μὲν οὖν ξυμφοραὶ τοιαῦται
σαρκός εἰσι, πᾶσαι δὲ κοιναὶ τοῦ ξυμφύοντος·[2] καὶ
τὰ μὲν τούτων ἐν ἑτέροις σημεῖα δεδήλωται, καὶ ᾗ
χρηστέον ἐστὶν ἐπιμελείᾳ, δι' ὧν δὲ τὸ ξυμφυόμενον
218 | διαλυθήσεται· καὶ τὸ πληρεύμενον, ἢ κοῖλον γινό-
μενον, ἢ τὴν εἰς πλάτος πορείαν ποιούμενον, προση-
κόντως περὶ τούτων ἐν ἄλλοις εἴρηται σημεῖα.

12. Περὶ δὲ καταπλασμάτων ὧδε· τῶν ἐπιτιθεμέ-
νων ὀθονίων ὅκου ἂν ἡ χρῆσις κατὰ τοῦ νοσεύματος
ἀκριβὴς εἶναι δοκέῃ, καὶ τῷ ἕλκει ἁρμόζου[3] τὸ ἐπι-
τιθέμενον ὀθόνιον, τῷ δὲ καταπλάσματι πρὸς τὸν
κύκλῳ τόπον τοῦ ἕλκεος χρῶ· χρῆσις γὰρ αὕτη
καταπλάσματός ἐστιν ἔντεχνός τε καὶ πλεῖστα ὠφε-
λεῖν δυναμένη· ἐδόκει γὰρ τῷ μὲν ἕλκει βοηθεῖν ἡ
τῶν περιτιθεμένων δύναμις, τὸ δ' ὀθόνιον φυλάσ-
σειν· τὰ δ' ἔξω μὲν τοῦ ἕλκεος τὸ κατάπλασμα ὠφελέει. τὴν μὲν οὖν χρῆσιν αὐτῶν εἶναι δεῖ τοιαύτην.

13. Περὶ δὲ καιρῶν, ὁκότε τούτοις ἑκάστοις χρη-

---

[1] Froben: ἐκπέμψαι V.   [2] Littré: -φέροντος V.

open beforehand, nor allow it to rupture spontaneously. Things capable of bringing about an even maturation have been discussed elsewhere.

11. Growths appear to have four modes of advance. First, downward: these are the fistulous growths, and such as are beneath the surface of the tissue, and hollowed out from inside. Second are those which form an excess of tissue upwards. Third are ones which grow laterally, and these are called herpetic (i.e. creeping). There is a fourth mode, and it alone seems to be a natural progression.[a] These then are the kinds of things that happen to tissue, and they all share the same process of condensation. Their signs have been pointed out elsewhere, and the kind of treatment to be used, and the means by which what has condensed will be resolved. The signs of those which are over-grown, or hollowed out, or making an excursion to the side are adequately recorded elsewhere.

12. Cataplasms. When the use of linen bandages with medications applied to them seems indicated in an ailment, fit the medicated linen to the lesion, and employ the cataplasm around the area of the lesion in a circle. For this use of the cataplasm is in accordance with the art, and can bring benefit in very many cases. For the strength of what is placed around the outside seems to benefit the lesion, while the linen bandage protects it, and the cataplasm helps the area outside the lesion. The method of their application, then, must be such.

13. As for the times when it is opportune to use each

---

[a] See Daremberg pp. 71f., note 32.

[3] Aldina: -ον V.

στέον ἐστί, καὶ τὰς δυνάμιας ὡς χρὴ τῶν γεγραμμένων καταμανθάνειν, παραλέλειπται δὲ τὰ τοιαῦτα,
ἐπεὶ[1] πλείω προῆκται τῆς κατ' ἰητρικὴν ἐπιμελείας
καὶ πόρρω τοῦ τῆς τέχνης ἤδη προεληλυθότος ἐστίν.

14. Ἐχόμενον δὲ τούτων ἐστὶ καὶ κατὰ στρατιὴν
γινομένων[2] τρωμάτων χειρουργίη περὶ τὴν ἐξαίρεσιν
τῶν βελέων. ἐν τῆσι κατὰ πτόλιν διατριβῇσι βρα-
χεῖά τίς ἐστι τούτων ἡ χρῆσις· ὀλιγάκις γὰρ ἐν
παντὶ τῷ χρόνῳ γίνονται πολιτικαὶ στρατιαὶ καὶ
220   πολεμικαί· | ξυμβαίνει δὲ τὰ τοιαῦτα πλειστάκις καὶ
ξυνεχέστατα περὶ τὰς ξενικὰς στρατιὰς γίνεσθαι.
τὸν μὲν οὖν μέλλοντα χειρουργεῖν στρατεύεσθαι δεῖ
καὶ παρηκολουθηκέναι στρατεύμασι ξενικοῖς· οὕτω
γὰρ ἂν εἴη γεγυμνασμένος πρὸς ταύτην τὴν χρείαν.
ὃ δὲ εἶναι δοκεῖ περὶ ταῦτα τεχνικώτερον, εἰρῆσθαι·
τῶν γὰρ ὅπλων ἐνόντων καὶ σημεῖα πεπορίσθαι
τέχνης ἐστὶ πλεῖστον μέρος καὶ τῆς πρὸς ταῦτα χει-
ρουργίης· τούτου γὰρ ὑπάρξαντος, οὐκ ἂν παραλί-
ποιτο τρωματίας ἀγνοηθεὶς ὅταν χειρουργῆται μὴ
προσηκόντως· μόνος δ' ἂν ὁ τῶν σημείων ἔμπειρος
εἰκότως ἐπιχειροίη. περὶ δὲ τούτων ἁπάντων ἐν ἑτέροις γεγραμμένον ἐστίν.

[1] Froben after Calvus' *quoniam*: ἐπὶ V.
[2] Froben: -μένην V.

of these things, and how one should learn the properties of the things that have been described, such things have been omitted, since they go further into the practice of medicine and are appropriate to the student who has already made considerable progress in the art.

14. Related to this is the surgery of wounds arising in military service, which concerns the extraction of missiles. In city practice experience of these is but little, for very rarely even in a whole lifetime are there civil or military combats. In fact such things occur most frequently and continuously in armies abroad. Thus, the person intending to practice this kind of surgery must serve in the army, and accompany it on expeditions abroad; for in this way he would become experienced in this practice. What seems to be the more technical side of these matters has been stated: to know the signs of weapons lodged in the body is the greatest part of the art and of the surgery concerning these things. For if this be known, the wounded man will not go unrecognized when he has not been suitably operated upon. Only someone who is knowledgeable about these signs can properly put his hand to the task. All this has been described in other writings.

USE OF LIQUIDS

INTRODUCTION

In spite of the tenuousness of its transmission—the text is present in only one mediaeval manuscript, A of the eleventh century[1]—and its apparently unfinished and/or mutilated state, the Hippocratic ancestry of this treatise is firmly established by several pieces of evidence. Extensive textual borrowings exist between *Use of Liquids* and *Aphorisms* 5, 16–25, although which work is the borrower and which the source has not been firmly established.

Furthermore, Erotian includes seven words from *Use of Liquids* in his *Glossary*,[2] although the name of the treatise is never given in these articles. In his preface, however, the title *On Waters* (περὶ ὑδάτων) appears among the dietetic works,[3] a title confirmed by Athenaeus, *Deipnosophistae* 2. 46b, where a paraphrase of the opening of *Use of Liquids* is ascribed to "Hippocrates in the περὶ ὑδάτων." In one gloss Erotian indicates that the work was known to Bacchius of Tanagra

---

[1] A second manuscript, Parisinus Graecus 2255 (= E) is, as Hans Diller demonstrated, at this point copied from the Froben edition: H. Diller, *Die Überlieferung der hippokratischen Schrift* ΠΕΡΙ ΑΕΡΩΝ ΤΔΑΤΩΝ ΤΟΠΩΝ, Leipzig, 1932, pp. 13ff.

[2] See Nachmanson pp. 329f.

[3] Erotian p. 9.

(third century B.C.),[4] and in another to Glaucias of Tarentum (second century B.C.).[5]

Galen, too, includes words unique to this treatise in his Hippocratic *Glossary*.[6] He also refers to the treatise twice by the title *Use of Liquids*, in his commentaries on *Aphorisms* 5, 16[7] and *Epidemics VI* 3,6,[8] and once quotes from it, albeit with an incorrect attribution to *Epidemics VI*, in his commentary to that work.[9]

True to its title, *Use of Liquids* instructs its reader on the external application of liquids—fresh water, salt water, vinegar, wines—to warm, cool, clean, soften, moisten, dry, and soothe the body.

This edition of *Use of Liquids* is based on Heiberg's CMG edition and on my collation from microfilm of the manuscript A. The resulting text is often obscure, requiring the translator to divine a meaning where none is clearly apparent.

[4] S.v. A 48.
[5] S.v. A 49.
[6] E.g. Galen vol. 19, 71 *αἰθόλικες*.
[7] Galen vol. 17B, 801.
[8] Galen vol. 17B, 36 = CMG V 10,2,2, p. 142.
[9] Galen vol. 17A, 891 = CMG V 10,2,2, p. 57.

ΠΕΡΙ ΥΓΡΩΝ ΧΡΗΣΙΟΣ

1. Ὕδωρ ποτόν· ἁλμυρὸν θάλασσα.

Ποτὸν μὲν κατ᾽ ἰητρεῖον κράτιστον· καὶ γὰρ
σιδηρίοισι καὶ χαλκείοισι κράτιστον, καὶ φαρμάκοισι τοῖσι πλείστοισι παλαιουμένοισι κοινότατον.
ἐς δὲ χρῶτα γνῶναι δεῖ, ὅτι τοῦτο ἢ τέγξει ἢ ψύξει
ἢ θέρμῃ, ἄλλῳ δὲ οὐδενὶ ὠφελέει ἢ βλάπτει πο-
τόν. ὅπου ὀλίγου ποτοῦ, σπόγγῳ· χρῆσις ἄριστον
ὀφθαλμοῖσιν. <εἰ>[1] αὐτὸ τὸ δέρμα ἐφήλκωται,
θερμῷ, αἰόνησις· πυρίη τοῦ σώματος ἅπαντος ἢ μέ-
ρεος, δέρματος σκληροῦ μάλθαξις, συντεταμένου
χάλασις νεύρων, σαρκῶν ἐκχύμωσις, ἱδρῶτος
ἄφοδος· ὑγρῆναι, προκλύσαι[2] οἷον ῥίνας, κύστιν,
φύσας· σαρκῶσαι, ἁπαλῦναι, τῆξαι, μινυθῆσαι,
χροιὴν ἀνακαλέσαι, χροιὴν ἀνασκεδάσαι. ὑπνικὸν
καὶ κατὰ κεφαλῆς καὶ ἄλλων· σπασμῶν, τετάνων
παρηγορικόν· ὀδύνας κωφοῖ ὠτός, ὀφθαλμῶν, ὅσα
τοιαῦτα. τὰ ψυχρὰ θερμῆναι, οἷον πίσσα, ἕλκεσι,
πλὴν τοῖσιν αἱμορραγεῦσιν ἢ μέλλουσι, κατήγμα-

---

[1] Littré.
[2] Foes in note after Cornarius' *prolutione*: -κλήσῃ A.

USE OF LIQUIDS

1. Drinking water: salty sea-water.

Drinking water is the best for use in the surgery. It is the best for instruments of iron and bronze, and the most commonly used for most medications that are conserved. You must know that this drinking water, when used on the skin, helps or harms by moistening, by cooling, or by heating, but in no other way. Where a little drinking water is being applied, do it with a sponge; this usage is the best for the eyes. If the skin itself is ulcerated, you will moisten it with warm water. A fomentation of the whole body or of part of it effects a softening of hard skin, a relaxation of cords when the body is rigid, a removal of fluid from the tissues, and an evacuation of sweat. Warm water is useful to moisten, to wash out beforehand,[a] as for example the nostrils, bladder, or flatus from the cavity; to promote the growth of tissue, to make plump; to reduce, to diminish; to restore colour, to dissipate colour. Poured over the head and other parts of the body, it promotes sleep; it is soothing to convulsions and spasms; it blunts pains of the ear, of the eyes, and such like. It warms what is cold, as pitch does, in sores other than those that are

[a] With A's reading: "provoke".

σιν, ἐκπτώμασι, τοῖσιν ἄλλοισι, οἷσιν ἂν ὀθόνια
120 ἰητρός· καρηβαρίη. τὸ μέτριον | ἑκάστῳ, μὴ πρόσω,
οἴδαμεν βασανίζειν, οἷον τὸ θερμὸν τῷ χρωτί, ἐξ
ὑπερβολῆς ἐφ᾽ ἑκάτερα, ὡς ἀμφοῖν μὴ ἁμαρτάνειν,[1]
σημαινόμενος ταῖς βλάβαις ἢ οὐκ ὠφελείαις, οἷον
χλιαροῦ· δεῖ γὰρ τῇσι βλάβῃσιν ἢ τῇσιν ὠφελείῃ-
σιν κἂν ὦσι, χρῆσθαι μέχρι τοῦ ὠφελέοντος ἢ
μέχρι τοῦ βλάπτοντος.

Τέγξις μὲν οὖν, ἀσθενές· ψύξις δὲ καὶ θάλψις,
ἰσχυρόν, ὡς ἐξ ἡλίου· τὸ δὲ ψυχρόν, θερμὸν ἐὸν ὡς
ποτὸν ἀσθενέοντι· ἀλλὰ τὸ μὲν θερμὸν μὴ πρόσω
καίειν, κρίνει δ᾽ αὐτός, πλὴν τοῖσιν ἀφώνοισιν, ἢ
παραπληγικοῖσιν, ἢ νεναρκωμένοισιν, ἢ οἷα ἐπὶ
τρώμασι κατεψυγμένοισιν ἢ ὑπερωδύνοισι, τούτοισι
δὲ ἀναίσθητα· λάθοις γὰρ ἂν κατακαύσας· καὶ τὰ
ἐκπτώματα δὲ τὰ βαθέα καὶ τὰ μεγάλα· ἤδη καὶ
πόδες ἀπέπεσον, καταψυχθέντες, ἐκ καταχύσιος θερ-
μοῦ· ἀλλὰ τούτοισιν ὁ τοῦ καταχέοντος χρώς, κρι-
τής. καὶ ψυχροῦ δὲ ὡσαύτως. τούτων δ᾽ αὐτῶν τὸ
ὀλίγον ἑκατέρου, ἀσθενές· τὸ δὲ πολύ, ἰσχυρόν·
ἀλλὰ μὴν ἐὰν, μέχρι γένηται οὗ ἕνεκα ποιέεται· τὸ
ἔσχατον προπαύειν πρὶν γενέσθαι· τούτων δὲ ἑκάτερον[2] βλάπτει. βλάπτει δὲ ταῦτα τὸ θερμὸν πλέον ἢ
πλεονάκις[3] χρεομένοισι, σαρκῶν ἐκθήλυνσιν, νεύρων
ἀκράτειαν, γνώμης νάρκωσιν, αἱμορραγίας, λειπο-
θυμίας, ταῦτα ἐς θάνατον· τὸ δὲ ψυχρόν, σπασμούς,

[1] Heiberg: -νει A.   [2] Zwinger: -ων A.
[3] ἢ πλεονάκις deleted by A[2].

bleeding or about to, in fractures, in dislocations and other conditions in which the physician uses linen bandages, and in headache. The right measure in each case—not to go too far—we know how to discern, as for example what is hot on the skin, from the evidence of an excess in either direction; in order not to make a mistake in the one direction or the other, we conjecture from seeing that we are doing harm, or from seeing that we are not doing any good, as for example in the application of lukewarm water. For you must draw conclusions from when harm or help becomes evident, and so continue to employ the measure until some benefit or harm results.

Moistening, then, is a weak measure, while cooling and heating have strong effects, as for example from the sun. Cold water, heated as a drink for the patient, is good; but do not heat it too much. Let the patient himself judge, except in cases of loss of speech, paralysis, numbness, or anaesthesia in wounds that are chilled or very painful. For these patients you could severely burn without it being noticed. Also dislocations that are profound and severe: there have been cases where frozen feet have been lost after being treated with hot affusions. In these cases, the touch of the person who is making the affusions is the judge, also of cold ones in the same way. A small amount of either of these is weak, a large amount strong. Let the treatment continue until that occurs which it was meant to achieve, but stop before any excess; each of these does damage. They harm those who use too much hot water or use it too often by causing a softening of the tissues, powerlessness of the cords, paralysis of judgement, haemorrhages, or loss of consciousness—even to the point of death. Cold water provokes convulsion,

τετάνους, μελασμούς, ῥίγεα πυρετώδεα. αἱ μὲν
μετριότητες ἐκ τούτων. τὰ δ' ἄλλα βλάπτει καὶ
ὠφελέει τὰ εἰρημένα ἡδονῆσι καὶ εὐφορίησι καὶ
ἀχθηδόσι καὶ δυσφορίησιν, αἳ καθ' ἓν ἕκαστον
αὐτῶν ὁμολογοῦσαι φαίνονται.

122   2. Ὑγιὲς σῶμα, τὸ μὲν ἐν σκέπῃ εἰθισμένον, ὅτι
ἄηθες, ὅτι προσωτάτω τοῦ οἰκείου θάλπεος, ἐγγύτατα δὲ τοῦ ἀλλοτρίου ψύχεος, διὰ ταῦτ' ἄρα τῷ
θερμῷ ἥδεται καὶ φέρειν δύναται. ἐγκέφαλος καὶ
ὅσα ἀπὸ τοιούτων, ψυχρῷ μὲν ἄχθεται, θερμῷ δ'
ἥδεται, καὶ ἢν ἄρα ᾖ ψυχρότερον καὶ στερεώτερον
φύσει. καὶ πρόσω τοῦ[1] οἰκείου θάλπεος, καὶ
πλεῖστα αὐτῶν· διὰ τοῦτο ὀστέοισιν, ὀδοῦσι, νεύροισι τὸ ψυχρὸν πολέμιον, τὸ δὲ θερμὸν φίλιον, ὅτι
ἀπὸ τούτων σπασμοί, τέτανοι, ῥίγεα πυρετώδεα, ἃ
τὸ μὲν ψυχρὸν ποιέει, τὸ δὲ θερμὸν παύει. διὰ τοῦτο
καὶ γονῇ τὸ θερμὸν ἡδοναὶ καὶ προκλήσιες, ἀπὸ δὲ
τοῦ ψυχροῦ ἀλγηδόνες καὶ ἀποτρέψιες· διὰ τοῦτο
ὀσφύος, στήθους, νώτου, ὑποχονδρίου[2] μᾶλλον ψυχρῷ
ἄχθεται,[3] θερμῷ δ' ἥδεται καὶ φέρειν δύναται· διὰ
τοῦτ' ἄρα ὀσφύς, νῶτον, στῆθος, ὑποχόνδριον τὰ
ἐναντία πέπονθεν, ὅτι ἐναντίαἰ· διὰ τοῦτ' ἄρα τούτῳ
τῷ χωρίῳ ἀσῶδες ἐνταῦθα χρωμένοισι τὸ θερμόν,[4]

[1] Zwinger in margin: πρὸς ὅτου A.
[2] A: ὀσφύς, στῆθος, νῶτον, ὑποχόνδρια A[2] and editors.
[3] A[2] in margin: ἄλλοστε A.
[4] A: τῷ θερμῷ A[2].

spasms, darkening of the flesh and febrile chills. Moderate usages follow from these things. In general, the measures discussed harm and help in relationship to the states of pleasure, well-being, annoyance and discomfort which appear to coincide with each of them.

2. A healthy body, that part of it that is usually sheltered—because heat is unfamiliar to it because it is furthest from the body's natural heat and closest to the alien cold—enjoys heat and can bear it, no doubt for those very reasons. The brain and other parts of similar material are vexed by cold and take pleasure in heat, even though they themselves are by nature colder and solider, and most of them are also located far from the body's own heat. For this reason cold is hostile to bones, teeth and cords, and heat is congenial to them, because convulsions, spasms and febrile chills arise from them, and these are things which are provoked by cold and stopped by heat. For the same reason to the parts of generation heat is pleasure and stimulation, whereas from cold they feel pain and aversion. For the same reason these parts are vexed more by cold than the loins, chest, back and hypochondrium,[a] and take pleasure in heat and can bear it; for the same reason, no doubt, loin, chest, back and hypochondrium sense the opposites since they are themselves opposite.[b] And for the same reason in this region (sc. the hypochondrium) heat is nauseating for those who employ it there,

[a] With the reading of A^2: "so also the loins, chest, back and hypochondria are vexed by cold. . . ."

[b] I.e. they are warm; with A^2's reading they would be cold. That in fact they are considered warm is proven by the sentence that follows.

τὸ δὲ ψυχρὸν παύει· διὰ τοῦτ' ἄρα ποτῷ ψυχρῷ ἥδεται, διὰ τοῦτ' ἄρα βρώμασι τοῖσι θερμοῖσιν ἥδεται.
διὰ τοῦτ' ἄρα καὶ κατὰ ἀκρέων ἐν λειποθυμίαις τὸ
ψυχρὸν καταχεόμενον ὠφελέει. ὅτι δὲ τὰ ὄπισθεν
τῶν ἔμπροσθεν τὸ θερμὸν μᾶλλον ἀνέχεται, τὰ εἰρημένα αἴτια· ὅτι δὲ καὶ τὰ ψυχρά, ὀρθῶς· τά τε γὰρ
ἄκρεα ἔμπροσθεν καὶ οὐκ ἐν σκέπῃ εἰθισμένα, |
ὥσπερ οἱ ἔνδον τοῦ ἔξω. μνηστέον δὲ καὶ ὅτι ἑκάτερον ἐφ' ἑκατέρου τοῦ σώματος κρέσσον τοῦ δέρματος τοῦ ἔξω, ὅτι συνεχές τε ἑωυτῷ καὶ νεύρῳ ἐναίμῳ·
διὰ τὸ ἔξω τοῦ οἰκείου θερμοῦ ἐν τῷ ἔξω ψυχρῷ[1]
εἶναι, ἐπ' ἀμφοῖν πυκνὰ κρατέεται, καὶ ἑκατέρων
πυκνὰ δεῖται, πυκνὰ δὲ μᾶλλον θερμοῦ ἐς ἡδονήν.
πάσχει δὲ καὶ ἄκρεα τοιοῦτο, ὥστε ταχέως ἐνακούειν
πολλῶν· βραδέως δὲ πρῶτον ἐπαείρεται, καὶ φλεβῶν
δῆλον, αἱ πρότερον καὶ ὕστερον· οὕτως πάντα φαίνεται, ὅκου τε ψύχεται ἄκρεα, ὅκου τε θερμαίνεται, ἐν
κενεαγγικοῖσιν, ἐν λειποθυμικοῖσι, καὶ κατὰ λόγον·
εἰκότως ἄρα φλέβας, καὶ τὰ ἀπὸ τούτων ἕπεται, καὶ
θερμαίνει πρῶτον τὸ θερμόν, οἷον τῶν χειρῶν τὰ
εἴσω. τὰ δ' ἕλκεα θερμῷ ἥδεται ὁμολογουμένως, ὅτι
ἐν σκέπῃ εἴθισται, εἰκότως ἄρα τῷ ἑτέρῳ ἄχθεται.
εἰκότως ἄρα καὶ αὐταὶ αἱ φλέβες, ὅτι ἐν θερμῷ εἰσιν.
εἰκότως ἄρα καὶ αὐτὸς ὁ θώρηξ, καὶ αὐτὴ ἡ κοιλίη
κρατουμένη ὑπὸ τοῦ ψυχροῦ μάλιστα ἀγανακτεῖ καὶ
θανατοῖ, ὅτι μάλιστα ἀήθεα·[2] ἀλλὰ πλεῖστον ἀπέχει

---

[1] Zwinger in margin: -χρᾶ A.   [2] Aldina: -έας A.

and cold stops the nausea. For the same reason again cold drinks provide pleasure, as do warm foods. For the same reason, where there is a loss of consciousness, cold water poured over the extremities is beneficial. What has been said is the reason why the posterior parts bear heat better than the anterior ones; so, too, do the colder parts; for the extremities are anterior and not accustomed to being sheltered as the internal parts are from the exterior. You must also remember that either one of them (sc. heat or cold) applied on either side of the body (sc. anterior or posterior) is more powerful than the external skin, which is continuous with itself and with the sanguineous cord. Because the skin is outside the body's own heat and is in the external cold, it is often dominated by them both, and often needs one or other of them, but more often heat, for pleasure. The extremities are affected in the same way, and so respond quickly to many influences; first they slowly swell up, and this becomes visible in the vessels— in some sooner, in others later. This is how everything occurs when the extremities are cooled or warmed in persons with empty vessels, in losses of consciousness, and so on. Naturally the heat pursues the vessels and heats them and the parts growing out of them, like the inner parts of the hands. Lesions, as is commonly agreed, take pleasure in heat, since they are accustomed to be sheltered; naturally they are vexed by the opposite. And naturally also the vessels themselves, since they are in a warm place. Naturally also both the chest itself and the cavity itself, on being overpowered by cold, are most violently irritated and mortified, since they are most unaccustomed to it.

τοῦ παθεῖν· τούτου γὰρ ἐγγύτατα [τοῦτο]¹ τὸ δεῖ-
σθαι· εἰκότως ἄρα ποτῷ ψυχρῷ ἥδεται· οὕτως ὁμο-
λογέει ταῦτα πάντα. ὅτι δὲ μάλιστα πάντα τὰ ἀποσύρματα καὶ τὰ ἐπικαύματα ἐπιπολῆς, οὐ μάλιστα
εἰθισμένα ἐν σκέπῃ, μάλιστα ἐν ψυχρῷ ἀγανακτεῖ
εἰκότως· τάχιστα γὰρ κρατέεται, καὶ τὰ βαθύτατα εἰ
κρατοῖτο, μάλιστ᾽ ἂν ἄχθοιτο· ἔπειτα καὶ τῆς
φύσιος τῆς νευρώδους μετέχει. ὅτι ὑπογά|στριον
ἥδεσθαι δοκέει θερμῷ, σκεπτέον τὸ χωρίον, καὶ μετ-
έχει· καὶ ἄκρεα, καὶ κύστις, καὶ γονή, καὶ ἄλλως²
γόνος ὁ γυμνός· ἔστι τε φύσει ψυχρότερος ἢ ὥς τις
οἴεται· ἄνω γάρ, οὐ κάτω θερμὸν ἀΐσσει· διὰ ταῦτα
ἥδεται. ὅτι μετὰ τὸ θερμὸν ψύχεται τὸ σῶμα δια-
χυθὲν³ μᾶλλον, μετὰ δὲ τὸ ψυχρὸν ἀναθερμαίνεται
μᾶλλον συσταλέν· οἷον καὶ τὰ ὕδατα, ψυκτέα, θερ-
μαντέα, διὰ λεπτότητα· ὅτι μετὰ τὸ θερμὸν σκλη-
ρύνεται μᾶλλον ἐπιξηρανθέν, οἷον ὀφθαλμοὶ μετὰ τὸ
ψυχρόν· τὸ μὲν γὰρ ὅμοιον τῷ περιέχοντι, τὸ δὲ οὔ.

3. Θάλασσα δέ, τοῖσι κνησμώδεσι καὶ δακνώ-
δεσι, καὶ λούειν καὶ πυριῆν θερμῇ· τοῖσι μὲν ἀνθή-
μασιν⁴ ὑπολιπαίνονται, πρὸς ἕλκεα δὲ πυρίκαυτα
καὶ ἀποσύρματα καὶ ὅσα τοιαῦτα πολέμιον, ἐπιτή-
δειον δὲ τοῖσι καθαροῖσιν, ἀγαθὸν καὶ ἰσχναίνειν εὖ,
ὡς τὰ τῶν ἁλιέων ἕλκεα· ταῦτα γὰρ οὐδ᾽ ἐκπυεῖ, ἢν
μὴ ψαύῃ· καὶ πρὸς ὑποδεσμίδας· καὶ τὰ νεμόμενα
παύει καὶ ἵστησιν, ὡς ἅλες καὶ ἁλμυρίδες καὶ

¹ Del. Heiberg.   ² Zwinger in margin: -ος A.

But the person is very far from sensing it, since the need of cold is nearest; naturally he takes pleasure in a cold drink. Thus all these things agree. That particularly all abrasions and surface burns not occurring in locations very accustomed to be sheltered are especially vexed by the cold, is natural, for they are very quickly overpowered by the cold; also if the deepest parts should be overpowered, they would be most vexed, since they also partake of a cord-like structure. Since the hypogastrium seems to take pleasure in heat, you must pay attention to that area; it also shares a cord-like structure. Also the extremities, the bladder, and the parts of generation, especially the exposed ones, are colder in nature than what one thinks; for heat moves upward, not downward. For this reason they take pleasure in heat. Note that after a warm affusion the body cools off, because it is more dilated, and after a cold affusion warms up, because it is more contracted, just as waters that are to be cooled or heated do, on account of their fineness. After an affusion of hot water the body becomes harder because it is more dried out, just as the eyes do after a cold affusion. For the body is like what is around it, but the eyes are not.

3. Sea-water is useful for itching and irritated skin: wash and foment with it warm. On eruptions it is anointed in a small amount. To lesions arising from burns and abrasions and the like it is hostile, but suitable for clean ones. It benefits and dries up well lesions such as those of fishermen; for these do not suppurate, if they are not touched. It is also used on bandages applied next the skin. It stops and limits lesions that are spreading, as do

---

[3] Littré after Triller: -ψυχθὲν A. [4] Potter: ἀήθεσιν A.

λίτρον· πάντα δὲ ταῦτα σμικρῷ μὲν χρωμένῳ ἐρεθι-
στικά, προσνικῶντι δὲ ἀγαθόν· βέλτιον δὲ θέρμη
πρὸς τὰ πλεῖστα.

4. Ὄξος δὲ χρωτὶ μὲν καὶ ἄρθροισι παραπλή-
σιον θαλάσσῃ καὶ δυνατώτερον καταχέειν καὶ
πυριᾶν· καὶ ἕλκεσι τοῖσι νεοτρώτοισι, θρόμβοισιν,
οὗ μέλασμα αἰδοίων, καῦσις οὐάτων ἢ καὶ ὀδόντων·
128 | θερμῷ δὲ ταῦτα, τά τε ἄλλα· καὶ τῇ ὥρῃ συν-
τεκμαίρεσθαι ἐκ τήξεως ἁλός·[1] καὶ πρὸς ἄλλα δὲ
ὅσα λειχῆσι, λέπρησιν, ἀλφοῖσι, συντείνει παχυν-
θὲν ἐν ἡλίῳ θερμῷ, μάλιστα δὲ ὄνυξι λεπροῖσι, κρα-
ταιοῖ γὰρ μετὰ χρόνον. μυρμηκίας ἀπαλύνει, καὶ
τοὺς ἐν ὠσὶν [ἐώσῃ][2] ῥύπους, μαλάσσει δὲ καὶ
χρῶτα, πολλαχῇ δὲ καὶ ἄλλη, εἰ μὴ ὀδμὴ ἔβλαπτε,
καὶ μάλιστα γυναῖκας· ἐδύνατο δ᾽ ἂν καὶ ποδαγρίῃ,
εἰ μὴ ὁ χρὼς ἐτιτρώσκετο. ταῦτα καὶ τρὺξ ὄξους
ποιέει.

5. Οἶνος δὲ γλυκύς, ὅσα χρόνια τρώματα, συνε-
χέως χρωμένῳ αὔταρκες, ἀτὰρ καὶ ἐς φαρμακοπο-
σίην. αὐστηρὸς δὲ ὁ λευκὸς καὶ μέλας οἶνος ψυχρὸς
ἐπὶ τὰ ἕλκεα ἐνδέχεται, ψυχρὸς διὰ τὴν θέρμην.
ὅρια δέ, ὅσα μὲν ψύξιος εἵνεκα ἢ καταχεῖται ἢ ἐνίε-
ται ἢ ἐμβάπτεται,[3] ὡς ὕδωρ ψυχρότατον· ὅσα δὲ
στύψιος,[4] ὁ μέλας οἶνος, καὶ εἴρια καταρρῆναι, οἷον
καὶ φύλλα τευτλίων ἢ ὀθόνια βάπτεται ἐπὶ τὰ
πλεῖστα· ὅσα δ᾽ ἔτι[5] στύψιος, οἷον κισσοῦ φύλλα, ὁ

salt, brine and soda; all these are irritant if used in a small amount, but when used over time beneficial. Sea-water is better used hot in most cases.

4. Vinegar has about the same effect on the skin and joints as sea-water; it is more powerful when applied in affusions and vapour-baths. It is also useful for recent lesions and thromboses, or where there is a darkening of the parts of generation, or a burning in the ears or teeth. Use it warm in these and the other cases, and take into account the time from the melting of the salt. It also exerts an effect on other conditions, such as lichen, lepra and alphos, when it has thickened in the hot sun, and especially leprous nails, for in time it gains the upper hand. It mollifies warts, and filth in the ears, softens the skin, and would find use in a good number of other instances, if its odour were not offensive, especially to women. It would also be effective in gout, if the skin is not broken. Wine lees, too, do the same.

5. Sweet wine, applied continuously, is adequate in itself for wounds that are chronic, but is also used in medications that are drunk. Dry white or red wine can be used cold on lesions, cold because the lesions are hot. Rules: for the sake of cooling, let wine as cold as the coldest water be applied by affusion, infusion or bathing; as astringent, red wine, and wool soaked in it, as beet leaves and linen cloths are often immersed; another astringent,

---

[1] Heiberg after Foes' *ex eliquatione salis*: ἅλες A.

[2] Del. Heiberg.

[3] Mack: ἐνβλάπτεται A: λ erased by A[2].

[4] Zwinger in margin: ψύξιος A.

[5] Zwinger: ἔνι A.

λευκός, καὶ ὅσα στρυφνότερα ἢ ψαθυρώτερα, οἷον
130 κίστος[1] τε καὶ βάτος, ῥοῦς | σκυτοδεψική, ἐλελίσφα-
κος, καὶ ὅσα μαλθάσσειν δεῖ, οἷον ἄλητον ἐφθόν.

6. Τὸ ψυχρὸν ὠφελέει τὰ ἐρυθρά, οἷα ἄλλη καὶ
ἄλλη ἐκθύει ὑποπλάτεα, οἷα τοῖσι[2] τοὺς σπλῆνας
ὀγκηροὺς ἀνίσχουσιν, εὐσάρκοισι δὲ καὶ ἀπαλο-
σάρκοισιν ὑπέρυθρα, μέλασι δὲ καὶ οἷαι στρογγύ-
λαι ἐοῦσαι, αἰθόλικας[3] λέγουσι, καὶ ἐν αὐτῇσι τῇσι
θερμολουσίῃσιν ἀνίσταται γενόμενα, καὶ γυναιξὶν
ἐν[4] γυναικείων ἀναδρομῇ ὑπὸ χρῶτα, καὶ ὑπὸ ἐρε-
θισμῶν[5] δέρματος, ἢ ἱματίων τρηχείων ἀήθει φορή-
σει, ἢ ἐξ ἱδρώτων ἐξόδου, ἢ ἐκ ψύχεος ἐξαίφνης
πρὸς πῦρ ἐλθόντι ἢ λουτρὰ θερμά, ταῦτα ἢν ὕστερον
ποιήσῃ, ὅτι οὐδαμῶς ἐκθύει. ἐπεὶ ὁκόσα γε ἐκ
132 ψύχεος γίνεται, ἢ ὁκόσα τρηχύνεται κερ|χνώδεα,[6]
εἶτ' ἐφελκοῦται, τὸ μὲν ψυχρὸν βλάπτει, τὸ δὲ θερμὸν ὠφελέει. ἃ δὲ ἄμφω ὠφελέει, τὰ ἐν ἄρθροισιν
οἰδήματα, καὶ ἄνευ ἕλκεος ποδαγρικά, καὶ σπάσ-
ματα[7] πλεῖστα, τούτων[8] ψυχρὸν πολλὸν καταχεόμενον ἱδρῶσιν ἰσχναίνει καὶ ὀδύνην ναρκοῖ, νάρκη δὲ
μετρίη ὀδύνης ληκτικόν· καὶ τὸ θερμὸν ἰσχναίνει καὶ
μαλθάσσει. τοῖσι δὲ ποδαγρικοῖσι, παρέσεσι, τετά-
νοισι, σπασμοῖσι, τὰ τοιαῦτα· συντάσιες, τρόμοι,
παραπληγίαι, τὰ τοιαῦτα· χαλάσιες [ἢ χωλώσιες],[9]
νάρκαι, ἀναυδίαι, τὰ τοιαῦτα· κάτωθεν ἀπολήμψιες·

---

[1] Zwinger in margin: κισσός A.
[2] τοῖσι Joly: οἱ A.   [3] Mack: θόλικας A.
[4] Joly: ἐκ A.   [5] Heiberg: -ὸν A.

like ivy leaves, is white wine; also use white wine in place of the more astringent and friable leaves such as rock-rose and bramble, the sumach used in tanning, and the salvia, and also for agents that should soften, like boiled meal.

6. Cold applications are of benefit for extended red eruptions that break out in different places, as for example in patients with swollen spleens, for reddish lesions in soft and fleshy bodies, and in dark bodies also for the roundish lesions they call pustules. Also for lesions that arise in warm baths, that occur in women when the menses are directed under the skin, or that arise from irritation of the skin either through the unaccustomed wearing of a rough piece of clothing, or the secretion of sweat, or in a person who suddenly comes out of the cold and approaches a fire or enters a hot bath—if he does this later, there is no eruption at all. Any lesions that arise from the cold, or that become rough like millet, and then ulcerate, are harmed by cold and benefited by warmth. Things that are benefited by both are swellings in the joints, gout without ulceration, most spasms. Copious cold affusions over them dry up the sweating and numb the pain; moderate numbness resolves pain. Heat too dries and softens. In gouts, paralyses, convulsions, spasms, use the like; contractions, trembling, palsies, the like; lamenesses, numbnesses, losses of speech, the like. Stoppages from below: take heed in the employment

---

⁶ Aldina: ψερχνώδεα A.

⁷ Zwinger in margin after Cornarius' *convulsa*: σώματι A.

⁸ Heiberg: τοῦτον A.   ⁹ Littré, correcting χαλάσιες from A's χαλῶσι, and following Zwinger in deleting what appears to be an intruded marginal gloss.

φυλάσσεσθαι δὲ ἐν τῇ ψυχρᾷ χρήσει, ψυχροῖσι
χρεόμενος μᾶλλον ἢ τἀναντία. τὰ δὲ ἐς τὰ ἄρθρα
ἐσκληρυσμένα ἢ ὑπὸ φλεγμονῆς ὁκότε γενομένης ἢ
ἀγκύλης, προὔργου[1] ἰᾶσασθαι, ἐς ἀσκίον θερμοὺς
ἐγχέοντας, τὴν χεῖρα ἐναποδῆσαι. καὶ ὄμματα,
δακρύου παρηγορικὸν καταλείφοντα, πρὸς τὰ δρι-
μέα λίπος προσηνές, ὥστε μὴ ἅπτεσθαι τὸ ἁλμῶδες,
καὶ τοῖσι βοθρίοισι διάνιψις καὶ πλήρωσις ἐς φύσιν
ἄγουσα. ὀφθαλμοῖσι θερμὸν ὀδύνῃσιν, ἐμπυήσεσι,
δακρύων δακνωδέων, ξηροῖσιν ἅπασιν· τὸ ψυχρόν,
ἀνωδύνοισιν,[2] ἐξερύθροισι· τοῖσι δὲ εἰθισμένοισι
134 συστροφὰς κατὰ φλέ|βας ποιέει, οἷα χοιραδώδεα,
κατὰ θώρηκα, καὶ ἄλλα σκληρά· ἀρχῷ δὲ καὶ ὑστέ-
ρῃσιν οὐ πάνυ ἐνδέχεται, αἷμα ἐν ψύχει οὐρέουσιν.
ἕλκεσι τὸ μὲν ψυχρὸν δακνῶδες, δέρμα περισκληρύ-
νει, ὀδυνώδεα ἀνεκπύητα ποιέει, πελιαίνει, μελαίνει,
ῥίγεα πυρετώδεα, σπασμούς, τετάνους. ἔστι δὲ ὅκου
ἐπὶ τετάνῳ ἄνευ ἕλκεος νέῳ εὐσάρκῳ, θέρεος μέσου,
ψυχροῦ πολλοῦ κατάχυσις θέρμης ἐπανάκλησιν
ποιέει· θέρμη δὲ ταῦτα ῥύεται, τοῖσι δὲ ἐν κεφαλῇ
καὶ καρηβαρίας. τὸ θερμὸν ἐκπυητικόν, οὐκ ἐπὶ
παντὶ ἕλκει, μέγιστον σημεῖον ἐς ἀσφάλειαν, δέρμα
μαλθάσσει, ἰσχναίνει, ἀνώδυνον, ῥιγέων, σπασμῶν,
τετάνων παρηγορικόν· τὸ δ' ἐν κεφαλῇ, καὶ καρηβα-
ρίαν λύει· πλεῖστον δὲ διαφέρει ὀστέων κατήγμασι,
μᾶλλον δὲ τοῖσιν ἐψιλωμένοισι, τούτων δὲ μάλιστα

[1] Linden: -ργ' A.

of cold, if you use cold more than its opposite. Useful to heal indurated joints that have arisen either through the occurrence of inflammation or from stiffness is to pour hot fluids into a wine-skin and bind the arm to it. It soothes the tearing of the eyes to smear on a gentle fat substance acting against what is acrid, so that the salty fluid does not touch them; for ulceration let there be a washing and filling up to the natural state. In the eyes use warm agents against pains, against suppurations that have arisen because of irritating tears, and against all forms of dryness; cold agents for painless disorders or for erythemas. In those that are accustomed to it, cold provokes convolutions in the vessels, something like scrofulas, in the chest and other rigid structures. For the anus and the uterus, cold is not very applicable, nor in persons who pass blood in their urines in cold weather. On sores cold is an irritant, it makes the skin all around hard, it causes pains without suppuration, it makes them livid or dark, and it causes febrile chills, convulsions and spasms. Sometimes, in a tetanus without a wound occurring in the middle of summer in a robust young man, a copious affusion of cold water effects a restoration of heat; and heat relieves these things, as well as symptoms in the head, and its heaviness. Heat, which provokes suppuration, although not in every wound, is the best sign of recovery; it softens the skin, reduces swelling, removes pain, and soothes rigors, convulsions and spasms. Also in the head it relieves heaviness. It is also of the greatest value for fractures of bones, especially if they are denuded, and

---

[2] Zwinger in margin: -δύνους A.

τοῖσιν[1] ἐν κεφαλῇ τρώμασιν ἑκουσίοισι καὶ ἀκουσίοισι, καὶ ὅσα ὑπὸ ψύχεος ἢ θνήσκει ἢ ἑλκοῦται· ἔτι[2]
ἑλκώμασιν δ᾽ ἑκουσίοισί τε καὶ ἀκουσίοισι, ἀποσύρ-
μασι, ἕρπησι, ἐσθιομένοισι, μελαινομένοισι ἐν νού-
σοισι, ἢ ἐν ἀκοῇ, ἢ ἐν ἕδρῃ ἢ ὑστέρῃ, τούτοισι πᾶσι
136   τὸ θερμὸν φίλιον | καὶ κρῖνον, τὸ δὲ ψυχρὸν πολέ-
μιον καὶ κτεῖνον, πλὴν ὁκόσα αἱμορραγέειν ἐλπίς.

7. Οὕτω κατάχυσις ὑγρῶν, ἐπίχρισις[3] ἀλειπτῶν,
ἐπίθεσις φύλλων ἢ ὀθονίων, κατάπλασις, ὁκόσα ἢ
ψῦξις ἢ θέρμη ὠφελέει ἢ βλάπτει.

---

[1] Linden: τῶν Α.
[2] Froben: ἐπὶ Α.
[3] Littré: -χρήσας Α.

most particularly in injuries, either intentional or accidental, in the head, and wherever, because of cold, tissue dies or is ulcerated; also for intentional or accidental ulcerations, for abrasions, shingles, corrosions and darkening of the skin in diseases, or for the ear, anus or uterus—in all these heat is favourable and promotes the crisis, whereas cold is hostile and mortal, except in cases where there is an expectation of bleeding.

7. Thus it is with the affusion of liquids, the anointing of ointments, the binding on of leaves or bandages, and the use of cataplasms whenever either cold or heat benefits or harms.

ULCERS

INTRODUCTION

This treatise is known under this name to both Erotian and Galen. Erotian lists it among the therapeutic works in his preface,[1] and includes about a dozen words from it in his *Glossary*.[2] Galen, besides including words from it in his *Glossary*,[3] uses the treatise's account as the basis of his discussion of ulcers in *Methodus Medendi* 4, quoting frequently and extensively from the text.[4] In his *Commentary to Articulations*, he even refers to a now lost commentary he wrote to *Ulcers*.[5]

Ulcers presents a thorough and well-ordered account of surface lesions, arranged under the heads pathology (ch. 1–10), general therapy (ch. 11–17), and special therapy (ch. 18–27). The treatise is convincing testimony to its author's extensive clinical experience of the signs of superficial disorders, their modes of healing, and their

[1] Erotian p. 9.

[2] See Nachmanson pp. 358–61.

[3] Galen vol. 19, 108 Καρικόν; 19, 121 μέλανι φαρμάκῳ.

[4] Galen vol. 10, 274–95.

[5] Galen vol. 18A, 693. J. Ilberg ("Über die Schriftstellerei des Klaudios Galenos," *Rheinisches Museum* N.F. 44 (1889), p. 230) sets this commentary in the time of Galen's second visit to Rome, during the reign of Marcus Aurelius; cf. Galen vol. 18B, 538.

complications, and draws on a rich store of pharmacological and surgical treatments.

Besides appearing in all the collected Hippocratic editions and translations, *Ulcers* is included in the surgical collections of Vidius, Manialdus and Petrequin. The treatise's textual transmission receives a careful investigation in D. Raupach, *Die handschriftliche Überlieferung der hippokratischen Schrift "De ulceribus,"* Diss. Göttingen, 1965. On the basis of the conclusions reached by Dr. Raupach, I have collated the manuscript M from microfilm, as our sole independent witness for the text of *Ulcers*.

An English translation of *Ulcers* appeared in Adams vol. 2, 794–809.

ΠΕΡΙ ΕΛΚΩΝ

VI 400
Littré

1. Ἕλκεα ξύμπαντα οὐ χρὴ τέγγειν, πλὴν οἴνῳ, ἢν μὴ ἐν ἄρθρῳ ἔῃ τὸ ἕλκος· τὸ γὰρ ξηρὸν τοῦ ὑγιέος ἐγγυτέρω ἐστί, τὸ δὲ ὑγρὸν τοῦ μὴ ὑγιέος· τὸ γὰρ ἕλκος ὑγρόν ἐστι, τὸ δὲ ὑγιὲς ξηρόν. ἀνεπίδετον δὲ ἐᾶν ἄμεινόν ἐστιν, ὅ τι γε μὴ καταπλάσσεται· οὐδὲ καταπλάσσειν ἐνδεχόμενόν ἐστιν ἔνια τῶν ἑλκέων, μᾶλλον δὲ τὰ νεότρωτα τῶν παλαιοτέρων, καὶ τὰ ἐν τοῖσιν ἄρθροισιν. ὀλιγοσιτέειν τε ὡς μάλιστα καὶ ὕδωρ ξυμφέρει πᾶσι τοῖσιν ἕλκεσι τῶν παλαιοτέρων, καὶ ὅ τι φλεγμαίνει ἕλκος ἢ μέλλει, καὶ ὅ τι σφακελίσαι κίνδυνος, καὶ τοῖσιν ἐν τοῖσιν ἄρθροισι ἕλκεσι καὶ φλέγμασι, καὶ ὅκου σπασμὸν κίνδυνος ἐπιγενέσθαι, καὶ τοῖσιν ἐν κοιλίῃ τρώμασι, παντῶν δὲ μάλιστα τοῖσιν ἐν τῇ κεφαλῇ καὶ μηρῷ κατεαγέντι, καὶ ἄλλῳ ὅ τι ἂν κατεαγῇ. ἑστάναι δ' ἕλκεσι ἥκιστα ξυμφέρει, καὶ ἄλλως ἢν ἐν σκέλεϊ ἔχῃ τὸ ἕλκος, οὐδὲ καθῆσθαι οὐδὲ πορεύεσθαι· ἀλλ' ἡσυχίη καὶ ἀτρεμίη ξυμφέρει.

Τὰ δὲ νεότρωτα ἕλκεα πάντα ἥκιστα ἂν φλεγμήναιεν αὐτά τε καὶ τὰ περιέχοντα, εἴ τις διαπυήσειε

ULCERS[a]

1. Surface lesions should generally not be moistened, except with wine, unless the lesion is at a joint; for dryness is nearer to health, and moistness to unhealthiness, since a lesion is moist, but healthy tissue dry. It is better to leave lesions without any bandage, unless a plaster is being applied; and plasters should not be applied to certain lesions, in particular recent injuries as opposed to older ones, and those at joints. It is beneficial to eat as little food as possible and to drink water in all more long-standing lesions, and in any case where the lesion is swollen or is about to swell, or where there is a danger of sphacelus, or in lesions and swellings at the joints, or when there is a danger of convulsion, or in wounds of the cavity, or most especially of the head or in a fractured thigh, or wherever else a fracture occurs. For the patient to stand up is least beneficial for these lesions, especially if the lesion is in a leg, nor does it help to sit or to walk; on the contrary it is rest and quiet that help.

All fresh lesions, both themselves and the area around them, swell least if someone brings about suppuration

[a] See note on terminology following the general introduction to this volume.

ὡς[1] τάχιστα, καὶ τὸ πῦον μὴ ἀπολαμβανόμενον ἀπὸ
τοῦ ἕλκεος τοῦ στόματος ἴσχοιτο, ἢ εἴ τις ἀποτρέποι
ὅκως μηδὲ μελλήσῃ διαπυῆσαι πλὴν τοῦ ἀναγκαίου
πύου ὀλιγίστου, ἀλλὰ ξηρὸν εἶναι ὡς μάλιστα φαρμάκῳ μὴ περισκελέῒ. πυρῶδες γὰρ γίνεται, ἐπὴν
402   φρίκη ἐγγένηται καὶ σφυ|γμός· φλεγμαίνει γὰρ τὰ
ἕλκεα τότε, ὅταν διαπυῆσαι μέλλῃ· διαπυεῖ δέ,
ἀλλοιουμένου τοῦ αἵματος καὶ θερμανθέντος, ἕως
σαπὲν πῦον γένηται. τῶν τοιούτων ἑλκέων, ὅταν
δοκέῃ δεῖσθαι καταπλάσιος, οὐ χρὴ αὐτὸ τὸ ἕλκος
καταπλάσσειν, ἀλλὰ τὰ περιέχοντα, ὅκως τὸ πῦον
ἀποχωρέῃ, καὶ τὰ σκληρυνόμενα μαλαχθῇ. τῶν δὲ
ἑλκέων ὅ τι μὲν ἂν ὀξέῒ βέλεῒ διατμηθῇ ἢ διακοπῇ,
ἐνδέχεται ἔναιμον φάρμακον καὶ τὸ κωλῦον διαπυεῖν
ἀναξηραῖνόν τι. ἥ τις δ' ὑπὸ τοῦ βέλεος ἐφλάσθη
καὶ ἐκόπη σάρξ, ταύτην δὲ ἰητρεύειν, ὅκως διάπυος
ὡς τάχιστα γένηται· ἧσσόν τε γὰρ φλεγμαίνει· καὶ
ἀνάγκη τὰς σάρκας τὰς φλασθείσας καὶ κοπείσας
σαπείσας καὶ πῦον γενομένας ἐκτακῆναι, ἔπειτα
βλαστάνειν νέας σάρκας.

2. Ἕλκεϊ νεοτρώτῳ παντί, πλὴν ἐν κοιλίῃ, ξυμφέρει <ἐκ τοῦ τρώματος αἷμα ῥυῆναι αὐτίκα πλέον ἢ
ἔλασσον· φλεγμαίνει γὰρ ἧσσον αὐτὸ τὸ ἕλκος καὶ
τὰ περιέχοντα. καὶ ἀπὸ τῶν πεπαλαιωμένων ἑλκέων
ξυμφέρει>[2] αἷμα ποιέειν ἀπορρέειν πυκνά, ὅκως ἂν
δοκέῃ καιρὸς εἶναι, καὶ ἀπ' αὐτῶν τῶν ἑλκέων
<καὶ>[3] τῶν περιεχόντων τὸ ἕλκος, ἄλλως τε καὶ ἢν

as quickly as possible and the pus that issues from the mouth of the lesion is not held back, or if one diverts the fluid so that the lesion does not suppurate, except for the unavoidable minimum of pus, but is dried as much as possible through the action of a non-irritant medication. A lesion becomes purulent when chills and throbbing occur, for lesions become inflamed when they are about to suppurate, and they suppurate when the blood in them is altered and heated until it putrefies and becomes pus. In the case of such lesions, when they seem to require plasters you should not apply the plaster to the lesion itself, but to the area around it, so that the pus can escape, and the part that is indurated become soft. Any lesion that has been produced by a sharp missile or receiving a gash lends itself to the application of a styptic which prevents suppuration by drying. Any tissue that was crushed or severed by the missile treat so that it suppurates as soon as possible, since then it will swell less; it is inevitable for crushed or severed tissues to dissolve and ooze out after they have putrefied and turned to pus, and then for new tissues to grow.

2. In every lesion that results from a recent injury, except in the cavity, it is beneficial <for more or less blood to flow immediately from the wound, for thus the lesion itself and the area around it swell less. It is also beneficial> to make blood flow frequently from long standing lesions, whenever it seems opportune, both from the lesions themselves and from the area around them, espe-

---

[1] M[2]: -ση ἕως M.
[2] Vidius and Foes from Galen vol. 10, 293.
[3] Linden.

ἐν κνήμῃ ἔῃ τὸ ἕλκος ἢ <ἐν>¹ δακτύλῳ ποδὸς ἢ
χειρός, μᾶλλον ἤ κου ἄλλοθι τοῦ σώματος· γίνεται
γάρ, ἀπορρέοντος τοῦ αἵματος, ξηρότερα καὶ μείονα
ἰσχναινόμενα· κωλύει γὰρ μάλιστα μὲν τὰ τοιαῦτα
ἕλκεα ὑγιαίνεσθαι,² ἔπειτα δὲ καὶ τὰ ξύμπαντα,
αἵματος σηπεδών, ἢ³ ἐξ αἵματος μεταστάσιος
γένῃ|ται. ξυμφέρει δὲ μετὰ τὴν τοῦ αἵματος ἀπορ-
ροὴν ἐπὶ τῶν τοιούτων ἑλκέων καὶ σπόγγον ἐπιδεῖν
πυκνὸν καὶ μαλθακόν, τετμημένον, ξηρότερον ἢ
ὑγρότερον, καὶ ἐπὶ τῷ σπόγγῳ ἄνωθεν φύλλα ἰσχνά.
ἔλαιον δὲ καὶ ὅσα μαλθακώδεα ἢ ἐλαιώδεά ἐστι
φάρμακα οὐ ξυμφέρει τοῖσι τοιούτοισιν ἕλκεσιν, ἢν
μὴ πάνυ ἤδη πρὸς ὑγιείην τείνῃ. οὐδὲ τοῖσι νεοτρώτοισιν ἕλκεσι ξυμφέρει ἔλαιον, οὐδὲ μαλθακώδεα
οὐδὲ στεατώδεα φάρμακα, ἄλλως τε καὶ ὅ τι ἂν δέηται ἕλκος πλείονος καθάρσιος· τὸ δὲ ξύμπαν εἰπεῖν,
ἐλαίῳ τὴν χρῆσιν ποιέεσθαι καὶ ἐν θέρει καὶ ἐν χειμῶνι, πρὸς ἃ τῶν τοιούτων φαρμάκων δεόμεθα.

3. Ὑποκάθαρσις τῆς κάτω κοιλίης ξυμφέρει
τοῖσι πλείστοισι τῶν ἑλκέων καὶ ἐν τρώμασιν ἐν
κεφαλῇ ἐοῦσι, καὶ ἐν κοιλίῃ, καὶ ἐν ἄρθροισι, καὶ
ὅσα σφακελίσαι κίνδυνος, καὶ ὅσα ῥαπτά, καὶ
τοῖσιν ἐσθιομένοισι καὶ ἑρπυστικοῖσι, καὶ τοῖσιν
ἄλλως πεπαλαιωμένοισιν ἕλκεσι, καὶ ὅκῃ ἂν μέλλῃ
ἐπιδεῖν.

4. Οὐ χρὴ οὐδ' ἐμπλάσσειν τὰ φάρμακα, πρὶν ἂν
πάνυ ξηρὸν ποιήσῃς τὸ ἕλκος· τότε δὲ προστιθέναι·
ἀνασπογγίζειν δὲ τὸ ἕλκος πολλάκις σπόγγῳ, καὶ

cially if the lesion is in the shin, or a toe or finger—more than when it is in some other part of the body; for, when blood flows out, lesions become drier and smaller as their swelling goes down. Particularly detrimental to the healing of such lesions, and then of the whole body, is putrefaction of the blood, which arises from its alteration. It is beneficial after the efflux of blood in such lesions to bandage on a dense soft sponge which has been cut to shape, drier rather than moister, and on top of the sponge dried leaves. Olive oil and medications that are softening and oily are not beneficial to such lesions, unless they are already tending strongly to recovery. Olive oil is also not beneficial to lesions from more recent injuries, nor are softening and fatty medications, especially to any lesion that requires more cleaning. In summary, olive oil is to be used in both summer and winter for those conditions which in our opinion require such medications.

3. Evacuation of the lower cavity benefits most surface lesions, including those arising in wounds of the head, cavity and joints, lesions in danger of sphacelus, lesions that have been sewn, decaying and spreading lesions, in particular lesions that are older, and when you are about to bandage.

4. You must not apply medications in a plaster before you have rendered the lesion very dry; then apply them: sponge off the lesion many times with a sponge, and wipe

---

[1] I.

[2] Foes in note after Galen vol. 10, 277: ὑγραίνεσθαι M.

[3] Potter: καὶ M.

αὖθις ὀθόνιον ξηρὸν καὶ καθαρὸν προσίσχων πολλάκις, οὕτω δὲ ἐπιθεὶς τὸ φάρμακον τὸ δοκέον ξυμφέρειν, ἐπιδεῖν ἢ μὴ ἐπιδεῖν.

5. Ἕλκεσι τοῖσι πλείστοισιν ὥρη ἡ θερμοτέρη ξυμφορωτέρη τοῦ χειμῶνος, πλὴν τοῖσιν ἐν κεφαλῇ καὶ κοιλίῃ, μάλιστα δὲ ἰσημερινή.

6. Τὰ ἕλκεα ὁκόσα ἂν κα|θαρθέντα καλῶς τε καὶ ἐς τὸ δέον, ἀεὶ ἐπὶ τὸ ξηρότερον ποιῆται τὴν βλάστησιν· ταῦτα δὲ οὐχ ὑπερσαρκέει ὡς ἐπὶ τὸ πολύ.[1]

7. Ἢν ὁκοθενοῦν ὀστέον ἀφιστῆται ἢ καυθὲν ἢ πρισθὲν ἢ ἄλλῳ τῳ τρόπῳ, τῶν ἑλκέων τούτων αἱ οὐλαὶ κοιλότεραι γίνονται.

8. Ἕλκεα οὐ κεκαθαρμένα οὐκ ἐθέλει ξυνιέναι ξυναγόμενα, οὐδ᾽ αὐτόματα ξυνέρχεται. ὧν τὰ περιέχοντα φλεγμαίνει τοῦ ἕλκεος, ἔστ᾽ ἂν μὴ παύσηται τῆς φλεγμασίης, οὐκ ἐθέλει ξυνιέναι· οὐδὲ ὧν τὰ περιέχοντα τοῦ ἕλκεος μελανθῇ αἵματος σηπεδόνι, ἢ καὶ κιρσοῦ παρέχοντος τὴν ἐπιρροὴν τοῦ αἵματος, οὐδὲ ταῦτα ἐθέλει ξυνιέναι, ἢν μὴ τὰ περιέχοντα τοῦ ἕλκεος ὑγιέα ποιήσῃς. τῶν ἑλκέων τὰ κυκλοτερέα ἢν ὑπόκοιλα ᾖ, ἐν κύκλῳ πάντη ἐπιτάμνειν χρὴ τὰ ἀφεστεῶτα, ἢ πάντα, ἢ τὰ ἡμίσεα τοῦ κύκλου, κατὰ μῆκος τῆς φύσιος τοῦ ἀνθρώπου.

9. Ἐπὶ παντὶ ἕλκεϊ ἐρυσιπέλατος ἐπιγενομένου, κάθαρσιν ποιέεσθαι τοῦ σώματος, ἐφ᾽ ὁπότερα ἂν ξυμφέρῃ τῷ ἕλκεϊ, εἴτε ἄνω, εἴτε κάτω.

---

[1] Vidius, Foes, Littré, Ermerins and Petrequin expand the

it again many times with a dry clean piece of linen; then apply the medication that you think will help, and do or do not apply a bandage.

5. For most lesions—the exception being those in the head and cavity—the warmer season is more favourable than winter, but the equinoctial season is most favourable.

6. Lesions that are cleaned thoroughly and as they should be, always form tissue on the dry side; these do not generally grow excessive flesh.

7. If a bone exfoliates at any point, either as the result of being burnt or sawn or in any other way, the scars of these lesions are quite hollow.

8. Lesions that are not cleaned refuse to unite when they are joined, nor do they come together spontaneously. Lesions whose peripheries are inflamed refuse to unite as long as the inflammation persists. Those also whose peripheries turn dark with putrid blood, or if they have a varix that produces an afflux of blood, refuse to unite unless you heal their peripheries. Circular lesions, if they are slightly hollowed out, you must incise in the part that projects, all the way around the circle, either the whole circle or half of it, according to the size of the lesion's form in the person.

9. If erysipelas develops in a lesion, clean the body in whichever direction will benefit the lesion—up or down.

---

text of this chapter on the basis of Galen's quotation of the passage (vol. 10, 281): τὰ δὲ ἕλκεα ὅσα μὴ καλῶς καθαρθέντα ἐς τὸ δέον, ἀεὶ πρότερον ἄρξεται βλαστάνειν, ταῦτα ὑπερσαρκέει μάλιστα· ὁκοῖα δ᾽ ἂν καθαρθέντα καλῶς καὶ ἐς τὸ δέον ἀεί, ἐπὶ τὸ ξηρότερον θεραπεύεται, πλὴν εἰ θλασθῇ, ταῦτα οὐχ ὑπερσαρκέει ὡς ἐπιπολύ.

10. Ὅτῳ ἂν οἴδημα γένηται παρὰ τὸ ἕλκος, ἀφλεγμάντου ἐόντος τοῦ ἕλκεος, χρόνῳ ὕστερον πύου ὑπόστασιν ἴσχει τὸ οἴδημα. καὶ ὅ τι ἂν τῇ φλεγμασίῃ οἰδῆσαν μὴ καθιστῆται, τῶν ἄλλων καθισταμένων, ὅσα ἅμα ἤρξατο φλεγμαίνειν καὶ οἰδίσκεσθαι, καὶ τοῦτο κίνδυνος μηδ᾽ ἅμα ξυνιέναι. ὅσα δὲ πιπτόντων ἢ ἄλλῳ τῳ τρόπῳ διακόπτεται καὶ φλᾶται, καὶ ἀνοιδίσκεται τὰ περιέχοντα τὸ ἕλκος, |

408 καί, διαπυήσαντα, πῦον ἀπὸ τῶν οἰδημάτων ἀποχωρέει κατὰ τὸ ἕλκος, τῶν τοιούτων ὅ τι ἂν δοκέῃ δεῖσθαι καταπλάσιος, οὐ χρὴ αὐτὸ τὸ ἕλκος καταπλάσσειν, ἀλλὰ τὰ περιέχοντα, ὅκως τὸ πῦον ἀποχωρέῃ, καὶ τὰ σκληρυνόμενα λαπαχθῇ· ἐπειδὰν δὲ λαπαχθῇ, καὶ ἡ φλεγμασίη παύσηται, ἐπὶ τὰ[1] ἀφεστηκότα[2] σπόγγους ἐπιδέων προσιστάναι, ἀρχόμενος ἀπὸ τοῦ ὑγιέος ὀλίγον προσχωρέων· ἐπὶ δὲ τῷ σπόγγῳ ἄνωθεν φύλλα ἐπέστω ἰσχνά.

Ὅ τι δ᾽ ἂν μὴ δύνηται προσστῆναι, ἡ σὰρξ ὑγρὴ ἐοῦσα αἰτίη ἐστίν· ταύτην ἐκβάλλειν. ἢν ὑπὸ βαθείῃ σαρκὶ τὸ ἕλκος ἔῃ, κατ᾽ ἄμφω κἀκ τῆς ἐπιδέσιος κἀκ τοῦ προσπιέζοντος ὑποκιρσοῦται· τὸ δὲ τοιοῦτον ἤν τις τάμνῃ, πρὸς μήλην, ἢν ἐνδέχηται, εὔροον ἀπὸ τοῦ στόματος τὸ ἕλκος ἀνατάμνειν, ὅπῃ ἂν δοκέῃ καιρὸς εἶναι, καὶ οὕτως ἰητρείην προσφέρειν, ὁκοίης ἂν δοκέῃ προσδεῖσθαι. ὡς δὲ τὰ πολλὰ ἐπὶ παντὶ ἕλκει, ὅ τι ἂν κοιλίην ἔχῃ ἐς τὸ ἰθύ, καταφανέα ἰδεῖν, οἰδήματος μὴ προσεόντος· ἢν μὲν ἔῃ ἐν

10. In any case where swelling occurs beside a lesion which is itself not inflamed, at some later time the swelling forms a purulent abscess. In any lesion that swells together with an inflammation, but does not go down when the rest goes down, there is a danger—since they began to be inflamed and to swell together—that this too will not unite at the same time. Any lesion in persons who have fallen or that results in some other way from being gashed or crushed, and then the area around the lesion swells up and suppurates, and pus comes out of the area of swelling through the lesion—no matter which plaster seems to be required in such cases, you should not apply the plaster to the lesion itself, but to the area around it, so that the pus will come out, and what has become indurated soften; when it softens and the inflammation ceases, approximate the borders by bandaging sponges over the abscess, beginning the bandage by advancing a little into the healthy tissues; place dried leaves on top of the sponge.

When union cannot occur, it is because the tissue is moist: remove the tissue. If the lesion goes deep under the flesh, it forms varices for two reasons; from the bandaging, and from the tissue pressing against it. If you make an incision in such a case, make your cut against a probe, if the lesion will permit it, so as to insure free flow from the mouth of the lesion wherever it seems to be appropriate; afterwards apply whichever medication seems to be required. Generally, in every lesion that has a cavity in a straight line, this is clear to see, as long as swelling is not present. If there is suppuration in the

---

[1] Parisinus Graecus 2145: ἔπειτα M.   [2] I: -κότας M.

αὐτῷ σηπεδών, ἢ ἡ σὰρξ ὑπέῃ μυδῶσα καὶ σαπρή,
410 ἔσται | τοῦτο τὸ ἕλκος καὶ τὰ περιέχοντα τὸ ἕλκος
ἰδεῖν μέλανα ὑποπέλια· καὶ τῶν ἐσθιομένων ἑλκέων,
ὅπῃ ἂν φαγέδαινα ἐνέῃ, ἰσχυρότατά τε νέμηται καὶ
ἐσθίῃ, ταύτῃ τοῦ ἕλκεος τὸ περιέχον χροιὴν ἕξει
μέλαιναν ὑποπέλιον.

11. Καταπλάσματα οἰδημάτων καὶ φλεγμασίης
τῆς ἐν τοῖσι περιέχουσιν· ἡ ἑφθὴ φλόμος, καὶ τῆς
τριφύλλου τὰ φύλλα ὠμά, καὶ τοῦ ἐπιπέτρου τὰ
φύλλα ἑφθά, καὶ τὸ πόλιον· ἢν δὲ καὶ καθαίρεσθαι
δέῃ τὸ ἕλκος, παντὰ μὲν καὶ ταῦτα καθαίρει· ἀτὰρ
καὶ τῆς συκῆς τὰ φύλλα καὶ τῆς ἐλαίης, καὶ τὸ πρά-
σιον. ἕψειν δὲ ταῦτα πάντα, μάλιστα δὲ τούτων
ἕψειν τὸν ἄγνον, καὶ τὴν συκῆν, καὶ τὴν ἐλαίην, καὶ
τῆς σίδης τὰ φύλλα ὡσαύτως ἕψειν. ὠμοῖσι δὲ
τοῖσίδε χρέεσθαι, τῆς μαλάχης τὰ φύλλα τρίβων
ξὺν οἴνῳ, καὶ τοῦ πηγάνου τὰ φύλλα καὶ τῆς ὀριγά-
νου χλωρῆς· πᾶσι τούτοισι χρὴ τοῦ λίνου τὸν καρ-
πὸν φρύξαντα καὶ κόψαντα ὡς λειότατον μιγνύναι.
ὅκου δ' ἐρυσίπελας κίνδυνος ἐφ' ἕλκεσι γενέσθαι,
τῆς ἰσάτιδος τὰ φύλλα τρίβων ὠμὰ καταπλάσσειν
σὺν τῷ λίνῳ, ἢ τὸ λίνον[1] δεύων στρύχνου χυλῷ ἢ
ἰσάτιδος καταπλάσσειν. ὅταν δὲ τὸ ἕλκος καθαρὸν
μὲν ἔῃ, φλεγμαίνῃ δὲ τό τε ἕλκος καὶ τὰ περιέχοντα
τοῦ ἕλκεος, φακὸν ἐν οἴνῳ ἑψήσας καὶ τρίψας λεῖον,
ἐλαίῳ ὀλίγῳ φυρήσας, καταπλάσας, ἐπιδεῖν· καὶ τοῦ
κυνοσβάτου ἑψήσας τὰ φύλλα ἐν ὕδατι, τρίψας λεῖα,
καταπλάσσειν, ὀθόνιον ὑποτείνας λεπτὸν καθαρόν,

lesion, or the tissue is dripping and putrid, this lesion and
the area around it look darkish and somewhat livid; in
rodent ulcers, wherever the erosion is located, and feeds
and consumes most actively, at that point the periphery of
the lesion will have a darkish, livid colour.

11. Plasters for swellings and for inflammation in the
areas around lesions. Boiled mullein, raw leaves of clover,
boiled leaves of rock-plant, hulwort; if the lesion must
also be cleaned, all these clean as well. Also fig-leaves,
olive leaves and horehound. Boil all these, but especially
chaste-tree, fig and olive; also boil the leaves of the
pomegranate-tree in the same way. Apply the following
raw: pound mallow leaves together with wine, rue leaves
and leaves of green marjoram: with all these you must mix
linseed roasted and pounded very fine. Where there is a
danger of erysipelas arising in lesions, pound raw woad
leaves and apply as a plaster with flax, or apply a plaster of
flax steeped in nightshade juice or woad. When the lesion
is clean, but the lesion and the area around it are
inflamed, boil lentils in wine and grind fine, mix in a little
olive oil, apply as a plaster and hold in place with a bandage. Also boil white rose leaves in water, pound them
fine, apply as a plaster, and put under it a fine, clean linen

---

[1] Aldina after Calvus' *linteum*: τῷ λίνῳ M.

οἴνῳ καὶ ἐλαίῳ τέγξας· καὶ ὅταν ξυνάγειν βούλῃ,
τοῦ κυνοσβάτου τὰ φύλλα ὥσπερ τὸν φακὸν σκευά-
ζειν. σαυρίδιον, οἶνος καὶ λίνου καρπὸς παραμίγνυ-
ται | λεπτός· καὶ τόδε, ὁ τοῦ λίνου καρπός, καὶ ἄγνος
ὠμός, καὶ Μηλεία στυπτηρίη, ὄξει ταῦτα δευθέντα.

12. Ὄμφακα λευκὴν ἐς χαλκεῖον θλίψας ἐρυθρὸν
δι᾽ ἠθμοῦ, πρὸς ἥλιον τιθέναι τὰς ἡμέρας, τὰς δὲ
νύκτας αἴρειν, ὅκως μὴ δροσίζηται, ἀνατρίβειν δὲ
τῆς ἡμέρης ἀπαύστως, ὡς ὁμαλῶς ξηραίνηται, καὶ
ἀπὸ τοῦ χαλκείου ὡς ὅτι πλεῖστον ἀναλαμβάνῃ,
τιθέναι δὲ ἐς τὸν ἥλιον τοσοῦτον χρόνον, ἔστ᾽ ἂν
παχὺ γένηται ὥσπερ μέλι· ἔπειτα ἐς χύτρην χαλκῆν
ἐγχέαι, καὶ μέλι ὡς κάλλιστον, καὶ οἶνον γλυκύν,
ἐναφεψήσας πρότερον ῥητίνην τερμινθίνην, ἕψειν δὲ
τὴν ῥητίνην ἐν τῷ οἴνῳ, ἕως ἂν σκληρὴ γένηται
ὥσπερ μέλι ἐφθόν· ἔπειτα τὴν μὲν ῥητίνην ἐξελεῖν,
τὸν δὲ οἶνον ξυγχέαι· ἔστω δὲ πλεῖστος μὲν ὁ χυλὸς
τῆς ὄμφακος, δεύτερον δὲ ὁ οἶνος, τρίτον δὲ τὸ μέλι·
καὶ σμύρναν τὴν στακτὴν καὶ ἄλλως ὡς βελτίστην
τρίψας λείην, δίεσθαι τοῦ οἴνου τοῦ αὐτοῦ παρα-
χέοντα κατ᾽ ὀλίγον· ἔπειτα ἕψειν αὐτὴν ἐφ᾽ ἑωυτῆς
τὴν σμύρναν ξὺν τῷ οἴνῳ ἀνακινέοντα, ὅταν δὲ
δοκέῃ ἤδη καλῶς ἔχειν τὸ πάχος, ξυγχέαι ἐς τὸν
χυλὸν τῆς ὄμφακος, καὶ νίτρον ὡς ἄριστον φρύξας,
ἡσύχως μιγνύναι ἐς τὸ φάρμακον, καὶ ἄνθος χαλκοῦ
ἔλασσον τοῦ νίτρου· ταῦτα δὲ ἐπειδὰν μίξῃς, ἕψειν
μὴ ἔλασσον τριῶν ἡμερέων, ξύλοισι συκίνοισιν
ὡς ὀλίγιστον ὑποκαίοντα ἢ ἄνθραξιν, ὡς μὴ

cloth moistened in wine and olive oil. When you wish to draw the lesion together, prepare white rose leaves as you did the lentil. Nose-smart, wine and fine linseed are mixed together. Another: linseed, raw chaste-tree, Melean alum, these steeped in vinegar.

12. Press unripe white grapes through a strainer into a vessel of red copper; place in the sun by day, but remove by night so that it so that it will not be covered with dew; by day stir continuously so that the mixture dries evenly, and so that as much as possible is taken up from the copper of the vessel; continue to place in the sun until the juice becomes as thick as honey, then pour this into a copper pot with the finest honey, and sweet wine into which turpentine has been boiled: boil the resin in the wine until it becomes hard like boiled honey; then remove the resin and decant the wine. Let the unripe grape juice be greatest in quantity, the wine second, and the honey third. Also take virgin gum of myrrh,[a] otherwise of the finest kind, pound it smooth, and soak it by adding the same wine a little at a time. Then boil the myrrh by itself in the wine, stirring it continually, and when it seems to be of the right consistency pour it into the unripe grape juice; roast the finest soda, and mix it gently into the medication, and also a lesser quantity of flower of copper. When you have made this mixture, boil it for not less than three days, very gently burning fig wood or coals under it, so

[a] I.e. *stacte*, the gum that exudes naturally from the living tree.

φρύγηται· καὶ ἐμβαλλόμενα πάντα ἄνυδρα ἔστω,
καὶ τὰ ἕλκεα μὴ τεγγέσθω, ὅκῃ ἂν ἐπαλείφηται
τοῦτο τὸ φάρμακον· χρῆσθαι δὲ τούτῳ τῷ φαρμάκῳ
πρὸς τὰ πεπαλαιωμένα ἕλκεα, καὶ πρὸς τὰ νεότρωτα,
καὶ ἐς πόσθιον, καὶ ἐς κεφαλῆς ἕλκεα[1] καὶ ἐς οὖς.
τῶν αὐτῶν· χολὴ βοὸς ξηρή, μέλι ὡς κάλλιστον,

414 οἶνος | λευκός· ἐναφεψῆσαι δ' ἐν αὐτῷ λωτοῦ τορνεύ-
ματα· λιβανωτός, σμύρνα ἴση, κρόκος ἴσος, ἄνθος
χαλκοῦ· ὁμοίως δὲ ὑγρῶν, οἶνος πλεῖστος, μέλι δεύτερον, ὀλίγιστον ἡ χολή. ἕτερον· οἶνος, μέλι κέδρινον, ὀλίγον· τὰ δὲ ξηρά, ἄνθος χαλκοῦ, σμύρνα,
σίδιον αὖον. ἕτερον· ἄνθος χαλκοῦ ὀπτὸν ἡμιμοί-
ριον, σμύρνης δύο ἡμιμοίρια, κρόκου τρεῖς μοῖραι,
μέλι ὀλίγον, οἴνῳ[2] ἕψεται.

Ἔναιμον· λιβανωτοῦ μοῖρα, σμύρνης μοῖρα,
κηκίδος μοῖρα, κρόκου τρεῖς μοῖραι· τούτων ξηρὸν
ἕκαστον τρῖψαι ὡς λειότατον, ἔπειτα μίξας, τρίβειν
ἐν ἡλίῳ ὡς θερμοτάτῳ, παραχέων χυλὸν ὄμφακος ἕως
ἂν ἰξῶδες γένηται, ἐπὶ τρεῖς ἡμέρας· ἔπειτα οἴνῳ
αὐστηρῷ μέλανι εὐώδεϊ παραχέων κατ' ὀλίγον δίεσθαι.
ἔναιμον· ἐν οἴνῳ γλυκέϊ ἕψειν λευκῷ πρίνου ῥίζας·
ἐπειδὰν δὲ καλῶς δοκέῃ ἔχειν, ἀποχέας, τοῦ οἴνου
δύο μοίρας ποιῆσαι τοῦδε καὶ ἀμόργεως ἐλαίων ὡς
ἀνυδροτάτων μοῖραν μίαν, ἔπειτα ἕψειν, ἀνακινέων
ὡς μὴ φρυγῇ, μαλθακῷ πυρί, ἕως ἂν δοκέῃ τοῦ
πάχεος καλῶς ἔχειν. ἔναιμον· τὰ μὲν ἄλλα, τὰ
αὐτά, ἀντὶ δὲ τοῦ οἴνου, ὄξος ὡς ὀξύτατον ἔστω
λευκόν· ἐμβάψαι δὲ ἐς αὐτὸ εἴρια ὡς οἰσυπώδεα·

that it does not dry out. All the ingredients should be dry, and should not moisten the lesions where the medication is anointed. Use this medication against old lesions, as well as against new injuries, and for the foreskin, for lesions of the head, and for the ears. For the same: dried bull's gall, finest honey, white wine in which the shavings of nettle-tree wood have been boiled, frankincense, an equal amount of myrrh, an equal amount of saffron, and flower of copper: likewise with the liquids: most wine, second honey, very little bile. Another: wine, a little cedar honey: dry components: flower of copper, myrrh, dry pomegranate peel. Another: a half portion of burnt flower of copper, two half portions of myrrh, three portions of saffron, a little honey, boiled in wine.

Styptic: a portion each of frankincense, myrrh, and oak-gall, three portions of saffron: pound each of these very fine in the dry state; then mix them together, triturate in the very hot sun, and add juice of unripe grapes until after three days it becomes sticky; then dissolve it a little at a time into fragrant dry dark wine. Styptic: in sweet white wine boil roots of the holm-oak; when this seems to be done, pour the wine out. Make a mixture of two parts of this wine and one part very dry olive paste; then boil on a gentle fire, stirring in order to prevent burning, until you think the consistency is right. Styptic: otherwise the same, but instead of the wine use very acidic white vinegar; immerse into it greasy wool, then,

---

[1] Aldina after Calvus' *capitis . . . vulnera*: ἐν κεφαλῇ ἑλκέων M.

[2] H: οἶνος M.

κἄπειτα δεύσας τῇ ἀμόργῃ ἕψειν· καὶ ὀπὸν ἐρινεοῦ
ξυγχέαι, καὶ στυπτηρίην Μηλείην, καὶ νίτρον καὶ
ἄνθος χαλκοῦ μῖξαι ὀπτὰ ἀμφότερα. τοῦτο μᾶλλον
τοῦ προτέρου καθαίρει, ξηραίνει δὲ τὸ πρότερον οὐχ
ἧσσον. ἔναιμον· τὰ εἴρια βάψαι ὡς ἐν ὀλιγίστῳ
ὕδατι, ἔπειτα οἶνον ξυγχέας μέρος τρίτον, ἕψειν ἕως
416 ἂν καλῶς ἔχῃ τὸ πάχος. ἀπὸ τῶνδε | τὰ νεότρωτα
διαπυΐσκεται τάχιστα. ἄρον ξηρὸν ἐμπάσσειν, καὶ
στέλλειν. κράδης ἐν ὀπῷ φλοιὸν χλωρὸν τρίβων ἐν
οἴνῳ ἐνστέλλειν, καὶ ἄνευ οἴνου αὐτὸν καὶ ξὺν
μέλιτι. [τάδε μάλιστα ἀποτρέπει τὰ νεότρωτα· διαπυΐσκεται τά τε ἔναιμα.]¹ ὄξος, ἐναφεψῶν λωτοῦ
τορνεύματα, ἔστω δὲ λευκὸν τὸ ὄξος, κἄπειτα μῖξαι
ἀμόργην ἐλαίων καὶ ὀρὸν πίσσης, τοῦτο ὠμόν· καὶ
ἐπαλείφειν² καὶ καταστάζειν καὶ ἐπιδεῖν.

13. Ξηρὰ ἀποτρέπει τὰ νεότρωτα διαπυΐσκεσθαι,
ἢ ὄξει ἀπονίψας, ἢ οἴνῳ ἀποσπογγίσας. τὸν μόλι-
βον τὸν λεῖον ξὺν τῇ σποδῷ τῇ κυπρίῃ λεανθέντα
ἐπιπάσσειν· καὶ τοῦ λωτοῦ τὰ ἰχθυήματα ἐπιπάσσειν, καὶ τὴν λεπίδα τοῦ χαλκοῦ, καὶ τὴν στυπτηρίην, καὶ τὴν χαλκῖτιν μετὰ τοῦ χαλκοῦ, καὶ μόνην,
καὶ μετὰ τῶν τοῦ λωτοῦ ἰχθυημάτων. καὶ ἄλλως,
ὅταν δέηται, ξηροῖσι τοῖσι τοιούτοισι χρέεσθαι, καὶ
τῇ σποδῷ τῇ ἰλλυριώτιδι λείη μετὰ τῶν ἰχθυημάτων,
καὶ αὐτῇ μόνῃ, καὶ ἰχθυήμασι, καὶ ἄνθει ἀργύρου
μόνῳ ὡς λειοτάτῳ· καὶ τὴν ἀριστολοχείην ξύων <τε
καὶ>³ τρίβων λείην ἐπιπάσσειν.

14. Ἔναιμον· σμύρνα, λιβανωτός, κηκίς, ἰός,

mixing in the olive paste, boil; add wild fig juice, and Melean alum, and soda and flower of copper, both roasted. This medication cleans more than the preceding one, but the preceding one dries equally well. Styptic: immerse wool in a very little water, then add a third part wine, and boil until the consistency is right. With these medications new lesions suppurate most quickly. Sprinkle on dry cuckoo-pint and cover it. Grind into fig-juice green bark of fig-tree branches, and administer in wine; also the same with honey instead of wine. Boil the shavings of nettle-tree wood in vinegar—the vinegar should be white—and then add olive paste and the watery part of wood-tar (pitch whey), this unboiled: anoint, apply as drops, and hold in place with a bandage.

13. Dry things prevent new wounds from suppurating; apply them either after washing the wound off with vinegar or sponging it off with wine. Dust on fine lead powder triturated with copper oxide; also sprinkle on shavings of nettle-tree wood, also copper scales, also alum, also copper ore (*chalcitis*) together with copper, also copper ore alone or together with the shavings of nettle-tree wood. At all events, when they are necessary, employ dry substances like these; also fine tin oxide with the shavings, also the oxide alone, also the shavings, and flower of silver alone ground very fine; also shred aristolochia, grind it fine, and sprinkle it on.

14. Styptic: myrrh, frankincense, oak-gall, verdigris,

---

[1] Del. I.

[2] Froben: $\dot{\alpha}\pi$- M.

[3] I.

ἄνθος χαλκοῦ ὀπτόν, στυπτηρίη Αἰγυπτίη ὀπτή, οἰνάνθη, οἰσυπίδες, μολύβδαινα, τούτων ἴσον ἑκάστου, ἡ δίεσις οἴνῳ ὥσπερ τὸ πρότερον, καὶ ἄλλη ἐργασίη κατὰ τὰ αὐτά. ὄξος ὡς ὀξύτατον λευκόν, μέλι, στυπτηρίην Αἰγυπτίην, νίτρον ὡς ἄριστον 418 ἡσύχως φρύξας, χολῆς | ὀλίγῳ[1] ἕψεται· τοῦτο τὰ ὑπερσαρκέοντα καθαίρει καὶ κοιλαίνει. ποίη ἡ μικρόφυλλος, ᾗ οὔνομα παρθένιον τὸ μικρόφυλλον, ἢ τὰ θύμια τὰ ἀπὸ τοῦ ποσθίου ἀφαιρεῖ, καὶ στυπτηρίη ἡ χαλκῖτις· καὶ Μηλιάδος ὠμῆς· ἐλατήριον λεπτὸν ξηρὸν προστεῖλαι, καὶ τὸ σίδιον λεπτὸν ξηρὸν ὡσαύτως.

15. Πληροῖ δὲ μάλιστα τὰ κοῖλα τὰ καθαρά, ποίη, ᾗ λαγώπυρος οὔνομα· ἔστι δὲ πιτύροισιν ὁμοίη ὅταν αὐαίνηται, μικρὸν τὸ φύλλον, ὥσπερ καὶ τὸ τῆς ἐλαίης, καὶ μακρότερον· καὶ πρασίου τὸ φύλλον, σὺν ἐλαίῳ. ἔμπλαστρον· ἰσχάδος τὸ εἴσω, τὸ πῖαρ, τὸ μελιτοειδές, ὡς ξηροτάτης, ὕδατος δύο μοίρας, καὶ λίνου καρποῦ φρύξας μὴ σφόδρα ὡς λεπτοτάτου μοίραν. ἕτερον· τῆς ἰσχάδος, καὶ ἄνθος χαλκοῦ ὀλίγον λεπτόν, καὶ συκῆς ὀπόν. τὸ δ' ἐκ τῆς ἰσχάδος, χαμαιλέων μέλας, χολὴ βοὸς ξηρή· τὰ μὲν ἄλλα τὰ αὐτά. τὰ δὲ ξηρά. κάρδαμον λεπτόν, ὠμόν, ἐρύσιμον, ἑκατέρου ἴσον, τῆς ἰσχάδος δύο μέρη, λίνου καρποῦ δύο μοίρας, ὀπὸν συκῆς. ὅταν τούτων τινὶ χρέῃ τῶν φαρμάκων, σπλῆνας ἄνωθεν ὀξηροὺς ἐπιθείς, σπόγγον ἄνωθεν τῶν σπληνῶν ἐπίθες, καὶ ἐπίδει, καὶ προσπιέσαι ὀλίγῳ μᾶλλον· τὰ δὲ περι-

burnt flower of copper, burnt Egyptian alum, flower of the wild-vine, tufts of greasy wool, galena[a]: an equal amount of each of these; moisten in wine as above, and the rest of the preparation is just the same. Very concentrated white vinegar, honey, Egyptian alum, finest soda gently roasted, all boiled in a little bile. This removes excessive flesh, and hollows out. The grass with small leaves, called small-leaved fever-few, which removes large warts from the foreskin, astringent copper ore, unroasted Melean earth. Apply dry powdered squirting-cucumber, and dry powdered pomegranate peel likewise.

15. What best fills lesions that are hollow and clean is an herb whose name is hare's foot trefoil; it is like bran when it is dry and has a small leaf like that of the olive-tree, but longer.—Also horehound leaf, with olive oil. Emplastrum: the insides—the greasy part, like honey—of a very dry fig, two parts water, and one part lightly roasted very fine linseed. Another: dried fig paste, a little fine flower of copper, and fig juice. Dried fig paste, black chameleon,[b] dry bull's gall. Others the same: these are drying medications. Narrow-leaved cress, raw, and hedge-mustard, an equal amount of each, two parts dried fig paste, two parts linseed, fig juice. When you use one of these medications, bandage compresses soaked in vinegar over the medication, put a sponge on top of the compresses, bandage, and apply a fair amount of pres-

---

[a] Native lead sulphide.

[b] The plant *Cardopatium corymbosum*.

---

[1] I: ὀλίγον M.

ἔχοντα ἦν φλεγμαίνῃ, ὅ τι ἂν δοκέῃ ξυμφέρειν,
περιπλάσσειν.

16. Ἢν βούλῃ ὑγρῷ χρέεσθαι, καὶ τὸ Καρικὸν
φάρμακον ἐπαλείφειν, ἐπιδεῖν δ' ὥσπερ τὰ πρότερα
γέγραπται κατὰ τὸν αὐτὸν τρόπον. ἔστι δὲ ἐκ τῶνδε
τὸ φάρμακον ποιεύμενον· ἐλλεβόρου μέλανος, σανδαράχης, λεπίδος, μολίβου κεκαυμένου[1] σὺν πολλῷ
θείῳ,[2] | ἀρρενικοῦ, κανθαρίδος· τούτῳ ὁποίῳ δοκέει
συντεθέντι χρῆσθαι· ἡ δὲ δίεσις, κεδρίνῳ ἐλαίῳ·
ἐπειδὰν δὲ ἅλις ἔχῃ ἐπαλείφοντι, ἐκβάλλειν τὸ φάρμακον, ἐπιπάσσων ἄρον ἐφθὸν λεῖον ἢ τρίβων
ξηρὸν τῷ μέλιτι δεύων· καὶ ἢν ξηρῷ χρῇ τῷ Καρικῷ
τούτῳ, χρὴ ἀφιστάναι τὸ φάρμακον ἐπιπάσσων.[3]
ποιέει δὲ τὸ ξηρὸν ἀπὸ ἐλλεβόρου μόνον, καὶ τῆς
σανδαράχης.

17. Ἕτερον ὑγρόν· ποίη, ἧς τὸ φύλλον ὅμοιον
ἄρῳ τὴν φύσιν, λευκὸν δέ, χνοῶδες,[4] κατὰ κισσοῦ
φύλλον τὸ μέγεθος· αὕτη ἡ ποίη ξὺν οἴνῳ ἐπιπλάσ-
σεται. ἢ τοῦ πρίνου τὸ περὶ τὸ στέλεχος τρίψας ἐν
οἴνῳ, ἐπίπλασσε. ἕτερον· ὀμφάκου χυλός, ὄξος ὡς
ὀξύτατον, ἄνθος χαλκοῦ, νίτρον, ὀπὸς ἐρινεοῦ. ἐς
ὄμφακος χυλὸν στυπτηρίην ἐμβάλλειν ὡς λειοτά-
την, καὶ θεῖναι ἐν χαλκῷ ἐρυθρῷ ἐς ἥλιον, καὶ ἀνακινέειν, καὶ ἀνελεῖν ὅταν δοκέῃ καλῶς ἔχειν τὸ
πάχος. ξηρὰ τάδε· ἐλλέβορος μέλας ὡς λειότατος

---

[1] Littré after Calvus' *plumbo deusto*: κεκλυμένου M.
[2] Littré after Calvus' *cum multo sulphure*: θείου M.
[3] Littré after Calvus' *superspargito*: -πλάσσων M.

sure. If the area around swells up, apply whatever plaster seems likely to help.

16. If you wish to employ a moist treatment, anoint with the Carian medication, and apply a bandage in the same manner as was described for the medications above. This medication is to be made from the following: black hellebore, red arsenic, copper scales, burnt lead with copious sulfur, yellow orpiment, and blister-beetles: use these combined as you judge best; dissolve them in juniper-oil. When the lesion has been sufficiently anointed, remove the medication, and apply as plaster cuckoo-pint—either boiled until smooth or dry and ground—mixed in honey. If you use this Carian medication dry, you must remove it after sprinkling it on. Make a dry application from hellebore alone, and from red arsenic.[a]

17. Another moist one: the herb whose leaf is like cuckoo-pint in form, but white, downy, and the size of an ivy leaf; this herb is applied in a plaster with wine. Or grind material taken from around the trunk of a holm-oak tree, and apply as a plaster in wine. Another: juice of unripe grapes, very strong vinegar, flower of copper, soda, juice of the wild fig-tree. To unripe grape juice add the finest alum, place in a vessel of red copper in the sun, agitate, and remove when the consistency seems to be right. Drying medication: sprinkle on very fine black hellebore,

---

[a] As many editors have seen, this sentence belongs to the paragraph about the Carian medication, not to the next paragraph as Littré has it.

[4] H: γνο- M.

ἐπιπάσσεται,[1] ἕως ἄν τι τοῦ ὑγροῦ ἐνέῃ καὶ νεμομέ-
νου· ἐπίδεσις δὲ ἡ αὐτή, ἥπερ ἐπὶ τοῖσιν ἐμπλά-
στροισιν. ἄλλο· ἁλὸς χόνδρους ὡς ξηροτάτους ἐς
χυτρίδιον χάλκεον ἢ κεραμεοῦν καινὸν ἐμβάλλειν,
ἴσους ὡς μάλιστα τὸ μέγεθος, μὴ ἁδρούς· καὶ μέλι
ὡς κάλλιστον διπλάσιον τῶν ἁλῶν εἰκάσας ἐπιχέαι
ἐπὶ τοὺς ἅλας· ἔπειτα ἐπιθεῖναι ἐπὶ τοὺς ἄνθρακας
τὸ χυτρίδιον, καὶ ἐᾶν ἕως ἂν κατακαυθῇ πᾶν· ἔπειτα
ἀνασπογγίσας τὸ ἕλκος καὶ ἐκκαθήρας, ἐπιδῆσαι
ὥσπερ τὸ πρότερον, καὶ πιέσαι ὀλίγῳ μᾶλλον· τῇ δ'
422   ὑστεραίῃ, ὅπη ἂν μὴ λάβηται τὸ φάρμακον, | ἐπι-
πάσας προσπιέζειν καὶ ἐπιδεῖν· ὅταν δὲ βούλῃ ἀφιστάναι τὸ φάρμακον, ὄξος θερμὸν ἐπιχέειν, ἕως ἂν
ἀποστῇ, καὶ αὖθις τὰ αὐτὰ ποιέειν, ἢν δέηται, ἀνασπογγίσας. ἄλλο ξηρὸν δάκνον· ἀποσπογγίσας ὡς
οἰσυπωδέστατα εἴρια ἐπ' ὀστράκου κατακαῦσαι δαι-
δίῳ προσίσχων ἕως πάντα κατακαύσῃς· τοῦτο λεῖον
τρίβων, ἐπιπάσας, ἐπιδεῖν τὴν αὐτὴν ἐπίδεσιν.
ἄλλο δάκνον· μίσυος ὡς λειοτάτου ἐπιπάσσειν ἐπὶ
τὰ ὑγρὰ καὶ τὰ σαπρά, καὶ ἄνθος λεπτὸν μὴ παντελῶς λεῖον. τῶν αὐτῶν ἑλκέων· χαμαιλέων μέλας,
στυπτηρίη ἡ τῷ ὀπῷ τῆς συκῆς δεδευμένη, δεύειν δὲ
ὀπτήν, καὶ ἄγχουσαν μῖξαι. ἀναγαλλὶς καὶ στυπτηρίη Αἰγυπτίη ὀπτή, ἐπίπαστον Ὀρχομένιον ἐπι-
πάσαι.

18.   Πρὸς δὲ τὰς νόμας· στυπτηρίη, ἥ τε Αἰγυ-

---

[1] Aldina after Calvus' *superspargito*: -πλάσσεται M.

as long as moisture is still present in the lesion and it is eroded; the bandage should be the same as for the other plasters. Another: place lumps—as far as possible equal in size, and not large—of very dry salt in a small copper vessel or a new pottery one, and add to the salt the finest honey, estimating the amount to be twice that of the salt. Then place the little pot on the coals, and leave it there until everything is completely burnt; then sponge the lesion off clean, and attach the application with a bandage as above, but with a little more pressure. On the following day, wherever the medication has not been taken up, sprinkle it on again, apply pressure, and bind with a bandage. When you wish to remove the medication, pour warm vinegar over it until it comes off; repeat if required, after sponging clean. Another dry irritant medication to be applied after sponging the lesion clean: burn very greasy wool on a potsherd by applying pine splinters, until it is completely burnt up; grind this fine, sprinkle it on the lesion and bind with the same kind of bandage. Another irritant medication: sprinkle finest copper ore on to moist and putrid lesions, and fine but not completely smooth flower of copper. For the same lesions: black chameleon and alum steeped in fig juice—steep it after it has been roasted; also mix alkanet with this. Pimpernel and burnt Egyptian alum, Orchomenean powder[a]: sprinkle on.

18. Against rodent ulcers: alum, both the burnt Egyp-

[a] Perhaps powdered ashes of reeds growing in the Copaic lake.

πτίη ὀπτή, καὶ ἡ Μηλείη, πρότερον δὲ ἀπονιτρῶσαι
ὀπτῷ καὶ ἀνασπογγίσαι. καὶ ἡ χαλκῖτις στυπτηρίη
ὀπτή· ὀπτᾶν δὲ ἕως ἂν φλογοειδὴς γένηται.

19. Τῶν παλαιῶν ἑλκέων τῶν ἐν τοῖσιν ἀντι-
κνημίοισι γινομένων· αἱματώδεα δέ τοι γίνεται καὶ
μέλανα· μελιλώτου ἄνθος τρίψας, μέλιτι φυρῶν, ἐπι-
πλάστῳ χρῆσθαι.

424 20. Ἐπὶ νεῦρα δὲ διατμηθέντα ἐπιδεῖν μυρρίνης
ἀγρίης ῥίζας[1] κόψας καὶ διαττήσας, φυρήσας ἐλαίῳ.
καὶ τὴν ποίην τὴν πεντάφυλλον, λευκὴ δ' ἐστι καὶ
χνοώδης, καὶ ὑψηλοτέρη ἀπὸ τῆς γῆς ἢ τὸ μέλαν
πεντάφυλλον, ταύτην τρίψας ἐν ἐλαίῳ ἐπιδεῖν, ἀπο-
λύειν δὲ τριταῖον.

21. Μαλθακώδεα, τοῖσίδε χρὴ τοῖσι φαρμάκοισι
χρῆσθαι ἐν χειμῶνι μᾶλλον ἢ ἐν θέρει· μαλθακώδεα,
ἃ καὶ οὐλὰς καλὰς ποιέει· σκίλλης τὸ εἴσω τὸ μυξῶδες τρίψας, ἢ πεύκην σὺν ὑείῳ στέατι νέῳ, ὀλίγον
ἔλαιον καὶ ῥητίνης ὀλίγον, καὶ ψιμυθίου. καὶ στέαρ
χηνός, καὶ συὸς νέον, καὶ σκίλλα, καὶ ἔλαιον ὀλίγον. κηρὸς ὡς λευκότατος, στέαρ πρόσφατον καθαρόν. ἢ σκίλλαν, ἔλαιον λευκόν, ῥητίνης ὀλίγον.
κηρὸν, στέαρ συὸς παλαιὸν καὶ νέον, καὶ ἔλαιον, καὶ
ἰός, καὶ σκίλλα, καὶ ῥητίνη, ἔστω δὲ δύο μοῖραι τοῦ
παλαιοῦ στέατος, τῶν δὲ ἄλλων ὁκόσον δοκέει καιρὸς εἶναι. στέαρ συντήξας πρόσφατον, ἀποχέας ἐς
ἕτερον χυτρίδιον, καὶ τῆς μολυβδαίνη· τρίψας ὡς
λειότατον, διαττήσας, ξυμμίξας, ἕψειν, καὶ κυκᾶν τὸ
πρῶτον, ἕψεῖν δὲ ἕως ἂν ἐπὶ τῆς γῆς πηγνύηται ἐπι-

tian and the Melean: first rub the lesion off with burnt soda and sponge it clean. Also burnt astringent copper ore; burn it until it flames.

19. For old ulcers in the shins; these become bloody and dark: grind the flower of the melilot, mix with honey, and employ as a plaster.

20. On severed cords, apply and bandage roots of wild myrtle, chopped and sifted, and mixed with olive oil. Also the herb cinquefoil (it is white and downy, and higher above the ground than the black cinquefoil), pound this in olive oil and attach with a bandage; untie on the third day.

21. To soften you must employ the following medications—more so in winter than in summer—medications for softening, which also make the scars neat. Pound the internal pulpy material of the squill, or pine bark, with fresh lard, a little olive oil, a little resin, and white lead. Also goose grease, fresh lard, squill and a little olive oil. Whitest wax and fresh, clean grease. Or squill, white olive oil, and a little resin. Wax, aged and fresh lard, olive oil, verdigris, squill and resin: let there be two parts of the aged lard, and of the other components as much as seems appropriate. Melt fresh grease, pour it off into another cup, grind galena very fine, sift, mix the two together, and boil, stirring at first; boil until when dripped on to the

---

[1] I: ῥίζας ἀγρίας M.

σταχθέν, ἔπειτα καθελὼν ἀποχέαι τὸ ἄλλο πλὴν τῆς
λίθου τῆς ὑποστάσης, καὶ ἐμβάλλειν ῥητίνην καὶ
ἀνακινέειν, καὶ κέδρινον ἔλαιον ὀλίγον ξυμμῖξαι καὶ
τὸ ἀφῃρημένον. πᾶσι χρὴ τοῖσι μαλθακοῖσιν ὅκου
ἂν ῥητίνην ξυμμιγνύῃς, ἐπειδὰν ἀφέλῃς ἀπὸ τοῦ
426 πυρὸς | τὸ φάρμακον, ἐς θερμὸν ἔτι ἐὸν καθεὶς τὴν
ῥητίνην κυκᾶν. ἕτερον· στέαρ συὸς παλαιόν, καὶ
κηρός, καὶ ἔλαιον, τὰ δὲ ξηρά, ἰχθήματα λωτοῦ,
λιβανωτός, μολύβδαινα, λιβανωτοῦ μοῖρα, μολυβδαίνης μοῖρα, τοῦ ἰχθνήματος μοῖρα· ἔστω δὲ τοῦ
παλαιοῦ στέατος δύο μοῖραι, τοῦ δὲ κηροῦ μία, καὶ
τοῦ ἐλαίου[1] μία. στέαρ μόνον παλαιὸν ὕειον, σὺν
τῷδε στέαρ αἰγὸς πρόσφατον ὡς ἥκιστα ξὺν τῷ
ὑμένι, κατακαθήρας, μικρὰ τρίψας ἢ κατακόψας
λεῖα, ἔλαιον παραχέειν, καὶ παραπάσσειν τὸν μόλι-
βον ξὺν τῇ σποδῷ, καὶ λωτοῦ ἰχθυημάτων τὸ ἥμισυ.
ἕτερον· στέαρ αἰγός, σποδός, χαλκῖτις κυανέη,
ἔλαιον.

22. Πυρικαύτων· ἑψεῖν χρὴ πρίνου ῥίζας ἁπαλάς,
ὧν εἰ ὁ φλοιὸς παχύτατος καὶ χλωρότατος, κατα-
ταμὼν μικρά, οἶνον λευκὸν ἐπιχέας, μαλθακῷ πυρὶ
καθεψεῖν, ἕως ἂν δοκέῃ καλῶς ἔχειν τὸ πάχος, ὡς
ὑπάλειπτον, καὶ ἐν ὕδατι τὸν αὐτὸν τρόπον. τοῦτο
δάκνει οὐδέν· ὑὸς στέαρ μόνον παλαιὸν ὑπαλείφειν,
τήξας, ἄνωθεν δὲ[2] τῆς σκίλλης τὴν ῥίζαν διαιρῶν
καὶ προστιθεὶς καταδεῖν, τῇ δὲ ὑστεραίῃ ἐπαιονᾶν.
ἕτερον· τήξας ὑὸς στέαρ παλαιόν, καὶ κηρόν, καὶ
ἔλαιον συμμίξας <καὶ>[3] λιβανωτόν, καὶ λωτοῦ

earth it congeals; then, removing it from the fire, pour off everything except the stony mass that precipitates, add resin and agitate; also mix together a little cedar oil and what was set aside. Whenever you mix resin into softening medications you must, when you remove the medication from the fire, pour the resin in, and stir as long as the mixture is still hot. Another: aged lard, wax, olive oil; dry components: shavings of nettle-tree, frankincense, galena—one part each of frankincense, galena and the shavings; of the aged lard let there be two parts, of the wax one, and of the oil one. Aged lard alone,[a] and with this fresh goat's fat, with as few membranes as possible, that has been thoroughly cleaned and kneaded fine, or cut into small pieces; add olive oil, and sprinkle in lead with copper oxide, and nettle-tree shavings one half in amount. Another: goat's fat, oxide of copper, blue copper ore, olive oil.

22. For burns. You must boil tender roots of the holm-oak, and if their bark is very thick and green, cut them into small pieces; add white wine, and boil them down on a gentle fire until they seem to have the right consistency to spread like an ointment; boil them in water in the same way. This one irritates nothing: melt aged lard alone, and apply by anointing; then divide the root of a squill, place it on top of the lard, and bind it with a bandage; on the next day foment. Another: melt aged lard and wax, mix together with them olive oil, frankincense,

[a] I.e. without wax.

[1] Littré after Manialdus: νέον M.
[2] Littré after Manialdus: διὰ M. [3] I.

ἰχθυήματα, καὶ μίλτον, τούτῳ ὑπαλείψας, ἄρου
φύλλα ἐν οἴνῳ καὶ ἐλαίῳ ἑψήσας, προστιθεὶς καταδεῖν. ἕτερον· ἐπειδὰν τῷ συείῳ στέατι ὑπαλείψῃς
παλαιῷ, καταλείφειν ἀσφοδέλου ῥίζας ἐν οἴνῳ
τρίψας λεῖον. ἕτερον· τήξας στέαρ συὸς παλαιόν,
ξυμμίξας ῥητίνη καὶ | ἀσφάλτῳ, αὐτὸ ἐπαλείψας ἐς
ὀθόνιον, θερμήνας πρὸς πῦρ, ἐπιθεὶς ἐπιδεῖν.

428

23. Ὅταν ἐν τῷ νώτῳ ὑπὸ πληγέων ἢ ἄλλως
ἕλκος γένηται, τῇ σκίλλῃ διέφθῳ τρίψας ἐς ὀθόνιον
ἐπαλείψας ἐπιδεῖν, ὕστερον δὲ στέαρ αἰγός, καὶ συὸς
νέον, καὶ σποδόν, καὶ ἔλαιον, καὶ λιβανωτόν.

24. Οἰδήματα ἐν τοῖσι ποσὶ γινόμενα αὐτόματα,[1]
οὐδὲν ὑπὸ τῶν καταπλασμάτων καθιστάμενα, τά τε
οἰδήματα, καὶ ἡ φλεγμασίη, καὶ ἢν σπόγγους ἐπι-
δέῃ τις ἢ εἴρια ἤ τι ἄλλο ἐπὶ τὸ ὑγιές, ἔπειτα ἀνοιδί-
σκηται αὐτόματον καὶ ἀναφλεγμαίνῃ, κατὰ φλέβας
ἐπίρρους αἴτιόν ἐστιν αἵματος, ᾧτινι μὴ φλάσμα
αἴτιόν ἐστι, καὶ ἤν που ἄλλοθι τοῦ σώματος τοῦτο
γίνηται, ὁ αὐτὸς λόγος. ἀλλὰ τοῦ αἵματος χρὴ
ἀφίεναι, μάλιστα μὲν κατὰ φλέβας τὰς ἐπιρρεούσας, ἢν καταφανέες ἔωσιν· ἢν δὲ μή, κατακρούειν τὰ
οἰδήματα βαθύτερα καὶ πυκνότερα, καὶ ἄλλο πᾶν ὅ
τι ἂν κατακρούῃς, οὕτω χρὴ ποιέειν, καὶ ὡς ὀξυτά-
τοισι σιδηρίοισι καὶ λεπτοτάτοισι, καὶ ὅταν ἀφαι-
ρῇς τὸ αἷμα, τῇ μήλῃ μὴ κάρτα πιέζειν, ὡς μὴ
φλάσις προσγίνηται· ὄξει δὲ κατανίζειν, καὶ θρόμ-
βον αἵματος ἐν τοῖσι σχάσμασι μὴ ἐᾶν ἐγκαταλεί-

shavings of nettle-tree wood, and red ochre: anoint with
this, then boil leaves of cuckoo-pint in wine and olive oil,
apply them to the burn, and bind them in place with a
bandage. Another: when you have anointed with aged
lard, anoint again with asphodel roots ground fine in
wine. Another: melt aged lard, mix in resin and asphalt,
smear on to a linen cloth, warm at a fire, and apply and
bind on with a bandage.

23. When a lesion occurs in the back as the result of
blows or in some other way, boil squill thoroughly and
crush it, smear it on to a bandage, and bind it on. Later
apply goat's fat, fresh lard, copper oxide, olive oil and
frankincense.

24. Swellings that arise spontaneously in the feet, that
are not reduced at all by plasters either in their swelling
or in their inflammation, and in which, when someone
applies with a bandage sponges, or wool, or anything else
to a healthy part, that part swells up spontaneously and
becomes inflamed: in these the cause of the blood flow, if
no bruise is the cause, lies in the afferent vessels. If
this occurs in another part of the body, the reason is
the same. You must remove blood, especially from the
afferent vessels, if they are visible. If not, make narrow
incisions in the oedema quite deeply and close together—
also, anywhere else you make such narrow incisions, you
must carry out the operation this same way—with blades
as sharp and fine as possible; when you draw off the
blood, do not apply excessive pressure with your probe, in
order to prevent a bruise from arising. Wash well with
vinegar, and do not allow any clots of blood to remain in

---

[1] καὶ μὴ αὐτόματα added by M^2 in margin.

πεσθαι, καταχρίσας τῷ ἐναίμῳ φαρμάκῳ, εἴρια
οἰσυποῦντα κατεξασμένα μαλθακὰ ἐπιδῆσαι, ῥήνας
οἴνῳ καὶ ἐλαίῳ, καὶ ἐχέτω τὸ σχασθὲν ὅκως ἀνάρ-
ρους εἴη τοῦ αἵματος καὶ μὴ κατάρρους. καὶ μὴ
430 τεγγέτω, καὶ ὀλιγοσιτεέτω, καὶ ὕδωρ πινέτω · | ἢν δὲ
ἀπολύων εὑρίσκῃς τὰ σχάσματα φλεγμαίνοντα,
καταπλάσσειν τῷ ἐκ τοῦ ἀγνοῦ καὶ λίνου καρποῦ
καταπλάσματι · ἢν δὲ ἑλκωθῇ τὰ σχάσματα καὶ ξυρ-
ραγῇ, πρὸς χρῆμα ὀρέων, ἔπειτα προσφέρων ὅτου
ἂν δέῃ, τὰ λοιπὰ ἰητρεύειν.

25. Ὅκου δὲ κιρσὸς ἔνεστιν ἐπ' ἀντικνημίου ἢ
περιφανὴς ἢ κατὰ τῆς σαρκός, καί ἐστι μέλαν τὸ
ἀντικνήμιον, καὶ δοκέει δεῖσθαι αἷμα ἀπ' αὐτοῦ
ἀπορρυῆναι, οὐ χρὴ τὰ τοιαῦτα κατακρούειν οὐδαμῶς · ὡς γὰρ ἐπὶ τὸ πολὺ ἕλκεα μεγάλα γίνεται ἐκ
τῶν σχασμάτων διὰ τοῦ κιρσοῦ τὴν ἐπιρροήν · ἀλλὰ
χρὴ αὐτὸν τὸν κιρσὸν ἀποκεντέειν ἄλλοτε καὶ
ἄλλοτε, ὅπῃ ἂν δοκέῃ καιρὸς εἶναι.

26. Ὅταν δὲ φλέβα τάμῃς, ἐπειδὰν τοῦ αἵματος
ἀφῇς καὶ λύσῃς τὴν ταινίην, καὶ μὴ ἵσταται, τὰ
ἀντία ὅκως ἂν ὁ ῥοῦς γίνηται τοῦ αἵματος, ἐχέτω, ἤν
τε χεὶρ, ἤν τε σκέλος ᾖ, ὡσεὶ χωρέοντος τοῦ αἵματος
ὀπίσω, καὶ οὕτως ὑπομείνας χρόνον πλείω ἢ ἐλάσσω
κατακείμενος, ἔπειτα ἐπιδῆσαι αὐτὸν οὕτως ἔχοντα,
μὴ ἐνεόντος θρόμβου ἐν τῇ τομῇ, ἔπειτα σπληνίον
διπλόον προσθείς, τέγξας οἴνῳ, καὶ ἄνωθεν εἴριον
ἐλαιώσας καθαρόν · κἢν γὰρ ἐπίρρυσις τοῦ αἵματος
ἔῃ βιαίη, σχέσις γίνεται ἐπιρρέοντος · κἢν, θρόμβου

the incisions; anoint with a styptic, bandage on soft, well-carded greasy wool, sprinkled with wine and olive oil, and let the incised part be positioned so that it is upstream for the blood, rather than downstream. Let the patient avoid being moistened from without, eat little, and drink water. If on unbandaging you discover the incisions are inflamed, apply the plaster prepared from chaste-tree and linseed. If the incisions ulcerate and run together, observe their condition and then continue the treatment, employing whatever is required.

25. Where a varix is present in a shin, either conspicuous on the surface or under the flesh, the shin is darkened, and it appears that blood must be evacuated from it, you should in such instances certainly not make deep incisions—for generally large ulcerations will arise out of the incisions on account of the afflux of blood from the varix. Rather, when the time seems appropriate, you must pierce through the varix itself repeatedly.

26. If, in incising a vessel, when you have removed the blood and loosened the bandage, the blood does not stop, hold the limb, whether arm or leg, in the position opposite to how the blood is flowing, so that the blood will be running backwards, and, having the patient wait thus for a greater or lesser time lying down, then, if there is no clot in the incision, apply a bandage with the limb in this position; then put on a double linen compress moistened with wine, and over it a clean piece of oiled wool. For thus, if the afflux of blood is violent, it will be restrained; if a clot

ἐπὶ τῇ τομῇ γενομένου, οὕτω φλεγμήνῃ, διαπυΐσκεται. ἠριστηκότα δὲ χρὴ πλέον ἢ ἔλαττον καὶ πεπω-
κότα φλεβοτομέειν, καὶ ὑποτεθερμασμένον, καὶ ἡμέρης θερμοτέρης ἢ ψυχροτέρης.

27. Σικύην δὲ προσβάλλοντα χρή, ἢν ἐπιρρέῃ
τὸ αἷμα ἀφῃρημένης τῆς σικύης, κἢν πολὺ ῥέῃ, ἢ
ἰχὼρ ῥέῃ, αὐτοῖσι ταχέως, πρὶν πλησθῇ, αὖτις προσ-
βάλλων, ἐπεξέλκειν τὸ λειπόμενον· εἰ δὲ μή, θρόμ-
βοι ἐνεχόμενοι ἐν τοῖσι σχάσμασιν, ἔπειτα ἀνα-
432   φλεγμηλῆναντα ἕλκεα γίνεται ἐξ αὐτῶν. ὄξει δὲ χρὴ
πάντα τὰ τοιαῦτα κατανίζειν, καὶ ὕστερον μὴ τέγγειν, μηδὲ κατακέεσθαι ἐπὶ τὰ σχάσματα, τῶν δὲ
ἐναίμων τινὶ φαρμάκων καταχρῖσαι τὰ σχάσματα·
καὶ ὅταν κάτωθεν τοῦ γούνατος δέῃ προσβάλλειν[1] ἢ
πρὸς τὸ γόνυ, ἑστηκότι ὀρθῷ, ἢν δύνηται ἑστάναι.

[1] Littré: -βάλλον M.

forms in the incision and it becomes inflamed, suppuration occurs. You must phlebotomize when the person has taken a larger or smaller breakfast and had a drink, and has been heated a little—this on a warmer rather than a cooler day.

27. When you apply a cupping instrument, if blood keeps on flowing when you remove it and if much blood or serum flows out, you must quickly reapply the instrument to the same site before it fills up, and draw out what is left; if you do not do this, clots form in the incisions, and then inflamed lesions arise from them. You must bathe all such lesions in vinegar, but later not moisten them, nor should the patient lie on the incisions; anoint the incisions with some styptic. When you must apply the cupping below the knee or at the knee, do so with the person standing up, if he can.

HAEMORRHOIDS AND FISTULAS

INTRODUCTION

These two treatises, or parts of the same treatise, demonstrate similarities of language and content that strongly suggest common authorship. The order in which I present them is supported by the title under which they are recorded in Erotian's preface, περὶ αἱμορροΐδων καὶ συρίγγων,[1] as well as twice in Galen's Hippocratic *Glossary*.[2] This order is also given in the list of Hippocratic works at the head of the manuscript Vaticanus Graecus 276, and in the Bruxelles Hippocratic *Vita*: "item de emorroide post hunc de fistulis."[3]

Erotian includes one word from *Haemorrhoids* and three from *Fistulas*[4] in his glossary, and Celsus draws directly on both works in his own account of lesions of the anus.[5]

Haemorrhoids and *Fistulas* present a clear and well organized summary of the pathology and therapy of

[1] Erotian p. 9.

[2] Galen vol. 19, 130 πήρινα; 19, 141 στρυβλήν. See also 19, 110 κατοπτῆρι, which comes from *Haemorrhoids* 5 or *Fistulas* 3.

[3] J. R. Pinault, *Hippocratic Lives and Legends*, Leiden, 1992, p. 133.

[4] See Nachmanson p. 439.

[5] Celsus, *De medicina* 7, 30,1–2.

haemorrhoids, condylomas, fistula in ano, and their complications such as strangury and prolapse of the anus. Surgical treatments dominate: cautery, excision, scraping, application of putrefactant medications, and fomentation.

The treatises were edited by Robert Joly in his Budé edition, volume XIII, and it is on this work that my text for the most part depends. F. Adams published an English translation of the works in 1849.[6]

---

[6] Adams vol. 2, 816–822 and 826–830.

ΠΕΡΙ ΑΙΜΟΡΡΟΙΔΩΝ

1. Αἱμορροΐδων τὸ μὲν νόσημα ὧδε γίνεται· ἐπὴν χολὴ ἢ φλέγμα ἐς τὰς φλέβας τὰς ἐν τῷ ἀρχῷ καταστηρίξῃ, θερμαίνει τὸ αἷμα τὸ ἐν τοῖσι φλεβίοισι· θερμαινόμενα δὲ τὰ φλέβια ἐπισπᾶται ἐκ τῶν ἔγγιστα φλεβίων τὸ αἷμα, καὶ πληρούμενα ἐξογκέει τὸ ἐντὸς τοῦ ἀρχοῦ, καὶ ὑπερίσχουσιν αἱ κεφαλαὶ τῶν φλεβίων, καὶ ἅμα μὲν ὑπὸ τῆς κόπρου ἐξιούσης φλώμεναι, ἅμα δὲ ὑπὸ τοῦ αἵματος ἀθροιζομένου βιαζόμεναι, ἐξακοντίζουσιν αἷμα, μάλιστα μὲν ξὺν τῷ ἀποπάτῳ, ἐνίοτε δὲ χωρὶς τοῦ ἀποπάτου.

2. Θεραπεύειν δὲ δεῖ ὧδε· πρῶτον μὲν ὑπαρχέτω εἰδέναι ἐν οἵῳ χωρίῳ γίνονται. Ἀρχὸν γὰρ καὶ τάμνων, καὶ ἀποτάμνων, καὶ ἀναρράπτων, καὶ καίων,[1] καὶ ἀποσήπων, ταῦτα γὰρ δοκέει δεινότατα εἶναι, οὐδὲν ἂν σίνοιο. παρασκευάσασθαι δὲ κελεύω ἑπτὰ ἢ ὀκτὼ σιδήρια, σπιθαμαῖα τὸ μέγεθος, πάχος δὲ ὡσεὶ μήλης παχείης· ἐξ ἄκρου δὲ κατακάμψαι· καὶ ἐπὶ τῷ ἄκρῳ πλατὺ ἔστω ὡς ἐπὶ ὀβολοῦ[2] μικροῦ.

---

[1] Ermerins: δαίων I.
[2] I, Littré: ὀβελοῦ Linden, Ermerins, Joly.

HAEMORRHOIDS

1. The condition of haemorrhoids arises as follows. When bile or phlegm becomes fixed in the vessels of the anus, it heats the blood in them so that, being heated, they attract blood from their nearest neighbours. As the vessels fill up, the interior of the anus becomes prominent and the heads of the vessels are raised above its surface, where they are partly abraded by the faeces passing out, and partly overcome by the blood collected inside them, and so spurt out blood, usually during defecation, but occasionally at other times as well.

2. You must treat as follows. First, undertake to find out where the haemorrhoids are; for to incise the anus, to amputate from it, to lift it by sewing, to cauterize it, or to remove something from it by putrefaction—these seem to be dangerous, but in fact will do no harm. I bid you to prepare seven or eight irons, a span in length, and with the width of a wide probe; bend these at the end, and also make them flat at the end like a small obol.[a] Clean the site

---

[a] With Vander Linden's conjecture: "like a small spit."

ΠΕΡΙ ΑΙΜΟΡΡΟΙΔΩΝ

προκαθήρας δὴ φαρμάκῳ τῇ πρότερον, αὐτῇ δὲ ᾗ ἂν
ἐπιχειρέῃς καῦσαι, ἀνακλίνας τὸν ἄνθρωπον ὕπτιον,
καὶ προσκεφάλαιον ὑπὸ τὴν ὀσφὺν ὑποθείς, ἐξ-
αναγκάζειν ὡς μάλιστα τοῖσι δακτύλοισι τὴν ἕδρην
ἔξω, ποιέειν δὲ καὶ διαφανέα τὰ σιδήρια, καὶ καίειν
ἕως ἂν ἀποξηράνῃς, καὶ ὅκως μὴ ὑπαλείψῃς· καίειν
δὲ καὶ μηδεμίην ἐᾶσαι ἄκαυστον τῶν αἱμορροΐδων,
ἀλλὰ πάσας ἀποκαύσεις. γνώσει δὲ οὐ | χαλεπῶς
τὰς αἱμορροΐδας· ὑπερέχουσι γὰρ ἐς τὸ ἐντὸς τοῦ
ἀρχοῦ, οἷον ῥᾶγες πελιδναί, καὶ ἅμα ἐξαναγκαζομέ-
νου τοῦ ἀρχοῦ ἐξακοντίζουσιν αἷμα. κατεχόντων δ'
αὐτῶν, ὅταν καίηται, τῆς κεφαλῆς καὶ τὰς χεῖρας,
ὡς μὴ κινέηται, βοάτω καιόμενος· ὁ γὰρ ἀρχὸς μᾶλλον ἐξίσχει. ἐπὴν δὲ καύσῃς, φακοὺς καὶ ὀρόβους
ἑψήσας ἐν ὕδατι, τρίψας λείους, κατάπλασσε[1] πέντε
ἢ ἐξ ἡμέρας· τῇ δὲ ἑβδόμῃ σπόγγον μαλθακὸν
τάμνειν ὡς λεπτότατον, πλάτος δὲ εἶναι τοῦ σπόγγου ὅσον ἐξ δακτύλων πάντῃ· ἔπειτα ἐπιθεῖναι ἐπὶ
τὸν σπόγγον ὀθόνιον ἴσον τῷ σπόγγῳ λεπτὸν καὶ
λεῖον, ἀλείψας μέλιτι· ἔπειτα ὑποβαλὼν τῷ δακτύλῳ
τῷ λιχανῷ τῆς ἀριστερῆς χειρὸς μέσον τὸν σπόγγον, ὦσαι κάτω τῆς ἕδρης ὡς προσωτάτω· ἔπειτα
ἐπὶ τὸν σπόγγον προσθεῖναι εἴριον, ὡς ἂν ἐν τῇ
ἕδρῃ ἀτρεμίζῃ. διαζώσας δὲ ἐν τῇσι λαγόσι, καὶ
ὑφεὶς ταινίην ἐκ τοῦ ὄπισθεν, ἀναλαβὼν ἐκ τῶν σκελέων τὸν ἐπίδεσμον, ἀναδῆσαι ἐς τὸ διάζωσμα παρὰ
τὸν ὀμφαλόν. τὸ δὲ φάρμακον, ὃ εἶπον, ἐπίδει τὸ
πυκνὴν τὴν σάρκα ποιέον[2] καὶ ἰσχυρὴν φῦναι.

438

you are attempting to cauterize beforehand with a medication, have the person lie on his back, and place a pillow
beneath the loins. Force the anus out as far as possible
with your fingers; heat the irons red-hot, and burn until
you so dry the haemorrhoids out that you do not need to
anoint: burn them off completely, leaving nothing uncau-
terized. You will recognize haemorrhoids without difficulty, for they rise above the surface in the interior of the
anus like livid grapes, and when the anus is forced outwards, they spurt out blood. Let assistants hold the
patient down by his head and arms while he is being cauterized so that he does not move—but let him shout during the cautery, for that makes the anus stick out more.
After you have applied the cautery, boil lentils and chick-
peas in water, pound them smooth, and apply this as a
plaster for five or six days. On the seventh day, cut a soft
sponge as thin as possible—it should be six fingers broad
in every direction—place a piece of thin fine linen cloth
equal in size to the sponge on top of it, and smear with
honey. Then, placing the middle of the sponge over the
index finger of the left hand, press it into the anus as far as
you can, and insert a woollen plug up against the sponge,
to hold it in place in the anus. Tie a band around the body
at the flanks, run a strip down from it at the back, draw up
the end of this strip between the legs, and tie it to the circular bandage at the navel. Use this to bind on the medication which I mentioned, that will make the flesh grow

---

[1] Ermerins: -πασσε I.
[2] Littré: ποιέειν I.

ταῦτα δὲ δεῖ ἐπιδεῖν μὴ ἔλασσον ἡμερῶν εἴκοσι.
ῥοφέειν δὲ ἅπαξ τῆς ἡμέρης,¹ ἄλευρον, ἢ κέγχρον, ἢ
τὸ ἀπὸ τῶν πιτύρων, καὶ πίνειν ὕδωρ· ἢν δὲ ἐς ἄφοδον ἴζηται, ὕδατι θερμῷ διανίζειν· λούεσθαι δὲ διὰ
τρίτης ἡμέρης.

3. Ἑτέρη θεραπείη· ἐκβαλὼν τὴν ἕδρην ὡς
μάλιστα, αἰονᾶν ὕδατι θερμῷ, ἔπειτα ἀποτάμνειν
τῶν αἱμορροΐδων τὰ ἄκρα· φάρμακον δὲ προκατα-
σκευασθῆναι πρὸς τὴν τομὴν τόδε· οὐρήσας ἐς χαλκεῖον, ἐπίπασον ἐπὶ τὸ οὖρον χαλκοῦ ἄνθος ὀπτὸν
καὶ τετριμμένον λεῖον,¹ ἔπειτα διείς, καὶ κινήσας τὸ
χαλκεῖον, ξήρανον ἐν τῷ ἡλίῳ· ὅταν δὲ ξηρὸν
γένηται, συγχύσας τρῖψον λεῖον καὶ προστίθει τῷ
440   δα|κτύλῳ, καὶ σπληνία ἐλαιώσας προστίθει, καὶ
σπόγγον ἐπάνω ἐπίδει.

4. Προσφύεται πρὸς τῇ αἱματίτιδί τι κονδυλῶ-
δες² οἷον συκαμίνου καρπός· καὶ ἢν μὲν ἔξω σφόδρα
ᾖ ἡ κονδύλωσις, περιπέφυκεν αὐτῇ καλυπτὴρ ὁ τῆς
σαρκός. καθίσας οὖν τὸν ἄνθρωπον ὀκλὰξ ἐπὶ
ὅλμων δύο, σκόπει· εὑρήσεις γὰρ πεφυσημένα τὰ
μεσηγὺ τῶν γλουτῶν παρὰ τὴν ἕδρην, τὸ δὲ αἷμα
ἐκχωρέον ἔνδοθεν. ἢν γοῦν ἐνδιδοῖ ὑπὸ τῷ³ καλυ-
πτῆρι, [ἢ]⁴ τὸ κονδύλωμα τῷ δακτύλῳ ἀφελεῖν·
οὐδὲν γὰρ χαλεπώτερον ἤ περ προβάτου δειρομένου
τὸν δάκτυλον μεταξὺ τοῦ δέρματος καὶ τῆς σαρκὸς
περαίνειν· καὶ ταῦτα διαλεγόμενος ἅμα λάνθανε

---

¹ ὀπτὸν—τετριμμένον λεῖον Ermerins: ὀπτοῦ—τετριμμένου
λείου I.

dense and strong. The bandage must remain in place for at least twenty days. Have the patient drink gruel made from meal, millet or bran once a day, and also water. After he sits down to stool, wash him off thoroughly with warm water; have him bathe every other day.

3. Another therapy. Protruding the anus as much as possible, moisten it with warm water, and then excise the extremities of the haemorrhoids. Before the operation prepare the following medication: collect urine in a copper vessel, and sprinkle into it flower of copper that has been roasted and ground fine; then mix this together, agitate the vessel, and dry the mixture in the sun. When it is dry grate it up, grind it fine and insert some of it with your finger; moisten a linen compress with olive oil, apply it, place a sponge over this, and bind it with a bandage.

4. There can grow next the blood vessel of the anus a kind of knobbiness (condyloma) shaped like a mulberry. If this condyloma is especially protuberant, a covering of flesh grows around it. Sit the person down in a squatting position on two supports, and make an examination: for you will discover the area between the buttocks next the anus distended, and blood escaping inside. Now if the condyloma gives way when a speculum is applied, remove it with a finger, for this is no more difficult than for a person who is flaying an animal to get his finger between the skin and the flesh; talk to the person while you are doing

---

[2] Joly: τῇ κονδυλώσει I.

[3] Foes: τῇ I.

[4] Del. Foes in note.

ΠΕΡΙ ΑΙΜΟΡΡΟΙΔΩΝ

ποιέων. ἐπὴν δὲ ἀφέλῃς τὸ κονδύλωμα, ἀνάγκη λύε-
σθαι[1] δρόμους αἵματος ἀπὸ πάσης τῆς ἀφαιρέσιος·
ταῦτα χρὴ ἀποπλῦναι οἴνῳ αὐστηρῷ, κηκῖδας
ἐναποβρέξας· καὶ ἥ τε αἱματῖτις οἰχήσεται σὺν τῷ
κονδυλώματι, καὶ τὸ κάλυμμα καταστήσεται, καὶ
ὅσῳ ἂν παλαιότερον ᾖ, ῥηϊδίως ἔσται ἡ ἴησις.

5. Ἢν δὲ ἀνωτέρη ᾖ ἡ κονδύλωσις, τῷ κατοπτῆρι
σκέπτεσθαι, καὶ μὴ ἐξαπατᾶσθαι ὑπὸ τοῦ κατοπτῆ-
ρος· διοιγόμενος γὰρ ὁμαλύνει τὴν κονδύλωσιν, ξυν-
αγόμενος δὲ πάλιν δείκνυσιν ὀρθῶς. ἀφαιρέειν δὲ
χρή, ἐλλεβόρῳ μέλανι ὑπαλείφοντα τὸν δάκτυλον·
ἔπειτα τριταῖον οἴνῳ κλύζειν αὐστηρῷ. τὸ δὲ αἷμα,
ὅταν ἀφέλῃς τὴν κονδύλωσιν, ὅτι οὐ ῥέει, μὴ θαυμά-
ζειν· οὐδὲ γὰρ ἢν ἐν τοῖσιν ἄρθροισι διατάμῃς τὰς
χεῖρας ἢ τὰ σκέλεα, οὐ ῥεύσεται αἷμα· ἢν δ᾽ ἄνωθεν
ἢ κάτωθεν τάμῃς τῶν ἄρθρων, εὑρήσεις φλέβας
κοίλας καὶ αἱμόρρους, καὶ χαλεπῶς ἂν ἴσχοις εὐ-
πόρως. οὕτω καὶ τὴν ἐν τῇ ἕδρῃ αἱμορροΐδα, ἢν μὲν
ἄνωθεν ἢ κάτωθεν τάμῃς τῆς ἀφαιρέσιος τοῦ κονδυ-
λώματος, αἷμα ῥεύσεται· ἢν δὲ αὐτὴν ἀφέλῃς τὴν
κονδύλωσιν ἐν τῇ προσφύσει, οὐ ῥεύσεται. ἢν μὲν
οὖν οὕτω καθίσταται, καλῶς ἂν ἔχοι· ἢν δὲ μή, καῦσαι, φυλασσόμενος ὡς μὴ ἅψῃ τῷ σιδήρῳ, ἀλλ᾽
ἐγγὺς προσφέρων τὰ σιδήρια ἀποξηραίνειν, καὶ
προστιθέναι τὸ τοῦ χαλκοῦ ἄνθος τὸ ἐν τῷ οὔρῳ.

6. Καυτῆρα χρὴ ποιήσασθαι, οἷον καλαμίσκον

[1] Potter: κύεσθαι I.

this to distract his attention. When you have removed the condyloma, it is inevitable for streams of blood to be released from the whole amputation. These you must wash off with dry wine in which you have steeped oak-galls. The blood vessel will disappear along with the condyloma, and the covering will recede. The older the condition is, the easier healing will be.

5. If the condyloma is higher up, examine it by means of a speculum, but do not be deceived by the speculum, for when it is open it flattens the condyloma down, whereas when you close it again you see the condyloma correctly. Remove the condyloma after smearing a finger with black hellebore; then, on the next day but one, wash it with dry wine. That blood does not flow when you remove a condyloma is no cause for wonder: for blood will also not flow if you incise the arms or legs at the joints, whereas if you cut them above or below the joints, you will discover the vessels to be hollow and filled with blood, and only with difficulty would you prevent their free flow. Thus, too, if you incise a haemorrhoid in the anus above or below the amputation of a condyloma, blood will flow, but if you remove the condyloma itself at its attachment, none will flow. Now if the condition clears up with this, fine; if not, apply heat, taking care not to touch the flesh with your iron, but by bringing the iron close dry the condyloma out; also inject flower of copper dissolved in urine.

6. Make a cautery apparatus shaped like a hedge

φραγμίτην·[1] σιδήριον δὲ ἐναρμόσαι καλῶς ἁρμόζον·
ἔπειτα τὸν αὐλίσκον ἐνθεὶς ἐς τὴν ἕδρην, διαφαῖνον
τὸ σιδήριον καθιέναι, καὶ πυκνὰ ἐξαιρέειν, ἵνα μᾶλλον ἀνέχηται θερμαινόμενος· καὶ οὔτε ἕλκος ἕξει
ὑπὸ τῆς θερμασίης, ὑγιέα τε ξηρανθέντα τὰ φλέβια.

7. Ἢν δὲ βούλῃ μήτε καίειν, μήτε ἀποτάμνειν,
προκαταιονήσας[2] ὕδατι πολλῷ θερμῷ, καὶ ἐκτρέψας
τὴν ἕδρην, σμύρναν τρίψας λείην καὶ κηκίδα, καὶ
στυπτηρίην Αἰγυπτίην κατακαύσας, ἓν καὶ ἥμισυ
πρὸς τἄλλα, καὶ μελαντηρίης ἄλλο τοσοῦτον, τούτοισι ξηροῖσι χρῆσθαι· ἡ δὲ αἱμορροῒς τούτοισι
τοῖσι φαρμάκοισιν ἀποστήσεται, ὥσπερ σκῦτος
κατακεκαυμένον· ταῦτα ποιέειν μέχρις ἂν πάσας
ἀφανίσῃς. καὶ χαλκίτιδος ἥμισυ κεκαυμένης[3] τωὐτὸ
ἀπεργάζεται.

8. Ἢν δὲ βούλῃ βαλανίοισιν ἰῆσθαι, σηπίης
ὄστρακον, μολυβδαίνης τρίτον μέρος, ἄσφαλτον,
στυπτηρίην, ἄνθος ὀλίγον, κηκίδα, χαλκοῦ ἰὸν ὀλίγον, τούτων μέλι ἑφθὸν καταχέας, βάλανον ποιήσας
μακροτέρην, προστίθει, μέχρις ἂν ἀφανίσῃς.

444   9. Γυναικείην αἱμορροΐδα ὧδε θεραπεύειν· πολλῷ
ὕδατι θερμῷ αἰονήσας, σύνεψε δὲ ἐν τῷ θερμῷ τῶν
εὐωδέων, τρίψας μυρίκην, λιθάργυρον ὀπτὴν καὶ
κηκίδα, οἶνον λευκὸν παράχεε καὶ ἔλαιον καὶ χηνὸς
στέαρ, τρίψας πάντα ὁμοῦ, διδόναι, ὁκόταν αἰονηθῇ,
διαχρίσασθαι· αἰονᾶν δὲ καὶ τὴν ἕδρην ἐξώσας ὡς
μάλιστα.

reed,[a] and adapt the iron so that it fits easily into it. Then introduce the tube into the anus, heat the iron red-hot, and insert it into the tube: remove it frequently, in order that the patient will better withstand the heating; he will not have any ulceration as a result of the heat, and the vessels, being dried, will heal up.

7. If you wish to avoid both cautery and excision, moisten the haemorrhoids beforehand with copious warm water, and evert the anus; grind myrrh and oak galls into a smooth paste, and add one and a half times as much burnt Egyptian alum and an equal amount of black pigment: apply this medication dry. The haemorrhoid will be separated like a burnt hide by it; repeat this procedure until you have removed them all. Half this amount of burnt copper ore accomplishes the same thing.

8. If you wish to treat with suppositories: take bone of cuttle-fish, a third part of galena, asphalt, alum, a little flower of copper, oak gall, and a little verdigris, and over these pour boiled honey; form into a rather long suppository, and insert it until the haemorrhoids are removed.

9. Treat haemorrhoids in a woman as follows. Foment with copious warm water in which sweet-smelling substances have been boiled; grind tamarisk, burnt litharge, and oak gall, add white wine, olive oil and goose grease, pound these all smooth, and after she has been fomented give this to be anointed. Also foment the anus after forcing it out as far as possible.

---

[a] Cf. Dioscorides 1, 114.

---

[1] Foes after Cornarius' *arundinem sepiariam, phragmiten appellatam*: φαρμακίτην I.
[2] Froben: προσ- I.   [3] Ermerins: -μένον I.

ΠΕΡΙ ΣΤΡΙΓΓΩΝ

VI 448
Littré

1. Σύριγγες δὲ γίνονται μὲν ὑπὸ φλασμάτων καὶ φυμάτων, γίνονται δὲ καὶ ὑπὸ εἰρεσίης, καὶ ἱππασίης, ὅταν ἀθροισθῇ ἐν τῷ γλουτῷ αἷμα πλησίον τῆς ἕδρης· σηπόμενον γὰρ νέμεται ἐς τὰ μαλθακά, ἅτε ὑγροῦ ἐόντος τοῦ τε ἀρχοῦ, καὶ τῆς σαρκὸς μαλθακῆς, ἐν ᾗ νέμεται, ἔστ' ἂν τὸ φῦμα ῥήξῃ καὶ κάτω ἐς τὸν ἀρχὸν διασήψῃ. ἐπὴν δὲ τοῦτο γένηται, συριγγοῦται, καὶ ἰχὼρ ῥέει, καὶ κόπρος ῥεῖ δι' αὐτῆς[1] καὶ φῦσα καὶ βδελυγμίη πολλή. ὑπὸ μὲν οὖν τῶν φλασμάτων γίνεται, ὁκόταν τι τῶν περὶ τὸν ἀρχὸν χωρίων φλασθῇ ὑπὸ πληγῆς, ἢ ὑπὸ πτώματος, ἢ ὑπὸ τρώματος, ἢ ἱππασίης, ἢ εἰρεσίης, ἢ ὅσα τοιουτότροπά ἐστι· ξυνίσταται γὰρ αἷμα· σηπόμενον δὲ ἐκπυΐσκεται· ὑπὸ δὲ τοῦ ἐκπυϊσκομένου πάσχει ἅπερ ἐπὶ τῶν φυμάτων εἴρηται.

2. Πρῶτον μὲν οὖν ὅταν τι τοιοῦτον αἴσθῃ φυόμενον φῦμα, τάμνειν ὡς τάχιστα ὠμὸν πρὶν ἢ διαπυῆσαι ἐς τὸν ἀρχόν.

3. Ἢν δὲ νοσέοντα ἤδη τὴν σύριγγα παραλάβῃς, λαβὼν σκορόδου φύσιγγα νεαρήν, ἀνακλίνας

390

FISTULAS

1. Fistulas arise from contusions and tubercles, and also from rowing and horse-back riding, when blood collects in the buttock near the anus; for on putrefying, this blood migrates through the soft parts, the rectum being moist, and the intervening tissue soft, until the tubercle ruptures and suppurates its way down through into the rectum. When this happens, a fistula is formed and serum flows out through it, as well as faeces, wind and copious filth. Now a fistula arises from contusions when part of the area about the rectum is contused as the result of a blow, fall or wound, or from horse-back riding, or rowing, or things like these: for the blood collects, putrefies and suppurates, and from the suppuration of the blood the person suffers the same things that were described for tubercles.

2. In the first place, then, whenever you observe any such tubercle forming, incise it as soon as possible during the unconcocted stage before suppuration through into the rectum occurs.

3. But if you take on someone who already has a fistula, take a fresh stalk of garlic, lay the person down on his

---

[1] Littré: ἑωυτῆς I.

τὸν ἄνθρωπον ὕπτιον, τὰ σκέλεα διαγαγὼν τὸ μὲν
ἔνθα, τὸ δὲ ἔνθα, τὴν φύσιγγα καθιέναι ἔστ᾽ ἂν
προσκόψῃ, μετρῆσαί τε τὸ βάθος τῆς σύριγγος τῇ
φύσιγγι, καὶ σεσέλιος δὲ ῥίζαν κόψας ὡς λεπτοτά-
την, ὕδωρ ἐπιχέας, βρέχειν τέσσαρας ἡμέρας· καὶ
προνηστεύσας πινέτω μέλιτι παραμίσγων τὸ ὕδωρ
κατὰ τρεῖς κυάθους. [ἐν τούτῳ κάθαιρε καὶ τὰς
ἀσκαρίδας. ὁκόσοι δ᾽ ἂν καταλείφθωσιν ἀθεράπευ-
τοι, θνήσκουσιν.]¹ ἔπειτα ὀθόνιον βύσσινον τιθυ-
μάλλου ὀπῷ τοῦ μεγάλου δεύσας, καταπάσσων
ἄνθος χαλκοῦ ὀπτὸν τετριμμένον στροβίλην ποιή-
450 σας ἴσην τῇ σύριγγι τὸ μῆκος, | ῥάμμα διεὶς δι᾽
ἄκρας τῆς στροβίλης καὶ αὖθις διὰ τῆς φύσιγγος,
ὕπτιον κατακλίνας τὸν ἄνθρωπον, κατοπτῆρι κατι-
δὼν τὸ διαβεβρωμένον τοῦ ἀρχοῦ, ταύτῃ τὴν
φύσιγγα διεῖναι· καὶ ὁκόταν παρακύψῃ ἐς τὸν
ἀρχόν, ἐπιλαμβανόμενος ἕλκειν, ἄχρις οὗ ἡ στρο-
βίλη διωσθῇ καὶ ἰσωθῇ τῷ τε ἄνω καὶ τῷ κάτω·
ἐπὴν δὲ ἐσωθῇ, βάλανον ἐνθεὶς κερατίνην ἐς τὸν
ἀρχόν, γῇ διαχρίσας σμηκτρίδι, τὸν ἀρχὸν ἐᾶν·
ἐπὴν δὲ ἀποπατέῃ, ἐξαιρέειν, καὶ αὖθις προστιθέναι·
ἕως ἂν πεμπταίη γένηται· ἕκτῃ δὲ ἡμέρῃ ἐξαιρέειν,
ἕλκων τὴν στροβίλην ἔξω τῆς σαρκός· καὶ τρῖψαι
στυπτηρίην μετὰ ταῦτα, καὶ πλήσας τὴν βάλανον
καὶ ἐς τὸν ἀρχὸν ἐμβαλών, ἐὰν ἄχρις οὗ ἡ στυπτηρίη ὑγρὴ γένηται· τὸν δὲ ἀρχὸν σμύρνῃ ἀλείφειν,
ἄχρις οὗ ἂν δοκέῃ ξυμπεφυκέναι.

4. Ἑτέρη θεραπείη· ὠμόλινον λαβὼν ὡς λεπτό-

back, spread his legs one in each direction, insert the stalk until it touches the end, and so measure the depth of the fistula. Pound root of hartwort very fine, add water, and moisten the fistula with this for four days. Let the patient first fast, and then drink water mixed with honey, three cyathoi at a time. [This patient should also be cleaned of intestinal worms. Any patients that are left untreated die.] Then dip a linen cloth in the juice of great spurge,[a] sprinkle ground burnt flower of copper over it, form it into a plug equal in length to the fistula, and pass a thread through the end of the plug and also through the garlic stalk; lay the person on his back, inspect the eroded part of the rectum with a speculum, and pass the stalk through to there; when its end emerges into the rectum, take hold of it and pull until the plug is drawn in and resides in the middle of the fistula between its top and bottom. With the plug in place, insert into the anus a horn suppository smeared with fullers' earth, and leave it there. Whenever the patient wants to pass stools, let him remove the suppository and reinsert it afterwards; continue for five days. On the sixth day, remove the suppository and draw out the plug from the flesh. After that, grind up alum, apply it to the suppository, and insert into the rectum, leaving it in place until the alum becomes moist. Continue to anoint the rectum with myrrh until it appears to have united.

4. Another therapy. Take about a span of very fine

[a] Cf. Dioscorides 4, 164.

---

[1] Del. Ermerins.

τατον, συμβάλλειν ὅσον σπιθαμιαῖον πεντάπλουν,
καὶ ξυμπεριλαβεῖν ἱππείην τρίχα· ἔπειτα ποιησάμε-
νος μήλην κασσιτερίνην ἐπ' ἄκρου τετρημένην, ἐνεί-
ρας ἐς τὴν μήλην τὴν ἀρχὴν τοῦ ὠμολίνου συμβε-
βλημένου, καθιέναι τὴν μήλην ἐς τὴν σύριγγα, καὶ
ἅμα τῆς ἀριστερῆς χειρὸς τὸν δάκτυλον τὸν λιχανὸν
καθιέναι ἐς τὴν ἕδρην· ἐπὴν δὲ ψαύσῃ ἡ μήλη τοῦ
δακτύλου, ἄγειν ἔξω τῷ δακτύλῳ, ἀποκάμψας τῆς
μήλης τὸ ἄκρον καὶ τὴν ἀρχὴν τὴν ἐν τῇ μήλῃ· καὶ
τὴν μὲν μήλην πάλιν ἐξαιρέειν, τοῦ δὲ ὠμολίνου τὰς
ἀρχὰς ἀφάψαι δὶς ἢ τρίς· καὶ τὸ λοιπὸν τοῦ ὠμολίνου ἐπιστρέψας, ἐπιδῆσαι πρὸς τὸ ἅμμα· ἔπειτα
κελεύειν ἀπελθόντα διαπρήσσεσθαι τὰ ἑωυτοῦ. ὁκόσον δή, σηπομένης τῆς σύριγγος, χαλᾶται τοῦ ὠμολίνου, τοῦτο ἐπιτείνειν καὶ ἐπιστρέφειν ἀεὶ καθ'
ἑκάστην ἡμέρην· ἢν δέ σοι τὸ ὠμόλινον διασαπῇ
πρόσθεν ἢ τὴν σύριγγα διαβρωθῆναι, πρὸς τὴν
τρίχα προσάψας ἕτερον ὠμόλινον διεῖναι καὶ ἀφά-
452   ψαι (ἡ γὰρ θρὶξ διὰ τοῦτο παραβάλλεται | τῷ
ὠμολίνῳ ὅτι ἄσηπτός ἐστιν)· ἐπὴν δὲ διασαπῇ ἡ
σύριγξ, τάμνεσθαι χρὴ σπόγγον μαλθακὸν ὡς
λεπτότατον προστεθέντα· ἔπειτα ἐς μὲν τὴν σύριγγα
ἄνθος χαλκοῦ ὀπτὸν συχνὸν τῇ μήλῃ ἐνθεῖναι, τὸν
δὲ σπόγγον ἀλεῖψαι μέλιτι, καὶ ὑποβαλὼν μέσον[1]
τῷ λιχανῷ δακτύλῳ τῆς ἀριστερῆς χειρὸς ὦσαι
πρόσω, καὶ προσθεὶς ἕτερον σπόγγον ἀναδῆσαι τὸν
αὐτὸν[2] τρόπον, ὅν περ καὶ ἐπὶ τῇσιν αἱμορροῖσι· τῇ
δὲ αὔριον ἀπολύσας, περινίψαι[3] ὕδατι θερμῷ, καὶ

thread of raw linen, fold it in five, twist it together, and
wrap it around a horsehair. Then make a tin probe with
an eye at one end, thread the end of the twisted linen into
the probe, and, passing the probe into the fistula, at the
same time introduce the index finger of your left hand
into the anus. When the probe touches your finger, use
the finger to draw it out through the anus by bending
aside its end and the thread in it; remove the probe, and
tie together the ends of the thread with a double or treble
knot; turn back the rest of the thread, and tie it against
the knot with a bandage. Then let the person go about his
business. Any slackness in the thread that arises as the
fistula suppurates should be tightened up daily; again
turn back the ends. If you find that the thread dissolves
before the fistula is eaten through, attach another thread
to the horsehair, pull it through, and fasten it in the same
way (the horsehair is run along with the thread for the
reason that it is not liable to dissolution the way the
thread is). When the fistula suppurates through, you
must trim a soft, very fine piece of sponge and apply it.
Then introduce burnt flower of copper into the fistula
with a probe several times, anoint a sponge with honey,
place its centre over the index finger of the left hand, and
press it into place; apply another sponge and hold it in
place with the same kind of bandage used on haemorrhoids. On the following day, remove the bandage, wash
the area with warm water, and attempt to clean out

---

[1] Littré after Vidius' *spongiam mediam*: μέσῳ I.

[2] Littré: ἑωυτὸν I.

[3] Littré: περιθεῖναι I.

σπόγγῳ τῷ δακτύλῳ τῆς ἀριστερῆς χειρὸς πειρᾶν
διακαθαίρειν τὴν σύριγγα, καὶ αὖθις πάλιν τὸ ἄνθος
ἐπιδῆσαι· ταῦτα ποιέειν ἑπτὰ ἡμέρας, ἐν ταύτῃσι
γὰρ μάλιστα ὁ χιτὼν τῆς σύριγγος ἐκσήπεται· τὸ δὲ
λοιπόν, ἔστ' ἂν ὑγιανθῇ, τούτῳ ἐπιδεῖν· κατὰ γὰρ
τοῦτον τὸν τρόπον ὑπὸ τοῦ σπόγγου διαναγκαζο-
μένη καὶ ἀναπτυσσομένη ἡ σύριγξ οὔτε πάλιν ξυμ-
πέσοι ἂν οὔτε τὸ μὲν αὐτῆς ὑγιανθείη ἄν, τὸ δὲ
πάλιν ξυμπληρωθείη, ἀλλ' ἐν ἑωυτῇ πᾶσα ὑγιὴς
ἔσται. ἐν τῇ θεραπείῃ δὲ προσαιονᾶν ὕδατι πολλῷ
θερμῷ, καὶ λιμοκτονέειν.

5. Ἢν δὲ μὴ διαβεβρώκῃ ἡ σύριγξ, προμηλώ-
σας μήλῃ, τέμνε ἕως ἂν διέλθῃ, καὶ ἐπίπασσε ἄνθος
χαλκοῦ, καὶ ἐὰν πέντε ἡμέρας· κατάχεε δὲ ὕδωρ θερ-
μόν· καὶ ἐπάνω ὕδατι φυρῶν ἄλφιτον κατάπασσε,
καὶ φύλλα τεύτλων ἐπίδει· ἐπὴν δὲ ἐκπέσῃ τὸ ἄνθος
τοῦ χαλκοῦ, καὶ καθαρὸν ᾖ τὸ ἕλκος τῆς σύριγγος,
ἰῶ ὥσπερ τὴν ἔμπροσθεν.

6. Ἢν δὲ ἐν χωρίῳ ᾖ, ὃ μὴ οἷόν τε τάμνειν,
βαθείη δὲ καὶ ἡ σύριγξ, ἄνθει χαλκοῦ καὶ σμύρνῃ
καὶ λίτρῳ οὔρῳ διείς, κλύζειν, καὶ ἐς τὸ στόμα τῆς
σύριγγος μολύβδιον ἐντιθέναι, ὅπως μὴ ξυμφύηται·
κλύζειν δὲ πτεροῦ σύριγγα προσδήσας πρὸς κύστιν,
454 καὶ | καθεὶς ἐς τὴν σύριγγα, πρὸς τοῦτο διάγειν
κλύζων. ὑγιὴς δὲ οὐ γίνεται, ἢν μὴ τμηθῇ.

7. Ἢν ὁ ἀρχὸς φλεγμήνῃ, καὶ ὀδύνη ἔχῃ καὶ
πυρετός, καὶ ἐς ἄφοδον θαμινὰ καθίζῃ, καὶ μηδὲν
ὑποχωρέῃ, καὶ ὑπὸ τοῦ φλέγματος δοκέῃ ἐξιέναι ἡ

the fistula with a sponge on the finger of your left hand; apply flower of copper again and rebandage. Do this for seven days, for in that time the tunic of the fistula is usually suppurated away. Thereafter, until the fistula heals, bandage as indicated, for with this procedure the fistula is held open by the sponge and extended, so that it can neither relapse, nor fill up again in one part while another part is healing: instead it heals completely. In this method of treatment also douche the rectum with copious warm water, and limit food intake strictly.

5. If the fistula is not eaten through by this method, first examine it with a probe, and then make a cut down to the probe; sprinkle on flower of copper, and leave for five days. Bathe the wound with warm water, sprinkle it from above with meal mixed with water, and attach beet leaves with a bandage. When the flower of copper comes out and the wound of the fistula is clean, treat as above.

6. If the fistula is in a location that cannot be incised, and is deep, rinse it out with flower of copper, myrrh and soda dissolved in urine, and into its mouth introduce a lead probe to prevent it from growing together. Make this injection by tying the quill of a feather to a bladder and introducing it into the fistula. Continue to inject in this manner; the person does not recover unless he is incised.

7. If the rectum becomes inflamed, and pain and fever are present, if the person sits down often to stool but passes nothing, if the anus is seen to protrude on account

ἕδρη, καὶ ἐνίοτε στραγγουρίη ἐπιλαμβάνῃ, τοῦτο τὸ
νόσημα γίνεται, ὅταν φλέγμα ἐς τὸν ἀρχὸν καταστηρίξῃ ἐκ τοῦ σώματος. ξυμφέρει δὲ τὰ θερμά·
δύναται γὰρ τάδε προσφερόμενα λεπτύνειν καὶ ἐκτή-
κειν τὸ φλέγμα, καὶ ἅμα τῷ δριμεῖ τὸ ἁλμυρὸν ἐξυ-
δατοῦν, ὥστε μὴ εἶναί τι¹ καῦμα μηδὲ δῆξίν τινα ἐν
τῷ ἐντέρῳ.

Θεραπεύειν οὖν χρὴ ὧδε· καθίζειν ἐς ὕδωρ θερμόν, καὶ τρίψαντα τοῦ κόκκου τοῦ Κνιδίου ἑξήκοντα
κόκκους διεῖναι ἐν οἴνου κοτύλῃ καὶ ἐλαίου ἡμικοτυ-
λίῳ, χλιήνας, κλύσον· ἄγει δὲ ταῦτα φλέγμα καὶ
κόπρον. ὅταν δὲ μὴ ἐν τῷ ὕδατι καθίζῃ, ᾠὰ ἑψήσας
ἐν οἴνῳ μέλανι εὐώδει προστιθέναι πρὸς τὴν ἕδρην,
ὑποπετάσας τι κάτωθεν θερμόν, ἢ κύστιν ὕδατος
θερμοῦ πλήσας, ἢ λίνου σπέρμα πεφωσμένον ἀλέ-
σας, τρίψας καὶ μίξας ἴσον ἄλητον ἐν οἴνῳ μέλανι
καὶ εὐώδει καὶ ἐλαίῳ, καταπλάσσειν ὡς θερμότατον·
ἢ κριθὰς μίξας, ἢ στυπτηρίην Αἰγυπτίην τετριμμέ-
νην, καταπλάσσειν τε καὶ πυρίην· ἔπειτα πλάσας
βάλανον μακρήν, καὶ χλιαίνων πρὸς πῦρ, [καὶ]²
τοῖσι δακτύλοισι προσπλάσσειν· ἔπειτα ἀκροχλίη-
ρον ποιέων, ἐντιθέναι ἐς τὴν ἕδρην· τὰ ἔξωθεν δὲ
κηρωτῇ περιαλεῖψαι, καὶ καταπλάσσειν σκορόδοισιν
ἐφθοῖσιν ἐν οἴνῳ μέλανι κεκρημένῳ. ἐπὴν δὲ ἐξαι-
ρέῃς, ἐς ὕδωρ θερμὸν ἐφίζειν, καὶ συμμίξας χυλὸν
στρύχνου καὶ χηνὸς καὶ ὑὸς στέαρ καὶ χρυσοκόλ-
λαν καὶ ῥητίνην καὶ κηρὸν λευκόν, ἔπειτα διατήξας
ἐν τῷ αὐτῷ καὶ ξυμμίξας, τούτοισιν ἐγχρίειν,

of the collection of phlegm, and if strangury sometimes occurs, then the disease is arising because phlegm from the body is being deposited in the rectum. Warm medications are of benefit, since when these are applied they are able to thin and melt away the phlegm, and water down and expel the salty humour at the same time as the pungency, so that there will no longer be anything hot or corrosive in the intestine.

You must treat, then, as follows: sit the person down in warm water; grind sixty grains of Cnidian berry, dissolve in a cotyle of wine and half a cotyle of oil, warm and inject as a douche; this removes phlegm and faeces. If you do not have the person sit in water, boil eggs in fragrant dark wine and apply them to the anus; spread out something warm underneath, or fill a bladder with warm water. Or grind roasted linseed, crush and mix with an equal amount of flour in fragrant dark wine and oil, and apply this as a plaster as warm as possible. Or mix barley or ground Egyptian alum, apply this as a plaster and fomentation. Then shape a long suppository, warm it before a fire, giving it its final form with your fingers; then make it luke-warm, and introduce it into the anus. Then anoint all around the outside with wax-salves, and apply a plaster of garlic heads boiled in dilute dark wine. When you remove the suppository, have the person sit in warm water; mix together night-shade juice, goose grease, lard, chrysocolla,[a] resin and white wax, then melt it and mix it:

---

[a] Hydrous carbonate of copper used to solder gold.

---

[1] Ermerins: τὸ I.
[2] Del. Littré.

456 καὶ ἕως ἂν φλεγμαίνη, | καταπλάσσειν τοῖσι σκορό-
δοισι θερμοῖσι. καὶ ἢν μὲν πρὸς ταῦτα ἀπαλλάσ-
σηται τῆς ὀδύνης, ἀρκείτω· ἢν δὲ μή, πίσαι τὸ
μηκώνιον τὸ λευκόν· εἰ δὲ μή, ἄλλο ὅ τι φλέγμα
καθαίρει· διαιτᾶν δέ, ἕως ἂν φλεγμαίνη, ῥυφήμασι
κούφοισιν.

8. Ἡ δὲ στραγγουρίη ἐπιπίπτει ἐκ τῶνδε· θερ-
μαινομένη ἡ κύστις ἐκ τοῦ ἀρχοῦ προσάγεται τῇ
θερμότητι φλέγμα· ὑπὸ δὲ τοῦ φλέγματος στραγγουρίη γίνεται. ἢν μὲν οὖν ἅμα τῇ νούσῳ παύηται,
φιλέει γὰρ ὡς τὰ πολλὰ οὕτω γίνεσθαι· ἢν δὲ μή,
δίδου τῶν φαρμάκων τῶν στραγγουρικῶν.

9. Ἢν δὲ ὁ ἀρχὸς ἐκπίπτῃ, ἀνώσας σπόγγῳ
μαλθακῷ, καὶ καταχρίσας κοχλίῃ, τῶν χειρῶν
δήσας, ἐκκρέμασον ὀλίγον χρόνον, καὶ εἴσεισιν. ἢν
δὲ μεῖζον ἐκπέσῃ, καὶ μένῃ ἔνδον, διαζώσας ἐν τῇσι
λαγόσι, καὶ ὑφεὶς ὄπισθεν ἐκ τοῦ διαζώματος ταινίην, ὤσας ἔσω τὸν ἀρχόν, προσθεῖναι σπόγγον
μαλθακὸν βρέξας ὕδατι θερμῷ, ἐνεψήσας λωτοῦ
πρίσματα· καταχέαι δὲ καὶ κατὰ τοῦ ἀρχοῦ ἀπ'
αὐτοῦ τοῦ ὕδατος, τὸν δὲ σπόγγον ἐκπιέσαι· ἔπειτα
ὑποτείνας τὴν ταινίην διὰ μέσων τῶν σκελέων, ἀναδῆσαι περὶ τὸν ὀμφαλόν. ὅταν δὲ θέλῃ ἀφοδεύειν,
ἐπὶ λασάνοισιν[1] ὡς στενοτάτοισιν ἀφοδευέτω· ἢν δὲ

---

[1] Littré: ἔπλασσ- I.

[a] Petrequin (vol. 1, 374f., note 7) argues on surgical grounds
for the text of the manuscripts, rejecting the negative understood
by Cornarius (et non maneat intus) and introduced into the

anoint with this mixture, and as long as there is any inflammation apply the plasters of warm garlic. If, with these measures, the person is relieved of his pain, let that suffice; if not, give him white spurge to drink, or if not that, then anything else that cleans phlegm. He should follow a regimen of light gruels as long as the inflammation lasts.

8. Strangury attacks for the following reason: the bladder is heated from the rectum, and because of this heat attracts phlegm; on account of the phlegm strangury arises. Now if the strangury ceases together with the disease of the rectum—this is wont to happen in most cases—fine; if not, give medications specific for strangury.

9. If the rectum prolapses, support it with a soft sponge, anoint it with snail-medication, tie the person's hands and suspend him for a short time, and it will go in. If it prolapses further, but will still stay inside,[a] encircle the body at the flanks with a band, and run a strip down from the band at the back. Then press the rectum in, soak a soft sponge in warm water in which you have boiled saw-dust of lotus, and apply it; also make an infusion into the rectum with the same liquid. Squeeze out the sponge, and then run the strip down between the legs, and tie it up at the navel. When the person wishes to go to stool let him sit on as narrow a night-stool as

Greek by Littré (καὶ μὴ μένῃ ἔνδον). The author's classification of prolapse is, according to Petrequin, into three degrees: (a) mild prolapse; (b) more severe prolapse, but such that the rectum will still remain in place when reduced; (c) even more severe prolapse, such that the rectum will not remain in place when reduced.

401

παιδίον ᾖ, ἐπὶ γυναικὸς τῶν ποδῶν, πρὸς τὰ γούνατα προσκλιθείς. ὅταν δὲ ἀφοδεύῃ, τὰ σκέλεα ἐκτεινάτω· οὕτω γὰρ ἂν ἥκιστα ἐκπίπτοι ἡ ἕδρη. ἢν δὲ ὑγραίνηται ὁ ἀρχὸς καὶ ἰχὼρ ἀπορρέῃ, περινίψαι τρυγὶ κεκαυμένῃ καὶ ὕδατι ἀπὸ μυρσίνης, καὶ ἀδίαντον ξηρήνας καὶ κόψας, διασήσας, κατάπασσε. ἢν δὲ αἱμορροῇ, περινίψας τοῖσιν αὐτοῖσι, χαλκῖτιν καὶ πρίσμα κυπαρίσσου ἢ κέδρου | ἢ πίτυος [ἢ τερεβινθίνου]¹ ἢ τερμίνθου τρίψας, συμμίξας τῇ χαλκίτιδι ἴσον, καταπλάσσειν, τὰ ἔξωθεν δὲ κηρωτῇ παχείῃ περιαλείφειν.

Ὁκόταν ἀρχὸς ἐκπίπτῃ καὶ μὴ θέλῃ κατὰ χώρην μένειν, σίλφιον ὅτι ἄριστον καὶ πυκνότατον ξύσας λεπτὸν καταπλάσσειν. καὶ τοῦ πταρμικοῦ φαρμάκου πρὸς τὴν ῥῖνα προσθιθέναι καὶ παροξύνειν τὸν ἄνθρωπον. ἢ ὕδατι θερμῷ περιπλύνας σίδια, καὶ στυπτηρίην τρίψας ἐν οἴνῳ λευκῷ, καταχέαι τοῦ ἀρχοῦ, ἔπειτα ῥάκεα ἐμβαλεῖν, καὶ τοὺς μηροὺς ξυνδῆσαι ἡμέρας τρεῖς, καὶ νηστευέτω, οἶνον δὲ πινέτω γλυκύν. ἢν δὲ μηδὲ οὕτω διαμένῃ,² μίλτον μίξας ὁμοῦ μέλιτι διαχριέτω. ἀρχὸς ἢν ἐκπίπτῃ καὶ αἱμορροῇ· ἄρου ῥίζης περιελὼν τὸν φλοιὸν ἑψεῖν ἐν ὕδατι· ἔπειτα τρίβειν ἄλητον ξυμμίσγων, καὶ καταπλάσσειν θερμόν. ἄλλο· τῆς ἀμπέλου τῆς ἀγρίης, ἢν ἔνιοι καλέουσι ψιλώθριον, ταύτης τὰς ῥίζας τὰς ἀπαλωτάτας περιξέσαντα ἑψῆσαι ἐν οἴνῳ μέλανι

¹ Del. Aldina. ² Ermerins: διαχωρέῃ I.

possible; if it is a child, he should sit on a woman's feet leaning against her knees. During defecation let him extend his legs, since in this position the rectum is least likely to prolapse. If the prolapsed rectum becomes moist and serum runs out of it, wash it off all around with burnt wine lees[a] and myrtle juice, and sprinkle on powder made by drying, pounding and sifting maiden-hair. If the rectum haemorrhages, wash it with the same things, apply a plaster made by grinding copper ore and mixing with it an equal amount of the sawdust of cypress, juniper, pine or terebinth tree, and anoint the outside all around with thick wax-salves.

When the rectum prolapses and will not remain in place, grind up fine the best and densest silphium and apply it as a plaster. Also introduce a sternutatory into the nose to stimulate the person. Or wash pomegranate-peel in hot water, grind alum in white wine, and infuse into the rectum; then insert rags and bind the thighs together for three days; have the person fast but drink sweet wine. If not even then the rectum stays in place, mix red ochre together with honey and anoint. If the rectum prolapses and bleeds, strip the bark off cuckoo-pint root and boil it in water. Then grind it, mix it together with flour, and apply warm as a plaster. Another: the wild vine which some people call "hair-remover": scrape thoroughly the tenderest roots of this plant and boil them in neat dry

[a] I.e. acid potassium tartrate.

ἀκρήτῳ αὐστηρῷ· ἔπειτα τρίψαντα καταπλάσσειν
χλιηρόν· ξυμμίσγειν δὲ καὶ ἄλευρα, καὶ φυρῆν ἐν
οἴνῳ λευκῷ καὶ ἐλαίῳ χλιηρῶς. ἄλλο· κωνείου καρ-
πὸν τρίβοντα, παραστάζειν οἶνον λευκὸν εὐώδεα,
ἔπειτα καταπλάσσειν χλιηρόν. ἢν δὲ φλεγμαίνῃ,
κισσοῦ ῥίζαν ἑψήσας ἐν ὕδατι, τρίψας λεῖον, ἄλευρον
ξυμμίσγων ὡς κάλλιστον, ἐν οἴνῳ λευκῷ φυρήσας,
καταπλάσσειν, καὶ ἄλειφα πρὸς τούτοις ξυμμίξας.
ἄλλο· μανδραγόρου ῥίζαν μάλιστα μὲν χλωρήν, εἰ δὲ
μή, ξηρήν, τὴν μὲν οὖν χλωρὴν ἀποπλύναντα καὶ
ταμόντα, ἑψῆσαι ἐν οἴνῳ κεκρημένῳ, καὶ καταπλάσ-
σειν· τὴν δέ γε ξηρὴν τρίψαντα καταπλάσσειν
ὁμοίως. ἄλλο· σικύου πέπονος τὸ ἔνδον τρίψας λεῖον
καταπλάσσειν.

10. Ἢν δὲ γένηται ὀδύνη καὶ μὴ φλεγμαίνῃ,
460 λίτρον ὀπτήσας | ἐρυθρόν, καὶ τρίψας λεῖον, καὶ
στυπτηρίην καὶ ἅλας φώξας τοὺς ἴσους, καὶ τρίψας
λείους, συμμίξαι ἴσον ἑκάστου· εἶτα πίσσῃ ξυμμίξας ὡς βελτίστῃ, ἐς ῥάκος ἐναλείψας, ἐντιθέναι καὶ
καταδεῖν. ἄλλο· καππάριος φύλλα χλωρὰ τρίψας,
ἐς μαρσίπιον ἐμβαλών, προσκαταδεῖν· καὶ ἐπὴν
καίειν δοκέῃ, ἀφαιρέειν καὶ αὖθις προστιθέναι. ἢν
δὲ μὴ ᾖ φύλλα καππάριος, τὸν φλοιὸν τῆς ῥίζης
κόψας, φυρήσας οἴνῳ μέλανι, τὸν αὐτὸν τρόπον ἐπιδεῖν. τοῦτο καὶ πρὸς σπληνῶν ὀδύνην ἀγαθόν. τούτων τῶν καταπλασμάτων τὰ μὲν ψύχοντα κωλύει
ῥεῖν, τὰ δὲ μαλθάσσοντα καὶ θερμαίνοντα διαχέει,
τὰ δὲ ἐς ἑωυτὰ ἕλκοντα ξηραίνει καὶ ἰσχναίνει.[1]

dark wine; then crush them and apply warm as a plaster; also mix wheat-meal together with these in warm white wine and oil. Another: grind seed of hemlock, distil fragrant white wine on to it, and then apply warm as a plaster. If the rectum is inflamed, boil ivy root in water, grind it fine, mix with the finest quality wheat-meal, add white wine to form a paste, and apply as a plaster—also mix oil with this. Another: mandrake root, best fresh, but otherwise dried: the fresh you should wash thoroughly and cut, and then boil in diluted wine and apply as a plaster; the dry you grind and apply as a plaster in the same way. Another: pound the inside part of a melon fine and apply as a plaster.

10. If there is pain but no inflammation, burn red soda and grind it fine, roast and grind fine equal amounts of alum and salt, and mix together an equal amount of these two; then add pitch of the finest quality, smear the mixture on to a rag, insert this, bandage it in place. Another: pound green leaves of the caper-plant, make into a poultice, and bind over the anus; when this seems to be burning the person, remove it and apply anew. If you do not have caper leaves, cut off bark from the root, mix in dark wine, and bind on in the same way. This is also good against pain in the spleen. Of these plasters, the ones that cool prevent fluxes, the ones that soften and warm disperse, and the ones that attract dry and reduce swelling.

---

[1] κωλύει ... διαχέει ... ξηραίνει ... ἰσχναίνει Ermerins: -ειν I.

τοῦτο δὲ τὸ νόσημα γίνεται, ὅταν χολὴ καὶ φλέγμα
ἐς τοὺς τόπους καταστηρίξῃ. ἀρχοῦ δὲ φλεγμήναν-
τος, διαχρίειν τῷ φαρμάκῳ, ὅπῃ ἡ ῥητίνη καὶ τὸ
ἔλαιον καὶ ὁ κηρὸς καὶ ἡ μολύβδαινα καὶ τὸ στέαρ·
ὡς θερμότατα [διερρήθησαν][1] καταπλάττεσθαι.

[1] Del. Joly.

FISTULAS

This condition arises when bile and phlegm become fixed in these locations. When the rectum is inflamed, anoint it with a medication of resin, olive oil, bees-wax, galena and fat applied in a plaster as hot as possible.

INDEX

409

411

INDEX

INDEX

417

INDEX